Graphs in VLSI

Rassul Bairamkulov • Eby G. Friedman

Graphs in VLSI

Rassul Bairamkulov
University of Rochester
Rochester, NY, USA

Eby G. Friedman
University of Rochester
Rochester, NY, USA

ISBN 978-3-031-11046-7   ISBN 978-3-031-11047-4   (eBook)
https://doi.org/10.1007/978-3-031-11047-4

This Springer imprint is published by the registered company Springer Nature Switzerland AG
The registered company address is: Gewerbestrasse 11, 6330 Cham, Switzerland

To my dear Zhansaya
To my wife and companion in life, Laurie

Preface

Advances in semiconductor fabrication technology have produced explosive growth in the number of transistors within an integrated circuit (IC). Modern devices consist of dozens of components, thousands of modules, millions of registers, and many billions of transistors. Despite the capacity to fabricate exorbitant quantities of nanoscale devices, converting these transistors into functional products is a complex multifaceted challenge. Synchronization, power integrity, logic synthesis, and physical layout represent only a small portion of the many issues encountered during the very large scale integration (VLSI) design and analysis process. These issues are only expected to grow in complexity as microelectronic systems evolve due to three-dimensional integration, shrinking transistor dimensions, and emerging, beyond CMOS technologies.

Since every integrated system is fundamentally a network, solutions to many of the challenges inherent to VLSI can be resolved using graph theory. Many forms of graphs naturally occur at each level of the VLSI system design hierarchy. At the architectural level, register allocation is often viewed from a graph coloring perspective. Synchronization of the sequential logic is achieved by optimizing timing graphs. The electrical characteristics of a VLSI system are determined from the analysis of circuit graphs. Graph-based partitioning, floorplanning, placement, and routing are integral parts of the multi-tiered VLSI physical layout process.

The quality and complexity of ICs have been greatly enhanced by graph theoretic techniques and algorithms. In return, novel practical VLSI applications have revitalized certain subfields of graph theory. Classic graph theoretic problems, such as Steiner minimal trees, pathfinding, and graph partitioning, have been extensively studied and applied, in no small part, due to the practical effectiveness of graph theory to the design and analysis of VLSI systems. A virtuous cycle of theory and application has greatly advanced both graph theory and ever more powerful VLSI systems.

This book is based on the body of research produced by Rassul Bairamkulov during his doctoral studies from 2017 to 2022 at the University of Rochester under the supervision of Professor Eby G. Friedman. Two observations inspired this book:

- Despite the significance of graph theory to the design of VLSI circuits and systems, a comprehensive review of applications of graph theory in VLSI is currently missing from the literature. Books discussing the overall VLSI design process typically only provide a basic description of graph theory, while books focusing on graph theory contain few applications relating to the VLSI design process. Books discussing specialized topics in VLSI also exist, yet these books only cover a small subset of graph theoretic applications.
- Despite the apparent omnipresence of graphs in the VLSI system design process, the authors believe that the full potential of graph theory has yet to be fully realized in modern VLSI design tools. Many areas of the VLSI design process, such as system exploration and the integration of emerging technologies, will require novel design methodologies and algorithms, many of which rely on graph theory.

These observations are reflected in the organization of this book. The first half of the book is focused on existing applications of graph theory and algorithms to the design of integrated systems. After a brief description of the fundamental concepts of graph theory, common applications of graph theory at different levels of abstraction within the VLSI system design process are discussed. Individual chapters are dedicated to synchronization and circuit analysis, two particularly important issues in the VLSI design process which are deeply affected by graph theory.

The second half of the book is focused on three novel unorthodox applications of graph theory. The first application is the Infinity Mirror Technique (IMT) – a constant time mesh analysis algorithm accelerating the IR drop analysis process in practical on-chip power networks by several orders of magnitude. An IMT-based computationally efficient algorithm for on-chip voltage regulator distribution is also described. The second application is related to the exploration of system-level power delivery. The SPROUT – Smart Power ROUTing algorithm is presented to efficiently produce a prototype of a board-level power network, enabling efficient analysis with high-level architectural tradeoffs, such as the number of layers within the board or the position of discrete components. The second half of the book is completed with QuCTS – single flux Quantum Clock Tree Synthesis algorithm, which describes a graph-based algorithm to satisfy the stringent requirements for clock distribution networks in superconductive electronics.

Due to the focus on the VLSI design process, this book is expected to become a useful addition to the library of engineers, researchers, and students working in the areas of VLSI system design and computer science. For professionals working in the design of VLSI systems (typically electrical and computer engineers and computer scientists), this book provides a deeper insight into the theory behind many established design techniques based on graph theory, such as clock skew scheduling, system partitioning, circuit analysis and optimization, and interconnect routing. For mathematicians and computer scientists, the book elucidates the link between graph theory and the design and analysis of VLSI circuits and systems.

Acknowledgments

The authors would like to acknowledge the support of our many collaborators who facilitated the development of this book. The authors would like to thank Charles B. Glaser and Shabib Shaikh from Springer for their assistance in the publishing process. The authors are very grateful to Dr. Mikhail Popovich from Google, Dr. Kan Xu and Dr. Juan S. Ochoa from Apple, Dr. Abinash Roy from Intel, Mr. Mahalignam Nagarajan and Dr. Vaishnav Srinivas from Qualcomm Technologies, and Mr. Jamil Kawa from Synopsys, for their continued support and collaboration during the research projects that constitute a considerable part of this book. Special thanks are reserved for Tahereh Jabbari from the High Performance Integrated Circuit Design and Analysis Laboratory for sharing her expertise in single flux quantum circuit design.

This research is supported in part by the National Science Foundation under Grant Nos. CCF-1329374, CCF-1526466, CCF-1716091, Intelligence Advanced Research Projects Activity under Grant Nos. W911NF-14-C-0089 and W911NF-17-9-0001, American Institute for Manufacturing Integrated Photonics under Award No. 059447-007, the Intel Collaborative Research Institute for Computational Intelligence, Singapore Ministry of Education Tier 2 under Grant No. MOE2014-T2-2-105, and grants from Cisco Systems, Google, OeC, Qualcomm, and Synopsys.

Rochester, NY, USA
Rochester, NY, USA

Rassul Bairamkulov
Eby G. Friedman

Contents

About the Authors

 **Rassul Bairamkulov** was born in August 1994 in Karaganda, Kazakhstan. He received a Bachelor of Engineering in Electrical and Electronic Engineering degree from Nazarbayev University in Astana, Kazakhstan in 2016, and a Master of Science degree in Electrical Engineering from the University of Rochester in Rochester, New York in 2018. He completed the Ph.D. degree in electrical engineering from the University of Rochester under the supervision of Prof. Eby G. Friedman in 2022. In the summers of 2018 and 2020, he interned with the Power Design team at Qualcomm Inc. in San Diego, California. His research interests include graph theory, physical design of integrated circuits, and electronic design automation of conventional and emerging VLSI technologies.

 **Eby G. Friedman** received the B.S. degree in electrical engineering from the Lafayette College, Easton, Pennsylvania, and the M.S. and Ph.D. degrees in electrical engineering from the University of California at Irvine, Irvine, California. He was with Hughes Aircraft Company in southern California, from 1979 to 1991, rising to Manager of the Signal Processing Design and Test Department, where he was responsible for the design and test of high performance digital and analog ICs. He has been with the Department of Electrical and Computer Engineering, University of Rochester in Rochester, New York, since 1991, where he is a Distinguished Professor and the Director of the High Performance VLSI/IC Design and Analysis Laboratory.

He is also a Visiting Professor with the Technion–Israel Institute of Technology in Haifa, Israel. He has authored more than 600 articles and book chapters; authored or edited 20 books in the fields of high speed and low power CMOS design techniques, 3-D design methodologies, high speed interconnect, superconductive circuits, and the theory and application of synchronous clock and power distribution networks, and he holds 26 patents. His current research and teaching interests include high performance synchronous digital and mixed-signal circuit design and analysis with application to high speed portable processors, low power wireless communications, and server farms. Dr. Friedman is a recipient of the IEEE Circuits and Systems Mac Van Valkenburg Award, the IEEE Circuits and Systems Charles A. Desoer Technical Achievement Award, the University of Rochester Graduate Teaching Award, and the College of Engineering Teaching Excellence Award. He was the Editor-in-Chief (EIC) and Chair of the Steering Committee of the IEEE TRANSACTIONS ON VERY LARGE SCALE INTEGRATION (VLSI) SYSTEMS and EIC of the *Microelectronics Journal*, a Regional Editor of the JOURNAL OF CIRCUITS, SYSTEMS AND COMPUTERS, an editorial board member of numerous journals, and a program and technical chair of several IEEE conferences. He is a Senior Fulbright Fellow, a National Sun Yat-sen University Honorary Chair Professor, and an Inaugural Member of the UC Irvine Engineering Hall of Fame.

Chapter 1
Introduction

Considering the fundamental nature of graph theory and the importance of intercon-nected systems, the relative historical novelty of the field is somewhat surprising. Most fundamental areas of mathematics, such as arithmetic, geometry, and com-binatorics, emerged in ancient times. The earliest traces of combinatorics, for example, are found in the 6th century B.C. [1], whereas early forms of arithmetic date to 20,000 BC, older than the earliest writing system [2]. The history of graph theory however only dates back to the second quarter of the 18th century, when Leonhard Euler tackled the famous problem of the Seven Bridges of Königsberg [3]. This question is how to route a walk through the four land masses of Königsberg, divided by the Pregel River, such that each of the seven bridges is crossed exactly once (see Fig. 1.1a). Euler stated in this problem formulation that a path within a land mass is irrelevant. The solution is exclusively determined by the sequence of bridge crossings, enabling an abstract analysis of networks.

Using a graph notation, the Seven Bridges of Königsberg can be rephrased as: *Given the undirected multigraph G in Fig. 1.1b, determine a trail that contains all edges of G (an Eulerian trail).* In 1736, Leonhard Euler showed that no such trail exists by recognizing that the degree of all vertices in G should be even other than the start and finish nodes [4]. This proof of what today is known as the *handshaking lemma* is widely regarded as the earliest work in graph theory [5]. Despite the ingenuity of the solution and the generalization provided in Euler's work, the field of graph theory has been relatively dormant for more than a century. The practical significance of graph theory had not yet been recognized. Graph theoretic problems occasionally appeared in recreational mathematics. In 1805, for example, Louis Poinsot published a puzzle particularly relevant to graph theory:

Given some points situated at random in space, it is required to arrange a single flexible thread uniting them two by two in all possible ways, so that finally the two ends of the thread join up, and so that the total length is equal to the sum of all the mutual distances [5]. In graph theoretic terms, Poinsot asked a reader to find an Eulerian circuit within a complete graph of degree k, a task that is only possible

© The Author(s), under exclusive license to Springer Nature Switzerland AG 2023
R. Bairamkulov, E. G. Friedman, *Graphs in VLSI*,
https://doi.org/10.1007/978-3-031-11047-4_1

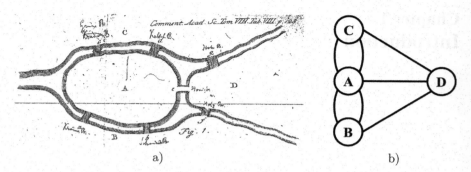

Fig. 1.1 Seven Bridges of Königsberg problem. a) Simplified map of Königsberg bridges as drawn by Euler in [4], and b) equivalent multigraph representation.

if k is odd. The first practical application of graph theory appeared in 1857 when Arthur Cayley discovered a subclass of graphs which he named *trees* and used this formulation as a tool to analyze nested operations [6]. The first use of graphs in chemistry is believed to be introduced by Alexander Crum Brown and published in 1866 by Edward Frankland [7]. In 'graphic notation,' each atom within a compound is represented by a circle with a letter denoting the element, where the lines represent the chemical bonds between atoms [7]. Notably, this system is still in common use today, often with minor modifications, such as the omission of circles or a different notation for benzene coils. A graphic notation quickly became popular and was crucial in explaining isomerism – the existence of substances with an identical composition but different properties (see Fig. 1.2). The term *graph* was proposed by James Joseph Sylvester in 1878 [8] based on an analogy with chemicograph, the visual representation of chemical bonds. Interest in graph theory gradually sprouted new branches during the late 19th century to early 20th century, such as algebraic graph theory, extremal graph theory, and random graph theory [9].

The advent of electronics and computers produced a variety of novel graph theory applications. Many known problems at the time, such as graph coloring [10] and partitioning [11], were successfully applied to computer design. The invention of the integrated circuit (IC) in 1958 [12] created new avenues for the application of graph theory. The demand for greater semiconductor integration motivated rapid advancements in algorithms for circuit partitioning, interconnect routing, and logic verification, all assisted in no small part by graph theory.

The primary purpose of this chapter is to introduce a graph theoretic perspective to the design of VLSI circuits and systems. In investigating the history of VLSI, the significance of graph theory to the design of integrated circuits is highlighted. Technology advancements preceding the birth of VLSI in the 1970's are discussed in Section 1.1. The growth of computer-aided design during the early years of VLSI is described in Section 1.2. The chapter is concluded with an outline of the book in Section 1.3.

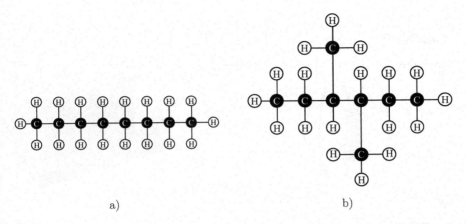

a)

b)

Fig. 1.2 Earliest application of graphs is in chemical notation. The figures illustrate the difference between C_8H_{18} isotopes [7], a) octane, and b) 3,4-dimethylhexane.

1.1 Precursors of VLSI

The initial stages of the IC design process were largely driven by advancements in materials science and semiconductor manufacturing. Prior to 1947, electronic applications, such as the radio, telephone, and telegraph, were based on vacuum tubes and relays. Relays are electrically controlled mechanical switches. The control electrode establishes a mechanical connection between terminals, as illustrated in Fig. 1.3. Relays were used in electronic engineering since the early 19th century [13] to amplify attenuated signals in telegraph lines. Vacuum tubes are similar to incandescent light bulbs due to the physical phenomenon used in both of these devices. A vacuum tube is depicted in Fig. 1.4. Electrons are emitted from a conductor heated to a sufficiently high temperature. This effect is called thermionic emission and was first documented by Frederick Guthrie in 1875 [14]. In vacuum tubes, electrons from a heated cathode filament travel towards the cold anode electrode, producing current. The current through the device can be controlled by adjusting the electrical potential at the cold and hot electrodes. Since electrons are not emitted by the cold conductor, the device behaves as a diode. An additional electrode (typically a grid) between the anode and cathode produces a triode, enabling more precise control of the output current.

The earliest electromechanical computers were manufactured using relays. For example, Z3 was an electromechanical programmable computer built in 1941 by Konrad Zuse, which utilized 2,600 electrical relays [17]. The Automatic Sequence Controlled Calculator built in 1944 used 3,500 relays [18]. Later, pre-semiconductor era computers primarily used vacuum tubes. The famous Electronic Numerical Integrator and Computer (ENIAC), shown in Fig. 1.5, was built in 1945 at the University of Pennsylvania [19] and contained 18,000 vacuum tubes. A major limitation of these systems was however soon exposed. Although vacuum tubes

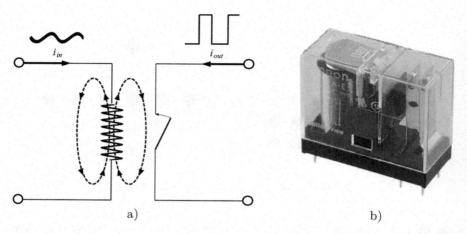

Fig. 1.3 Magnetic relay. a) Schematic diagram illustrating a normally open (NO) relay. The circuit is inactive if the input current is zero. The input current i_{in} passes through the input coil, producing a magnetic field that mechanically closes the switch. Output current i_{out} flows through the switch. Reducing the input current disables the switch, stopping the output current. A normally closed relay has a similar structure, but the switch is closed when i_{in} is zero and is open when input current is applied. b) Omron Electronics G2R-1A-AC240 NO relay [15]. Observe the coil under the metal plate. The metal plate is attached to one of the contacts pictured on the right part of the image. The current passing through the coil produces a magnetic field that displaces the metal plate, connecting the circuit.

and relays supported small scale electronics, the reliability of these devices was insufficient to simultaneously maintain stable operation of thousands of these devices. In the ENIAC, for example, on average, one vacuum tube burnt out every two days, necessitating regular preventive measures and thorough testing [19]. Furthermore, vacuum tubes required enormous power due to the need for heating.

The successful demonstration of the point contact transistor in 1947 [20] and bipolar junction transistor (BJT) in 1948 [21] revolutionized electronics and computers [22, 23]. As compared to relays and vacuum tubes, these novel devices dissipated much less power, were smaller in size, and cost considerably less to produce. Crucially, these early transistors were highly reliable, enabling thousands of hours of uninterrupted operation [24]. The first transistor-based computer was built in 1953 in Manchester, UK [25] and contained 92 point contact transistors [26], starting the era of computers based on semiconductor transistors. Although the computational performance of these early transistor-based computers was inferior to the vacuum tube computers of the time, a significant reduction in size, cost, and power was achieved [26]. A year later, in 1954, TRansistorized DIgital Computer (TRADIC), the first American transistor computer, bridged the performance gap, offering computational performance comparable to vacuum tube computers of the time [24]. Weighing 1,191 pounds and dissipating less than 100 watts of power, TRADIC was sufficiently small to install within a bomber aircraft, enabling use for navigation and targeting [27]. The small power and size enabled transistors to be

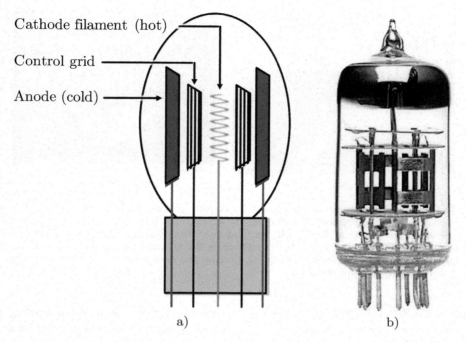

Cathode filament (hot)

Control grid

Anode (cold)

a)   b)

Fig. 1.4 Vacuum tube. a) Internal structure of a vacuum tube. The cathode filament is heated inducing thermionic emission [14]. The electrons emitted from the cathode are captured by the cold anode. The grid is placed between the anode and cathode to control the output current. Without the grid, the vacuum tube behaves as a diode. b) Solen Électronique SI-12AX7B vacuum tube [16]. The thick vertical wire at the center of the tube is a filament cathode. The filament is surrounded by several anode plates capturing electrons emitted by the filament.

used not only for computing, but also for data storage [28], radio [29], telephony [30], and a host of other novel applications.

The next decade is largely characterized by the many rapid advancements of semiconductor manufacturing technology. The first integrated circuit, built in 1958 by Jack Kilby, used multiple discrete components integrated onto a single substrate [31]. The next year, the first monolithic IC was invented by Robert Noyce based on a planar manufacturing process developed at Fairchild Semiconductor [32]. In the late 1950's, the research group lead by Mohamed M. Atalla developed a silicon oxidation process. This invention was an important precursor to the metal-oxide-semiconductor field effect transistor (MOSFET) developed in 1959 by Mohamed M. Atalla and Dawon Kahng [32]. Formation of p-n junctions based on ion implantation, described in 1965 by Manchester, Sibley, and Alton [33], was a crucial prerequisite of the self-aligned gate process proposed in the late 1960's by Bower, Dill, Aubuchon, and Thompson [34, 35]. The self-aligned gate in MOSFETs enabled highly accurate fabrication of MOSFETs, reliably operating at high frequencies due to the small parasitic capacitance [36]. These inventions were

a)   b)

Fig. 1.5 Electronic Numerical Integrator and Computer (ENIAC) built in 1945 using 18,000 vacuum tubes. The operating frequency of the ENIAC was 100 kilohertz, producing 5,000 operations per second. a) The ENIAC occupied more than 167 square meters and consumed more than 150 kilowatts [19]. b) The reliability of the ENIAC was poor due to the large number of vacuum tubes that occasionally burnt out. Locating and replacing the broken tubes required significant time and effort.

primary drivers of MOSFET technology, becoming a mainstay in IC manufacturing in the early 1980's, enabling the integration of large numbers of transistors within a single IC.

1.2   The rise of VLSI

The number of devices integrated within a single IC has rapidly increased, producing increasingly complex systems. In 1968, for example, the Rockwell '1502' large scale array contained 658 MOSFETs within approximately 10 mm^2 [37]. Only three years later, in 1971, the Intel 4004 microprocessor contained 2,250 transistors within 12 mm^2 [38], an almost threefold increase in density. The growing complexity of these systems required qualitatively new techniques to design these ICs, giving rise to the new field of large scale integration (LSI) and supporting fields like electronic design automation (EDA).

The issues induced by integrating a large number of transistors onto a single die were raised as early as 1968, when the excessive time and high error rates during the manual LSI design process were noted [39]. A sophisticated manufacturing technology alone was no longer sufficient for achieving high performance integrated circuits and systems. Similar progress in other aspects of the integrated system design process was required, including layout synthesis, computer architecture, and logic design. LSI technology during the 1970's was characterized by the proliferation of computer-aided design (CAD) and EDA, greatly assisted by advancements in graph theory in LSI engineering [40, 41].

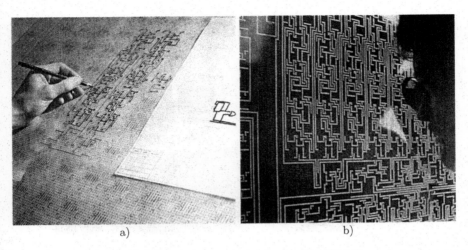

Fig. 1.6 Manual layout of an integrated system during the pre-CAD/EDA era. a) An interconnect pattern of a logic circuit scaled by a factor of 250, manually drawn with pencil on a routing grid [42]. b) Visual inspection of the resulting layout [42].

Prior to the advent of EDA, the physical layout of early ICs was drawn manually, as illustrated in Fig. 1.6, and therefore required significant time and labor. Systems consisting of hundreds to thousands of elements required enormous time to be schematically drawn and laid out. Early CAD tools for integrated systems were proposed in the late 1960's to assist the drawing of circuit schematics [45, 46]. A drafting tool for circuit layout was presented in [47] where the drawings are encoded using standardized blocks that enabled digital storage of the drawings for subsequent reproduction. Similar to most tools developed during the 1960's, these tools were intended for internal use within semiconductor manufacturing companies and were therefore used by only a few people. The market for commercial CAD tools for IC design and analysis emerged in the early 1970's. Many of the seminal CAD tools for LSI development were introduced into the marketplace during this time. These tools consisted of a computer equipped with a display and a graphic input device, such as a RAND tablet [48], as exemplified in Fig. 1.7. Among the early LSI graphic tools was the Calma Graphics Design System Integrated Circuit Mask maker (GDS-ICM) [43]. The GDS computer graphic format used in this tool evolved into the GDS-II format, today's industry standard for representing the physical layout of an IC [49]. Many tools were developed to verify the design of the transistor circuits and physical layout. For example, the Simulation Program with Integrated Circuit Emphasis, commonly known as SPICE, was developed in 1973 [50]. With SPICE, complex circuits are efficiently represented in text format, as illustrated in Fig. 1.8, allowing a circuit to be simulated to predict circuit behavior.

Despite the proliferation of CAD tools into the VLSI design process, errors frequently occurred, necessitating significant time and labor for corrections and modifications. The explosive growth in the complexity of these integrated circuits

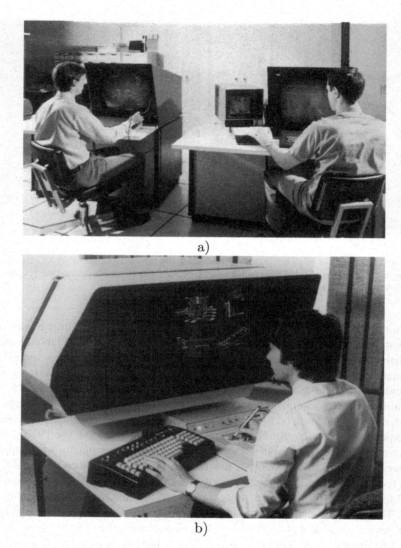

Fig. 1.7 Workstations used during the development of integrated systems during the early years of CAD/EDA. a) Two workstations used in Texas Instruments *circa* 1970 [43]. The left workstation, used for schematic design, is equipped with a large interactive screen and stylus. The right workstation is used for layout drawing and is equipped with a large interactive screen and a small additional display. b) IC design workstation during the mid-1980's used to generated the layout of an IC [44].

and systems motivated the automation of the labor intensive and error prone tasks. EDA tools were an upgrade over standard CAD tools so as to improve the speed and accuracy of the design process with minimal human intervention. Early EDA tools emerged in the late 1960's and targeted primarily IC layout. In 1969, one of the first automated layout synthesis tools was developed [51]. A topological layout graph

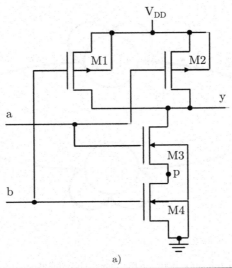

a)

M1	y	a	vdd	vdd	tp	L=0.6u	W=1.2u
M2	y	b	vdd	vdd	tp	L=0.6u	W=1.2u
M3	y	a	a	0	tn	L=0.6u	W=1.2u
M4	p	b	0	0	tn	L=0.6u	W=1.2u

b)

Fig. 1.8 Example of a two input NAND gate in SPICE using four transistors. a) Initial schematic representation. b) Circuit in SPICE format. Each line encodes the type and name of the component (e.g., M1 describes a transistor with name '1'), connection to other nodes (e.g., M1 is connected to nodes y, a, and vdd), and model parameters such as the channel dimensions (length and width).

was produced, specifying the relative position of the cells within an IC. These early EDA tools were primarily developed by IC manufacturers to accelerate the product development process. The widespread adoption of EDA was soon facilitated by the advent of commercial EDA companies and products. Three of the most prominent companies of the time were Daisy Systems, Mentor Graphics, and Valid Logic Systems, each of which produced a variety of EDA products in the areas of logic and circuit design and analysis [52, 53], automated placement and routing [54], and silicon compilation [55, 56]. The size of the EDA market grew rapidly, reaching $2.5 billion in 1994 [57], $5 billion in 2006 [58], $6.4 billion in 2013 [59], and over $9 billion in 2020 [60].

Graph theory plays an important role in enabling VLSI by managing the complexity of these systems. The characteristics of these objects are represented by graphs, intentionally omitting many design details to focus on only relevant

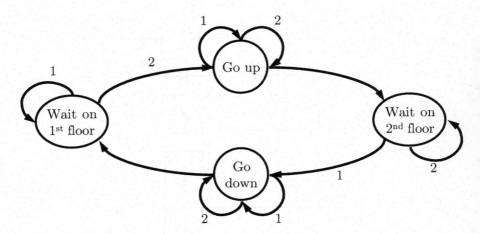

Fig. 1.9 An example of a finite state machine [62, 63] representing a two floor elevator. The nodes represent the four states of the system, *i.e.*, wait on the first or second floor, go up, or go down. The edges represent the transition between the states. The edge labels 1 and 2 denote the events of pressing button 1 and 2 in an elevator. These events trigger the transition of an elevator to the next state. The loops (edges starting and ending at the same node) indicate events that do not change the state of the system.

features. Many CAD/EDA tools heavily rely on graph theory [61]. Synchronization of sequential logical systems is one of the earliest applications of graph theory to the design of computing systems. Classic techniques include finite state machines (FSM), exemplified in Fig. 1.9. A FSM were first described in the mid-1950's by G. H. Mealy [62] and E. F. Moore [63], where a complex synchronous system is represented by a network of states. By applying abstraction, data flow within an integrated system could be efficiently modeled while disregarding the less significant details. SPICE is largely based on modified nodal analysis (MNA) [64] – a method utilizing a Laplacian matrix describing a circuit graph. Interconnect routing tools utilize graph-based methods for finding an optimal connection between terminals [61, 65, 66]. Graph theory continues to be used in modern IC design, including three-dimensional integration [67], hardware security [68, 69], circuit analysis [70–72], and networks-on-chip (NoC) [73, 74].

1.3 Outline of book

Applications of graph theory to the design of VLSI circuits and systems are the primary focus of this book. Fundamental concepts in graph theory frequently encountered in the IC design process are reviewed in Chapter 2. In addition to basic terminology, common graph theoretic problems are discussed, including graph traversal, construction of minimum spanning trees, and graph coloring.

Graph theory is introduced in Chapter 3 as an effective tool for managing the complexity of VLSI circuits and systems. Abstraction in the IC design process is explained, and four levels of abstraction are identified; namely register transfer level, gate level, circuit level, and physical level. Applying a hierarchical model, applications of graph theory at each level of the IC design process are reviewed, including task scheduling, logic verification, circuit analysis, and wire routing.

Synthesis of clock distribution networks in VLSI circuits are reviewed in Chapter 4. The process is composed of three parts; namely, clock skew scheduling, clock tree topology, and clock tree embedding. During clock skew scheduling, the arrival time of the clock signals is determined based on the topology of a synchronous logic circuit. The clock tree topology is generated based on target clock arrival times. The clock tree layout is produced during the embedding step.

Methods for electrical circuit analysis are reviewed in Chapter 5. Modified nodal analysis (MNA) based on the Laplacian matrix of a circuit graph is the standard method for analyzing electrical networks. Due to the large size of modern VLSI networks, different techniques for accelerating MNA are presented, such as domain decomposition, geometric multigrid, and hierarchical matrix. Alternative methods unrelated to MNA also exist, such as random walks and infinite mesh graph models.

A graph theoretic approach to modeling power networks in VLSI circuits and systems is presented in Chapter 6. Due to the large size of the power grids within modern ICs, an infinite lattice graph model can be used to evaluate IR drops. The nodes of the infinite lattice graph represent the vias, and the edges represent the horizontal and vertical interconnects connecting adjacent vias. The infinite grid model however exhibits poor accuracy near the boundaries of the grid. A novel image method, introduced in this chapter, restores the accuracy of the infinite lattice model near the corners and boundaries of the grid, reducing the worst case error from 40% to 4% [71].

This graph-based image method is extended to the analysis of finite rectangular grids in Chapter 7. Based on the infinity mirror technique, the IR drop is accurately determined at arbitrary points within a lattice without considering the entire grid [72]. The infinity mirror technique therefore greatly accelerates the IR drop analysis process in large grids by exclusively determining the electric potential at a few nodes of primary interest.

A general framework for distributing on-chip voltage regulators within a power grid is presented in Chapter 8. Based on the infinity mirror technique, a computationally efficient voltage drop analysis algorithm is proposed. The runtime of the analysis methodology does not depend upon the grid dimensions, enabling the efficient analysis of arbitrarily large power grids. By supplementing discrete particle swarm optimization with fast grid analysis, placement of the distributed on-chip voltage regulators to minimize parasitic voltage drops is efficiently determined. Practical scenarios, such as limited current capacity and restricted placement of the regulators, are considered. The framework is validated on a set of industrial power grid benchmark circuits.

Optimization of power delivery at the system level is the focus of Chapter 9. A system-level power network is often modeled as a linear circuit consisting of only passive elements (e.g., resistors, inductors, and capacitors), and current and voltage sources [75]. A simulation framework for linear electrical circuits based on a state-space model is developed using a Laplacian matrix of a circuit graph. This method is particularly effective for the repetitious simulation of circuits with a constant topology and perturbed parameters. A power network optimization tool is described based on this framework which significantly improves power integrity and lowers cost by adjusting the on-chip and off-chip decoupling capacitances.

The Smart Power ROUting Tool (SPROUT) for prototyping board-level power networks is presented in Chapter 10. Based on certain layout characteristics, such as physical design rules and the location of the terminals and obstacles, a prototype layout of a board-level power network is synthesized. The layout of a power rail is initially decomposed into small rectangular cells. Each cell is converted into a node within an equivalent layout graph, and the adjacent cells are connected by edges. The current density within the power rails is minimized by evaluating the layout graph, reinforcing those regions with the greatest current density. The layouts produced using SPROUT exhibit characteristics similar to manually designed layouts. SPROUT can therefore be used to generate multiple power network prototypes for efficient design exploration.

QuCTS – single flux Quantum Clock Tree Synthesis tool – is presented in Chapter 11. Clocked gates and datapaths within a sequential circuit are represented, respectively, by nodes and edges within a timing graph. The clock arrival time is determined for each clocked gate within a single flux quantum circuit by optimizing the timing graph. From the clock arrival times, a topological graph of a binary clock tree is generated using clustering. The clock tree is embedded into a physical layout using a proxy graph technique where the nodes and edges represent, respectively, the location of the cells within the layout and the distance between these locations.

In Chapter 12, the conclusions and closing comments of this book are provided. The research presented in this book is summarized. The significance of the research discussed in this book is discussed from the perspective of theoretical graph theory and practical VLSI design issues.

Chapter 2
Graph fundamentals

Before discussing graph theory in the context of VLSI, a review of graph theory is necessary. Despite more than 250 years of development, the terminology of graph theory is not completely standardized and many fundamental concepts have multiple names. The terminology used in this chapter is largely based on books by J. A. Bondy and U. S. R. Murty [76], and D. B. West [77], widely recognized as standard sources in the research community.

A *graph* is fundamentally an ordered triple $G = (V_G, E_G, \psi_G)$, where V_G is a set of *nodes (vertices)*, E_G is a set of *edges (arcs)*, and $\psi_G : E_G \rightarrow V_G \times V_G$ is an *incidence function* mapping each edge $e_i \in E_G$ to a pair of nodes in V_G,

$$\psi_G(e) = [u, v], \tag{2.1}$$

where $u, v \in V$ are the *endpoints* of e and are *incident to* (connected by) e. Similarly, edge e is incident to nodes u and v. The sets V_G and E_G are frequently referred to as, respectively, the *node set* and *edge set* of graph G. An example of a graph is G_1, as shown in Fig. 2.1a. The node set, edge set, and incidence function of G_1 are

$$V_1 = \{v_1, v_2, v_3, v_4, v_5, v_6\}, \tag{2.2}$$

$$E_1 = \{e_1, e_2, e_3, e_4, e_5, e_6, e_7, e_8\} \tag{2.3}$$

$$\psi_1 : E_1 \rightarrow V_1 \begin{cases} \psi_1(e_1) = [v_1, v_3], & \text{(2.4a)} \\ \psi_1(e_2) = [v_3, v_3], & \text{(2.4b)} \\ \psi_1(e_3) = [v_3, v_4], & \text{(2.4c)} \\ \psi_1(e_4) = [v_1, v_5], & \text{(2.4d)} \\ \psi_1(e_5) = [v_3, v_5], & \text{(2.4e)} \\ \psi_1(e_6) = [v_3, v_6], & \text{(2.4f)} \\ \psi_1(e_7) = [v_3, v_6], & \text{(2.4g)} \\ \psi_1(e_8) = [v_1, v_6]. & \text{(2.4h)} \end{cases}$$

© The Author(s), under exclusive license to Springer Nature Switzerland AG 2023
R. Bairamkulov, E. G. Friedman, *Graphs in VLSI*,
https://doi.org/10.1007/978-3-031-11047-4_2

In G_1, nodes v_3 and v_4 are incident to edge e_3, and e_6 and e_7 are edges incident to nodes v_3 and v_6. Edges e_1 and e_2, incident to the same pair of nodes, are called *parallel* or *multiple*. *Multiplicity* $\mu(u, v)$ of nodes u and v is the number of edges connecting u and v. For example, edges e_6 and e_7 are parallel. The multiplicity of nodes v_3 and v_6 is $\mu(v_3, v_6) = 2$. Nodes connected by an edge are called *neighbors* or *adjacent*, and the set of nodes connected to node v is called a *neighborhood* and is denoted as $N(v)$. The *degree* $d(v)$ of node v is the size of the neighborhood of the node, *i.e.*, the number of edges incident to node v. For example, every node u has degree $d(u) = 3$ in graph G_2, see Fig. 2.1b. The neighborhood of node 0 is set $N(0) = \{1, 10, 19\}$. A node with degree zero is called *isolated* (e.g., node v_2 in G_1).

An edge incident to two distinct nodes is called a *link*, while an edge connecting a single node with itself is called a *loop* or *self-loop*. Edge e_1 in G_1 is incident to

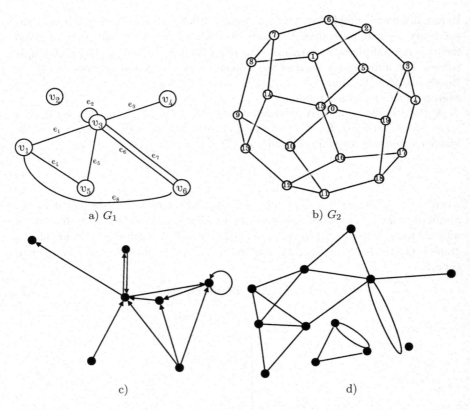

a) G_1 b) G_2

c)   d)

Fig. 2.1 Examples of graphs. a) A generic graph with order six and size eight. v_2 is an isolated node with degree zero. Edge e_1 is the link connecting two distinct nodes. Edge e_2 is the loop connecting node v_3 with itself. e_6 and e_7 are parallel edges connecting the same pair of nodes, v_3 and v_6. b) Dodecahedral graph. The nodes and edges represent, respectively, the vertices and edges of a regular dodecahedron (platonic solid). c) Directed graph with self-loops, and d) undirected multigraph.

nodes v_1 and v_3 and is therefore a link. Edge e_2 is an example of a loop connecting node v_3 with itself. The *order* of a graph G refers to the number of nodes within graph G, *i.e.*, the cardinality of node set $|V_G|$. Similarly, the *size* of G is the number of edges, *i.e.*, the cardinality of edge set $|E_G|$. Observe that

$$\sum_{v \in V_G} deg(v) = 2|E_G|, \tag{2.5}$$

since two nodes exist for every edge. Since the number $2|E_G|$ is even, the number of edges with an odd degree in a graph is inevitably even. This consequence of (2.5), called a *handshaking lemma*, was proven by Leonhard Euler in 1736 and is considered the first proof in graph theory [78].

2.1 Graph categories

There are several fundamental graph categories in graph theory that significantly affect the properties of a graph. A diagram classifying graphs into topological categories is shown in Fig. 2.2. Based on the edge properties, each category of graphs can be further subdivided into subcategories; namely, weighted graphs and directed graphs. In this section, the basic properties of graphs within each of these categories are described.

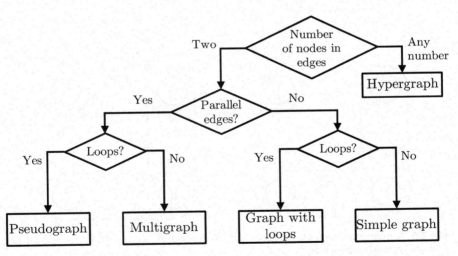

Fig. 2.2 Topological classification of a graph. A hypergraph is a superclass of graphs where the edges can connect an arbitrary number of nodes. Only two nodes (not necessarily distinct) can be connected with an edge in a graph. Depending on whether parallel edges and loops exist, a graph can be classified as a pseudograph, multigraph, graph with loops, or simple graph.

2.1.1 Hypergraph

A *hypergraph* is the superclass of a graph (*i.e.*, every graph is a hypergraph, but not every hypergraph is a graph). A hypergraph $H = (V_H, E_H, \psi_H)$ is an ordered triple, where V_H is the node set, E_H is the set of *hyperedges*, and

$$\psi_H : E_H \rightarrow \mathcal{P}(V_H) \setminus \{\varnothing\} \tag{2.6}$$

is the incidence function mapping each hyperedge $e \in E_H$ to a subset of nodes, and $\mathcal{P}(V_H)$ is the power set (set of all subsets) of node set V_H.

In a hypergraph, an arbitrary (nonzero) subset of nodes $V_e \subseteq V_H$ can be connected with a hyperedge e, as illustrated in Fig. 2.3. Hypergraphs are frequently found in modeling "multi-adic" relationships, where the relationship is not limited to only two objects [79]. In [80], for example, hypergraphs are used to model cellular mobile communications systems, where hyperedges represent the interference between cellular stations. Hypergraphs appear in the physical design of VLSI circuits, where multiple nodes are connected [81], as illustrated in Fig. 2.4. Hypergraphs are widely used in computational biology [82, 83], telecommunications [80, 84], image processing [85], and artificial intelligence [86].

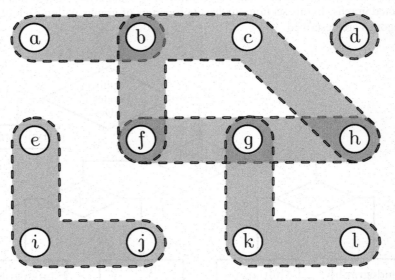

Fig. 2.3 A hypergraph with twelve nodes and six hyperedges, namely $\{a, b\}$, $\{b, c, f, h\}$, $\{d\}$, $\{e, i, j\}$, $\{f, g, h\}$, and $\{j, k, l\}$.

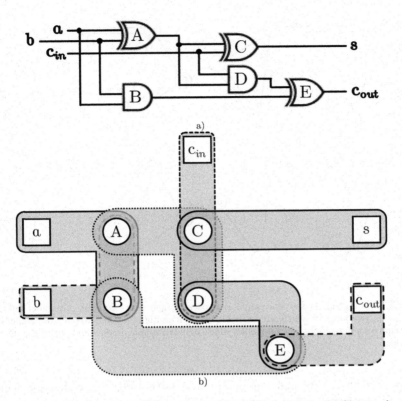

Fig. 2.4 A logic circuit converted into a hypergraph. a) The initial circuit. Observe the wires connecting more than two gates. b) Equivalent hypergraph. The nodes represent the gates and terminals. Eight hyperedges exist within the hypergraph, representing the wires connecting the gates.

2.1.2 Graphs with parallel edges

In a class of graphs, the edges are restricted to connecting only two nodes, not necessarily distinct nodes. Depending upon the existence or absence of parallel edges and self-loops, a graph can be a pseudograph, multigraph, simple graph, or a graph with self-loops. The *pseudograph* class is the least restrictive, permitting both parallel edges and loops. An example of a pseudograph is shown in Fig. 2.5a, where node v_5 contains two parallel loops, and two pairs of nodes are connected with parallel edges, namely, $[v_2, v_3]$, and $[v_4, v_6]$. Graph G_1, shown in Fig. 2.1a, is also a pseudograph, since loop e_2 and parallel edges e_6 and e_7 are found in the graph. Applications of pseudographs are used to model molecular structures of chemical compounds [87] and artificial intelligence [88].

Depending upon the application, self-loops may not occur in a graph. For example, self-loops frequently occur in finite state machines (FSM) [89], but are

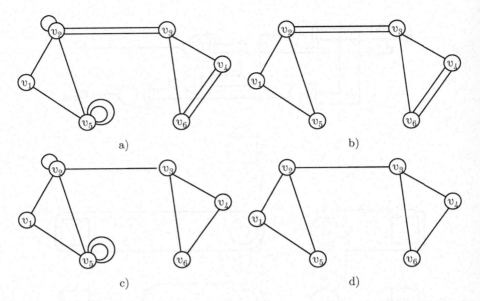

Fig. 2.5 Topological classes of a graph. a) A pseudograph with two pairs of parallel links, and three loops, two loops of which are parallel loops. b) a multigraph with two pairs of parallel links, c) a graph with three loops and no parallel edges, and d) a simple graph with no parallel edges and no loops.

rarely encountered in modeling automotive traffic [90]. A pseudograph without loops is commonly called a *multigraph*. The edges incident to the same pair of nodes are called *parallel* or *multiple* edges. An example of a multigraph is shown in Fig. 2.5b. Notably, the diagram of the Königsberg bridges [4], the first graph in the history of graph theory, is a multigraph, since multiple bridges connect the same pair of landmasses (see Figs. 1.1 and 2.6).

2.1.3   *Graphs without parallel edges*

Many applications do not permit multiple edges connecting the same nodes. Unlike multigraphs, however, pseudographs without parallel edges have not been assigned a common name. If no self-loops are permitted and any two edges are connected with at most one edge, the graph is called a *simple graph*. For example, the graph illustrated in Fig. 2.5d is simple, while the graph shown in Fig. 2.5c is not simple, since several edges form loops. An edge connecting nodes u and v within a simple graph can be unambiguously represented as a set of vertices $\{u, v\}$. A simple graph is therefore often defined as an ordered pair $G = (V_G, E_G)$, where $E_G \subseteq \binom{V_G}{2}$ and $\binom{V_G}{2}$ is the set of unordered pairs of elements of V_G. The maximum size of a simple graph G is

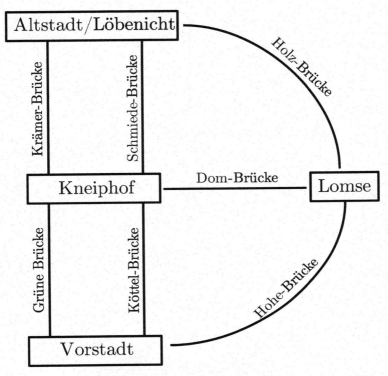

Fig. 2.6 A multigraph representing the seven bridges of Königsberg [4]. Four landmasses, represented by rectangles, are connected by seven bridges, represented by edges. Two pairs of parallel edges are formed by bridges, Krämer and Schmiede, and Grüne and Kötten.

$$\binom{|V_G|}{2} = \frac{|V_G|(|V_G| - 1)}{2}. \tag{2.7}$$

A simple graph with n nodes and a maximum number of edges is called a *complete* graph K_n. In a complete graph, every edge is connected to all other edges. A variety of examples of a complete graph is shown in Fig. 2.7.

2.1.4 Weighted graph

Graph systems modeling practical networks often require additional information describing objects and connections. A *weight* $w(e)$ of edge e is commonly used to quantitatively characterize a connection between nodes. In electrical circuits, for example, edge weights often represent a wire conductance. A graph with weighted edges is called a *weighted* graph or, more specifically, an *edge-weighted* graph.

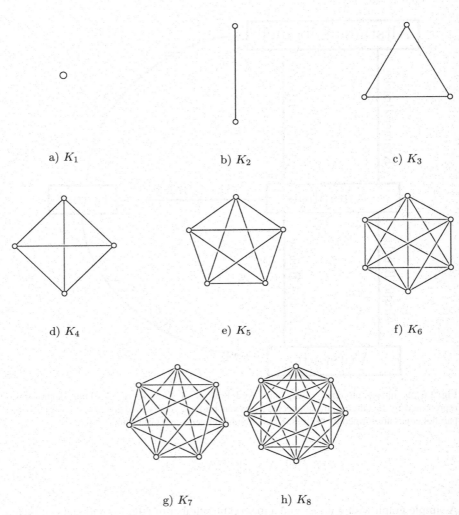

Fig. 2.7 Complete graphs K_n for $n \in [1, 8]$. a) A *trivial* graph K_1 with one node and no edges, b) a line graph K_2 with two nodes and one edge, c) a smallest cycle graph K_3 with three nodes and three edges, d) tetrahedral graph K_4 with four nodes and six edges, e) K_5 with five nodes and ten edges, f) K_6 with six nodes and 15 edges, g) K_7 with seven nodes and 21 edges, and h) K_8 with seven nodes and 28 edges.

Recall that the degree $d(v)$ of node v within an unweighted undirected graph $G_u = (V_u, E_u, \psi_u)$ is equal to the number of edges incident to a node. A similar measure is defined for a weighted graph $G_w = (V_w, E_w, \psi_w)$. The *strength* (or *weighted degree*) of node $u \in V_w$ is the sum of the weights of the edges incident to u,

$$s(u) = \sum_{e \in E_G | u \in \psi_G(e)} w(e). \tag{2.8}$$

Based on (2.8), an unweighted graph can be considered a weighted graph with all degrees equal to 1. A simple edge-weighted graph G is often defined as an ordered triple (V_G, E_G, w), where $w : E_G \rightarrow \mathbb{R}$ is the weight function assigning a weight to each edge within a network. A *node-weighted* graph is a graph whose nodes are assigned weights. Node-weighted graphs are less prevalent than edge-weighted graphs but are encountered in medical imaging [91], layout synthesis [65], routing in field programmable gate arrays (FPGA) [92], and cloud computing [93].

2.1.5   Directed graph

A graph whose edges are oriented is called a *directed* graph or *digraph*. An edge in a digraph is often represented by an ordered pair (u, v) such that $E_G \subseteq V_G \times V_G$. Note that $(u, v) \neq (v, u)$, since (u, v) and (v, u) are of opposite direction. An example of a directed graph is G_d, as shown in Fig. 2.8a. An edge (u, v) is incident *from* node u *to* node v. u and v are *consecutive nodes* and are called, respectively, the *tail* and *head* of an edge (u, v). Observe that nodes c and h in G_d are consecutive, since an edge is incident from c to h. u is a *direct predecessor* of v, and v is a *direct successor* of u. *Consecutive edges* are a pair of edges e_1 and e_2 sharing node v such that v is the head of e_1 and the tail of e_2. The number of edges incident to node u (*i.e.*, the number of edges for which node u is a head) is called an *indegree* $d_{in}(u)$. Similarly, *outdegree* $d_{out}(u)$ is the number of edges incident from u (*i.e.*, the number of edges for which node u is the tail). For example, the indegree of node g in G_d is $d_{in}(g) = 2$, since two edges are incident to g, a link from node e and a loop. The outdegree $d_{out}(g) = 1$, since only a single loop is incident from node g. Observe that

$$\sum_{v \in V_G} d_{in}(v) = \sum_{v \in V_G} d_{out}(v) = |E_G|, \tag{2.9}$$

since a tail exists for every head of an edge. The node with zero indegree is called a *source*. Similarly, a node with zero outdegree is called a *sink*. In Fig. 2.8a, for example, node c is a source, and node i is a sink.

2.2   Inter-graph relationships

A directed graph G is produced by *orienting* (assigning a direction) to each edge of an undirected graph G_u. G is therefore called the *orientation* of G_u. Conversely, G_u is called the *underlying graph* of G. Graphs G_d and G_u, depicted in Fig. 2.8,

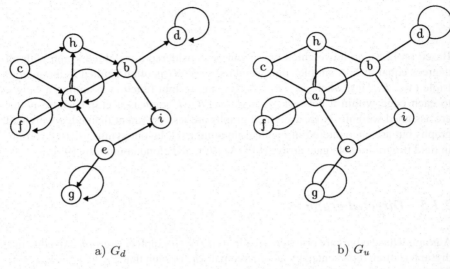

a) G_d b) G_u

Fig. 2.8 An example of a graph orientation and the underlying graph. a) Directed graph G_d, and b) underlying graph G_u.

are examples of, respectively, an orientation and an underlying graph. Similar to directed graphs, a multigraph with self-loops M can be converted into a simple graph G_u by removing the loops and replacing multiple edges with a single edge. For example, the simple graph shown in Fig. 2.5d is an underlying graph for the graphs shown in Figs. 2.5a to 2.5c.

Simple graphs G and H are *isomorphic* if there exists a bijection,

$$f : V_G \rightarrow V_H, \tag{2.10}$$

such that

$$(u, v) \in E_G \Longleftrightarrow (f(u), f(v)) \in E_H. \tag{2.11}$$

Map f is called *isomorphism*, and graphs G and H are called *isomorphic*, denoted as $G \cong H$. Consider graphs G and H depicted in Fig. 2.9. These graphs are isomorphic since there exists an isomorphism $f : V_G \rightarrow V_H$ such that edge $(u, v) \in E_G$ is mapped to edge $(f(u), f(v)) \in E_H$. Note that the direction of the edges is preserved in isomorphic directed graphs.

A graph $H = (V_H, E_H, \psi_H)$ is called a *subgraph* of $G = (V_G, E_G, \psi_G)$ if $V_H \subseteq V_G$, $E_H \subseteq E_G$, and $\psi_H(e) = \psi_G(e) \forall e \in E_H$. Conversely, graph G is called a *supergraph* of H. If edge set E_H of a subgraph includes all edges where both endpoints are in V_H, i.e.,

$$E_H = \{e | \psi_H(e) \subseteq V_H\}, \tag{2.12}$$

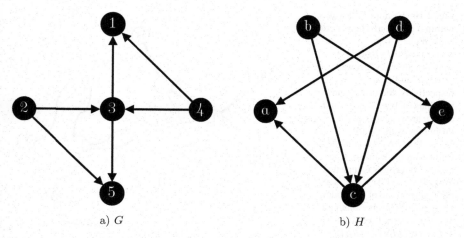

Fig. 2.9 Isomorphic graphs; a) G and b) H. Map $f : [1 \rightarrow a, \ldots, 5 \rightarrow e]$ is the isomorphism of graphs G and H, since for any edge $(u, v) \in E_G$, a unique edge $(f(u), f(v)) \in E_H$ exists. For example, edge $(4, 1) \in E_G$ is mapped to edge $(d, a) \in E_H$.

subgraph H is called an *induced subgraph* $G[V_H]$ or a subgraph induced by V_H. An induced subgraph is produced when a set of nodes V_r is removed from the node set such that $V_H = V_G \setminus V_r$. The edges incident to the nodes in V_r are removed from subgraph H.

2.3 Graph exploration

A sequence of alternating vertices and edges $W = [v_0, e_1, v_1, ..., e_k, v_k]$, where e_i is incident to v_{i-1} and v_i for $i \in [1 \ldots k]$, is called a *walk*. k denotes the number of edges within a walk and is called the *length* of a walk. The first node v_0 and the last node v_k in a walk are called, respectively, the *origin* and *terminus*. In a simple graph, a walk can be uniquely determined by the node sequence $[v_0, v_1, \ldots, v_{k-1}, v_k]$. A walk is called a *trail* if no edge occurs more than once. A trail is called a *simple path* if all of the nodes within a trail are distinct. Examples of a walk, trail, and path are depicted in Fig. 2.10a. Walk W_2 does not contain repeated edges, and is, therefore, a trail. Walk W_3 does not contain repeated nodes, and is therefore a path. A trail whose origin and terminus are the same node is called a *circuit*. A *cycle* is a type of circuit where no node occurs twice, not counting the origin. A cycle traversing an entire node set is called a *Hamiltonian cycle*. Example circuits are shown in Fig. 2.10b. Circuit W_4 is not a cycle, since node O occurs twice during the traversal (the origin is not counted).

Nodes u and v are *connected* if there exists a path from u to v. A graph where any pair of nodes is connected is called a *connected graph*. Conversely, a graph is

Fig. 2.10 Examples of walks within a graph. a) The generic walk $W_1 = YpOoUvTvUyBzE$ is shown with wavy arrows. Nodes Y and E are, respectively, the origin and terminus of W_1. Node U and edge v are repeated twice during the walk. Walk $W_2 = WaIbNcNdD$ is the trail (solid thick lines), since none of the edges is repeated. Walk $W_3 = SkArVwE$ is a path, since none of the nodes is repeated. Observe that path W_3 is also a trail. b) Walk $W_4 = OeDdNfOgRhFiO$ is a circuit since the origin and terminus of W_4 is the same node (O). Walk $W_5 = SjInTuHxEwVrAkS$ is a cycle since W_5 is a circuit with no repeated nodes.

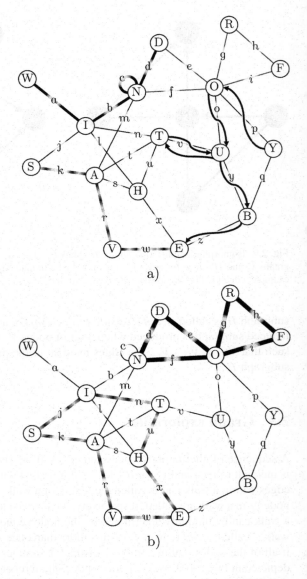

a)

b)

disconnected if a pair of disconnected nodes exists within the node set. The node set V_G of a disconnected graph G can be partitioned into multiple disjoint subsets $V_1, V_2, \ldots, V_n$ such that for $i, j \in [1, n]$, nodes $u \in V_i$ and $v \in V_j$ are connected if $i = j$ and disconnected otherwise. Subgraphs $G[V_1], G[V_2], \ldots, G[V_n]$ induced by these sets are called *connected components*.

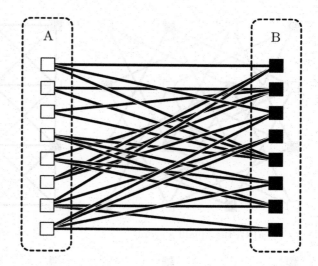

Fig. 2.11 A bipartite graph consisting of sets A and B. All edges include one endpoint in A and another endpoint in B. None of the edges connects the nodes within the same partition.

2.4   Bipartite graph

Graph G is called bipartite if the node set V_G can be split into two disjoint subsets $A \cap B = \varnothing$, $A \subset V_G$ and $B \subset V_G$, $A \cup B = V_G$, such that any edge has endpoints in both A and B, i.e., $(u, v) \in E_G$, $u \in A$, and $v \in B$. Sets A and B are called *bipartitions* of graph G, as depicted in Fig. 2.11. No nodes within the same partition are adjacent in bipartite graphs. Consider graphs G and H shown in Figs. 2.12a and 2.12b. Graph G is called a Knight's graph for a 4×4 chessboard, where the nodes represent squares on a chessboard and the edges represent the legal moves of a knight. Graph G is bipartite, since the knight's move always connects a black square with a white square, as illustrated in Fig. 2.12c. In fact, the graph shown in Fig. 2.11 is isomorphic to G, and sets A and B correspond to white and black squares on a 4×4 chessboard. Graph H is called a king's graph for a 4×4 chessboard, where the edges represent the legal moves of a king, as shown in Fig. 2.12d. The king's graph is not bipartite. Consider the cycle $[a, b, c]$. Any partition will contain edge $\{a, b\}$, $\{b, c\}$, or $\{a, c\}$ connecting nodes within the same partition. In general, a graph is bipartite if and only if no odd length cycle exists within the graph.

2.5   Directed acyclic graph

A graph G is called a *directed acyclic graph (DAG)* if no directed cycles exist within G. Consider directed graphs G_1 and G_2, as shown in Fig. 2.13a. Graph G_1 contains two directed cycles, namely, $[a, b, d, c]$ and $[e, c, d]$, and is therefore not a DAG. Reversing the edge (d, c) in graph G_1 produces graph G_2, as shown in Fig. 2.13b. No directed cycles exist within G_2, hence G_2 is a DAG. If a path from node u to

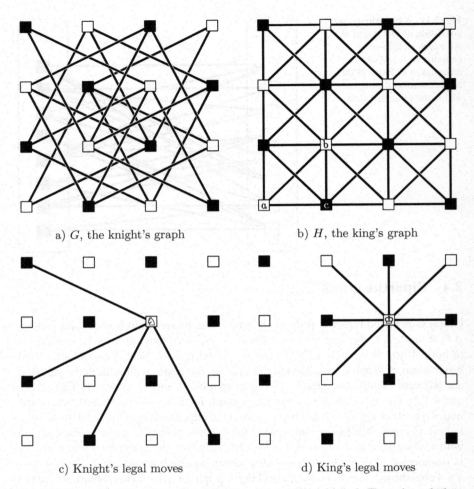

a) G, the knight's graph   b) H, the king's graph

c) Knight's legal moves   d) King's legal moves

Fig. 2.12 Examples of bipartite and non-bipartite graphs. a) Knight's graph. The nodes and edges represent, respectively, the chessboard squares and valid knight's moves. b) King's graph. The edges represent valid king's moves within the chessboard. c) Valid moves of a knight, and d) valid moves of a king.

node v exists in a DAG, u is an *ancestor* of v and v is a *descendant* of u. In G_2, for example, node d is a descendant of a and e, while every node except g is an ancestor of g.

The primary feature of a DAG is the existence of a mapping $f : V_G \rightarrow [1, \ldots, |V_G|]$, where

$$(u, v) \in E_G \iff f(u) < f(v). \tag{2.13}$$

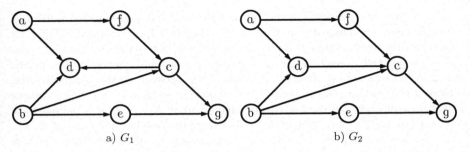

Fig. 2.13 Directed graphs with and without directed cycles. a) Directed graph G_1 with cycles $[a, b, d, c]$ and $[e, c, d]$, and b) directed acyclic graph (DAG) G_2.

Fig. 2.14 Topological orderings of DAG G_2 (see Fig. 2.13b). In topological ordering, any ancestor of node u in a DAG appears before u, while the descendants of u appear after u.

Topological ordering						
1	2	3	4	5	6	7
e	f	a	c	b	d	g
e	a	f	b	c	d	g
a	b	e	f	c	d	g
a	e	b	c	d	f	g

Mapping f is called a *topological sorting* or *topological ordering* of graph G. Topological ordering is generally not unique [94]. Any of the topological orderings of G_2 shown in Fig. 2.14 satisfy (2.13). DAGs naturally occur in systems that prohibit cyclic relationships, including combinatorial logic [95], artificial neural networks [96], task scheduling [97], and the analysis of influences in social networks [98].

2.6 Tree

A connected undirected simple graph with no cycles is called a *tree* $T = (V_T, E_T)$. The number of edges within a tree is always $|E_T| = |V_T| - 1$. A variety of examples of a tree is shown in Fig. 2.15. Any two nodes within a tree are connected by a unique path. Conversely, if more than one path exists between a pair of nodes, the graph is not a tree. A *forest* is a simple graph whose connected components are trees. Removing any single edge from a tree produces a forest (a disconnected graph with no cycles). Adding an edge $\{u, v\}$ to a tree produces a cycle containing this edge.

A *rooted tree* is the orientation of a tree, where one node is designated as the *root*, and the edges are directed from the root. A rooted tree T is illustrated in Fig. 2.16. Observe that the direction of each edge is uniquely determined by the root, since only a single path exists between a root and an arbitrary node. Several terms specific to a rooted tree exist to describe the relationship between nodes in a rooted tree. Node u is called a *parent* or *predecessor* of node v, and node v is called a *child* or *successor* of node u if there exists an edge $(u, v) \in E_T$. In T, a is the parent of c, and h is a child of d. Any node in V_T except the root has a single parent, *i.e.*, the indegree of any non-root node is 1. Nodes v_1 and v_2 are called *siblings* if these nodes, v_1 and v_2, have the same parent. For example, d and e are siblings since both of these nodes have the same parent b. Node u is a *leaf* if u has no children, otherwise u is called an *internal node*. T has six leaves, namely, g, h, i, j, k, and l.

Node u is called an *ancestor* of v, and node v is called a *descendant* of u if there exists a path connecting u to v. The number of ancestors of node v is called the *level* of v. Nodes b and d in T are both ancestors of node h and are both descendants of node a. The *level* of a node u in a rooted tree denotes the distance from the root to u and is equal to the number of ancestors of u. The root node is level 0. The maximum level of any node in V_T is called the *height h* of a tree. The height $h(T)$ of T is three, since the maximum level of a leaf in T is three. If the level of the leaves is either $h - 1$ or h, the tree is called *balanced*. T is balanced since the minimum level of a

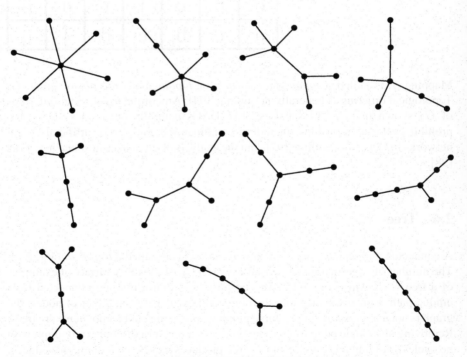

Fig. 2.15 Eleven possible non-isomorphic trees with seven nodes.

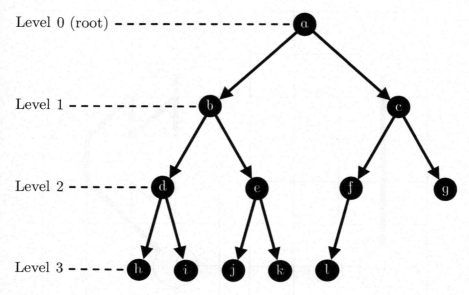

Fig. 2.16 An example of a complete balanced rooted tree with height $h = 3$. a is the root node. All edges are oriented away from the root.

leaf in T is two. If the maximum outdegree of a node within a rooted tree is m, the tree is called *m-ary*. A *full* m-ary tree is a tree whose internal nodes all have either 0 or m children. A *complete* m-ary tree is a balanced tree whose internal levels are all filled. The leaves in a complete m-ary tree are arranged to ensure that the leftmost node is filled first. T is a *binary* tree since the maximum number of children at any node is two. T is not full but complete, since all internal levels of T (levels 0 to 2) are filled, and the leaves within the last layer are arranged from left to right.

2.7   Common problems in graph theory

Graph theory is found in many practical applications in mathematics, physics, chemistry, and engineering. Different kinds of relationships between objects can be represented with nodes and edges. In telecommunication network models, for example, the edges represent physical routing channels, such as wired or wireless media. In graph-based register allocation, the edges represent the relationship between the data stored in the registers. Many of these problems, such as a Steiner minimum tree, exhibit high computational complexity [99], making the solution of these problems impractical if the graph size is sufficiently large.

Heuristic methods are commonly used to partially overcome this limitation. With heuristics, a solution to a computationally complex problem is approximated using a simpler method. For example, the shortest path between two nodes within a

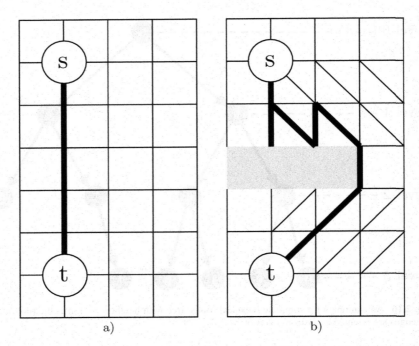

Fig. 2.17 Example of heuristics in graph pathfinding. a) Path finding in a grid graph. To determine the shortest path, the Euclidean distance from the target node is used as a heuristic. The shortest path between nodes s and t is efficiently determined by traversing only five edges. b) With the same heuristic, a suboptimal path is found due to the presence of an obstacle.

graph can be efficiently approximated using a heuristic, the distance to a target (see Fig. 2.17a). The use of heuristics, however, does not guarantee the optimal solution, as illustrated in Figs. 2.17b. Efficient and accurate heuristics are therefore a highly important objective to produce high quality solutions in practical time.

Several fundamental problems in graph theory are discussed in this section, namely, pathfinding, spanning tree construction, and graph coloring. Solutions of these problems are adapted to a wide range of practical applications. Pathfinding algorithms, for example, are often used to determine the fastest route through a communications network. In addition, the basic algorithms discussed here often form a basis for more complex algorithms. For example, a routing algorithm for wireless networks, described in [80], combines graph coloring and pathfinding algorithms. In [66], the pathfinding algorithms are used to determine the shortest balanced path between a splitter and clocked gate. Pathfinding algorithms are discussed in Subsection 2.7.1. Spanning trees and Steiner trees are introduced in Subsection 2.7.2. Graph coloring is described in Subsection 2.7.3. Topological sorting is described in Subsection 2.7.4.

2.7.1 Pathfinding

Finding paths within a network is one of the oldest problems in graph theory. The first work in graph theory, Euler's solution of Seven Bridges of Königsberg, is, to a great extent, a path finding problem. Graph *traversal* is the task of visiting every node within a node set and is widely used in path finding. Traversal algorithms are discussed in this subsection.

2.7.1.1 Depth-first search

The problem of finding a shortest path within a graph is commonly encountered in many applications, ranging from transportation networks to interconnect synthesis in microelectronic systems [100–102]. A depth-first search (DFS) is the oldest algorithm for path finding within a graph [103]. Application of the algorithm on an example graph is illustrated in Fig. 2.18a. An arbitrary node u is initially selected as a source and all other vertices within the graph are marked as not discovered. During each iteration, the DFS algorithm advances to the next node v selected among the undiscovered neighbors of current node u. If all neighbors of a current node are discovered, the algorithm returns to the predecessor node.

DFS was first published in the 19th century by Charles Pierre Trémaux [104]. A computer version of DFS was described by Tarjan in 1972 [105]. A stack data structure is commonly used in DFS [106]. The stack is a *Last-In, First-Out* (LIFO) structure [107]. The datum placed into a stack earliest is removed last. Any datum placed into a stack is placed on top of the other data. This operation is called push and is illustrated in the first two columns of Fig. 2.19. Similarly, retrieval of only the latest datum is possible, using a pop operation. A stack-based DFS is illustrated in Fig. 2.18b. The stack initially consists of only the root node. During each iteration, an unvisited neighbor of the top node is added to the stack. If all of the neighbors are visited, the top node is removed from the stack. In a finite connected graph, DFS is guaranteed to find a path from the source to an arbitrary node in $O(|V| + |E|)$ time [105]. The maximum worst case size of a stack is $|V|$. The path, however, is not guaranteed to be the shortest path. Furthermore, if the graph is infinite, DFS may fail to find a path even if the path exists [108].

2.7.1.2 Breadth-first search

Breadth-first search (BFS) was first published in 1959 by Edward F. Moore as a method for finding the shortest path out of a maze [109]. BFS is a fundamental algorithm for shortest path discovery within an unweighted graph. Those nodes closest to the source node are traversed first, ensuring that the first discovered path is the shortest path. A queue, another fundamental data structure, is commonly used in BFS. A queue is commonly referred to as a First In, First Out (FIFO) data structure,

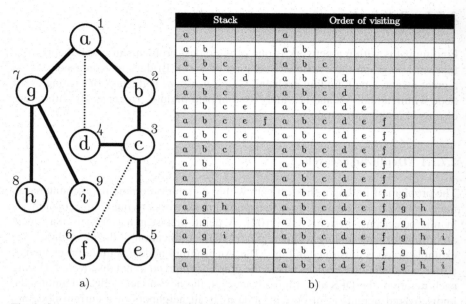

Stack					Order of visiting								
a					a								
a	b				a	b							
a	b	c			a	b	c						
a	b	c	d		a	b	c	d					
a	b	c			a	b	c	d					
a	b	c	e		a	b	c	d	e				
a	b	c	e	f	a	b	c	d	e	f			
a	b	c	e		a	b	c	d	e	f			
a	b	c			a	b	c	d	e	f			
a	b				a	b	c	d	e	f			
a					a	b	c	d	e	f			
a	g				a	b	c	d	e	f	g		
a	g	h			a	b	c	d	e	f	g	h	
a	g				a	b	c	d	e	f	g	h	
a	g	i			a	b	c	d	e	f	g	h	i
a	g				a	b	c	d	e	f	g	h	i
a					a	b	c	d	e	f	g	h	i

a)   b)

Fig. 2.18 An example of Depth-First Search (DFS). a) The traversed graph. The numbers indicate the order of traversal. The thick solid and thin dashed lines denote the traversed and non-traversed edges. b) DFS using a stack data structure. Unvisited neighbors of the top (rightmost) node in the stack are traversed. The new nodes are placed at the top of the stack (right), *i.e.*, the neighborhood of the nodes added last are traversed first. Once all of the neighbors of a node are visited, the node is removed from the stack. The algorithm is terminated after the last entry is removed.

Stack operations

Push	Push	Pop	Push	Push	Pop	Pop	Pop
				c			
	b			b	b	b	
a	a	a	a	a	a	a	

Fig. 2.19 Basic stack operations. An element is placed on top of the stack by using a push operation. The top element is removed from the stack by using a pop operation. This data structure is commonly called Last-In, First-Out (LIFO), where the last added element is removed first.

where the oldest entries are removed first [107]. Two queue operations are important in a BFS, namely, enqueue and dequeue. The entries are placed into a queue using the enqueue operation. A new entry becomes the latest (leftmost) in the queue, as illustrated in the first two columns in Fig. 2.20. Using the dequeue operation, the oldest entry within the queue can be removed while returning the value of the entry. A version of BFS using a queue is shown in Fig. 2.21b. The source node is initially pushed into the queue, and all nodes except the source node are

Queue operations							
E	E	D	E	E	D	D	D
				b			
	a			b	c	c	
a	b	b	c	d	d	d	

Fig. 2.20 Basic queue operations. An element is placed at the end of the queue with the enqueue operation, denoted here as E. The first (top) element is removed from the stack with the dequeue operation, denoted here as D. This data structure is commonly called First-In, First-Out (FIFO), where the last added element is removed last.

marked as unvisited. During each iteration, unvisited neighbors of the oldest node in the queue are pushed into the queue and marked as visited. Once all neighbors of the oldest node are marked, the node is removed from the queue. If a path to a specific target node is required, the algorithm continues until the target node is found. In an unweighted connected graph, a single-source shortest path, *i.e.*, the shortest path from the source node to all other nodes, can be discovered using BFS. This output is commonly called a shortest path tree, as illustrated in Fig. 2.21a. While finding the single source shortest path, the algorithm continues until the queue is empty, indicating that all nodes within the connected component of a graph are marked as visited.

The major advantage of BFS over DFS is the guaranteed discovery of a shortest path from the root to any other node in an unweighted graph [107]. Using BFS, if node v is located farther from the source node than node u, node v cannot be discovered before node u. For example, the length of the path from a to d is two when discovered using DFS (see Fig. 2.18a), and one when discovered using BFS (see Fig. 2.21a). Application of the queue algorithm to finding the shortest path within a finite graph requires at most $O(|V| + |E|)$ time [107], since every node and edge are checked while the size of the queue is at most $|V|$. Weighted graphs, however, require more advanced methods for shortest path discovery. Consider, for example, the graph shown in Fig. 2.22a. Using BFS, the shortest path from a to d is $[a, d]$ with total weight 9. A shorter path $[a, b, c, d]$ is however available with total weight 8. The Bellman-Ford [110–112] and Dijkstra's [113] algorithms are two of the oldest algorithms for finding the shortest path in a weighted graph.

2.7.1.3   Dijkstra's algorithm

The Dijkstra's algorithm can be viewed as a greedy expansion process, where the paths with the least cost are expanded. The algorithm was developed in 1956 by Edsger W. Dijkstra to identify the shortest path between two nodes within a weighted graph [113]. The algorithm is often extended to finding the single source

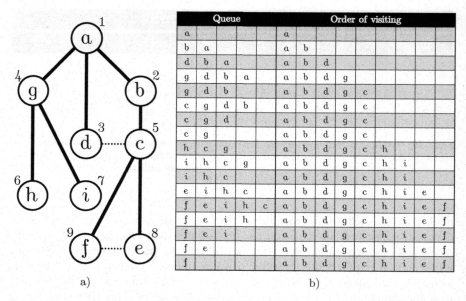

Queue				Order of visiting								
a				a								
b	a			a	b							
d	b	a		a	b	d						
g	d	b	a	a	b	d	g					
g	d	b		a	b	d	g	c				
c	g	d	b	a	b	d	g	c				
c	g	d		a	b	d	g	c				
c	g			a	b	d	g	c				
h	c	g		a	b	d	g	c	h			
i	h	c	g	a	b	d	g	c	h	i		
i	h	c		a	b	d	g	c	h	i		
e	i	h	c	a	b	d	g	c	h	i	e	
f	e	i	h	c	a	b	d	g	c	h	i	e	f
f	e	i	h	a	b	d	g	c	h	i	e	f
f	e	i		a	b	d	g	c	h	i	e	f
f	e			a	b	d	g	c	h	i	e	f
f				a	b	d	g	c	h	i	e	f

a)   b)

Fig. 2.21 An example of Breadth-First Search (BFS). a) The traversed graph. The numbers indicate the order of traversal. The thick solid and thin dashed lines denote the traversed and non-traversed edges. b) BFS using a queue data structure (First-In, First-Out, FIFO). Unvisited neighbors of the rightmost node in the queue are traversed. New nodes are placed at the (left) end of the queue. Once all of the neighbors of a node are visited, the node is removed from the queue. The algorithm is terminated after the last entry is removed.

shortest path. Each node u within a graph (except the source node) is assigned two attributes; namely, tentative cost and predecessor [114]. The tentative cost specifies the smallest known cost to reach node u starting from the source. The predecessor specifies node v preceding node u along the shortest known path.

An example illustrating the Dijkstra's algorithm is shown in Fig. 2.22b. An arbitrary source node s is initially selected as current node u. The set of unvisited nodes is set to $V \setminus \{s\}$. The cost of reaching s is set to zero, while the cost of reaching the other nodes is initially set to infinity. During each iteration, the unvisited neighbors of current node u are explored. If the cost of reaching node $v \in N(u)$ from current node u is smaller than the smallest known cost c_v of reaching node v, the cost is updated,

$$c_v \leftarrow \min (c_v, c_u + w_{uv}), \tag{2.14}$$

where c_u is the cost of reaching current node u (cost attribute), and w_{uv} is the weight of an edge connecting u and v. If cost c_v is updated, the predecessor attribute of node v is changed to u, indicating that the shortest path from s to v is composed of the shortest path from s to u followed by a transition from u to v. The shortest path is determined by reconstructing the path from the target node using the predecessor

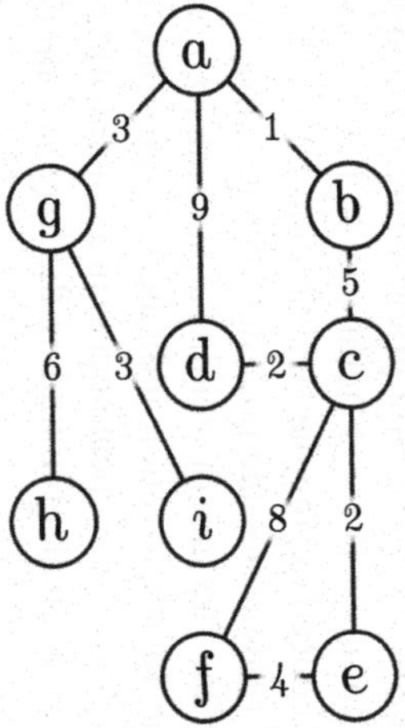

Iteration	Current node	a		b		c		d		e		f		g		h		i	
		Pred	Cost	Pred	Cost	Pred	Cost	Pred	Cost	Pred	Cost	Pred	Cost	Pred	Cost	Pred	Cost	Pred	Cost
1	a	None	0	?	∞	?	∞	?	∞	?	∞	?	∞	?	∞	?	∞	?	∞
2	b	None	0	a	1	?	∞	a	9	?	∞	?	∞	a	3	?	∞	?	∞
3	g	None	0	a	1	b	6	a	9	?	∞	?	∞	a	3	?	∞	?	∞
4	c	None	0	a	1	b	6	a	9	?	∞	?	∞	a	3	g	9	g	6
5	i	None	0	a	1	b	6	c	8	c	8	c	14	a	3	g	9	g	6
6	d	None	0	a	1	b	6	c	8	c	8	c	14	a	3	g	9	g	6
7	e	None	0	a	1	b	6	c	8	c	8	c	14	a	3	g	9	g	6
8	h	None	0	a	1	b	6	c	8	c	8	c	12	a	3	g	9	g	6
9	f	None	0	a	1	b	6	c	8	c	8	c	12	a	3	g	9	g	6
Result		None	0	a	1	b	6	c	8	c	8	c	12	a	3	g	9	g	6

b)

Fig. 2.22   An example of the Dijkstra's algorithm. a) The traversed graph. The numbers indicate the edge weights. b) Order of traversal. During each iteration, the neighborhood of a current node is explored. If a shorter path is determined, the cost and predecessor of a node are updated, as shown in light gray in the table. The node with the smallest cost is selected as the current node, and the cost of this node is not changed in subsequent iterations.

attribute. An unvisited node with the smallest tentative cost is selected as the current node and is marked as visited.

The performance of the Dijkstra's algorithm greatly depends upon the implementation. The original Dijkstra's algorithm requires time $(|V|^2)$ to find the shortest path to every node within a graph. Note that $(|V|^2)$ is also the worst case complexity for finding the shortest path to a single target. Using specialized data structures, such as heaps and priority queues, the algorithm can be accelerated to $O((|V|+|E|)\log|V|)$ [115] and $O(|E|+|V|\log|V|)$ [116].

2.7.1.4   Bellman-Ford

A major limitation of Dijkstra's algorithm is the inapplicability to directed graphs with negative-weight edges. Consider the example depicted in Fig. 2.23. The shortest path to node c estimated by the Dijkstra's algorithm is $[a, b, c]$ with cost 6. A shorter path $[a, d, c]$, however, exists with cost 0. The Bellman-Ford (BF) algorithm, developed independently by Shimbel in 1954 [110], Ford in 1956 [111], and Bellman in 1958 [112], utilizes an alternative approach that enables the analysis of graphs with negative edge weights.

The primary output of the BF algorithm is the shortest path to every node within a graph. An example of the BF algorithm is illustrated in Fig. 2.24. Similar to the Dijkstra's algorithm, nodes are assigned two attributes, namely cost and predecessor. A zero cost is assigned to the source node, while other nodes are assigned an infinite cost. During each iteration, the neighborhood of each node is evaluated. If a shorter path is identified, the cost and predecessor are updated by (2.14). The algorithm terminates if no improvement in cost for any of the nodes is achieved during an iteration. At most, $|V|-1$ iterations are required using the BF algorithm to determine the shortest path within a graph, where $|E|$ edges are traversed during each iteration. The time complexity of the BF algorithm is therefore $O(|V||E|)$ [107].

The BF algorithm successfully handles directed graphs with negative edge weights. Observe that the distinct nodes along a walk are not explicitly required in the BF algorithm. This limitation is exposed if the BF algorithm is applied to a graph with negative cycles, i.e., those cycles whose sum of weights is negative. If a graph has a negative cycle, a shortest path does not exist, since the cost of a walk can be made arbitrarily small by traveling along the negative cycle.

To mitigate this limitation, an additional iteration is incorporated into the BF algorithm to identify the negative cycles. In a graph without negative cycles, the shortest path is identified in at most $|V|-1$ iterations. In the absence of negative cycles, none of the paths is reduced during the $|V|^{\text{th}}$ iteration. Detecting a change in the cost at this stage therefore indicates the presence of a negative cycle. Consider the example shown in Fig. 2.25. The sum of weights along path $[b, c, d]$ is negative. A change in cost during the fourth iteration indicates the presence of a negative cycle. Finding the shortest *path* (i.e., no repeat nodes) in a graph with negative cycles is an $\mathcal{NP}$-hard problem [117], equivalent to finding the longest path in a graph.

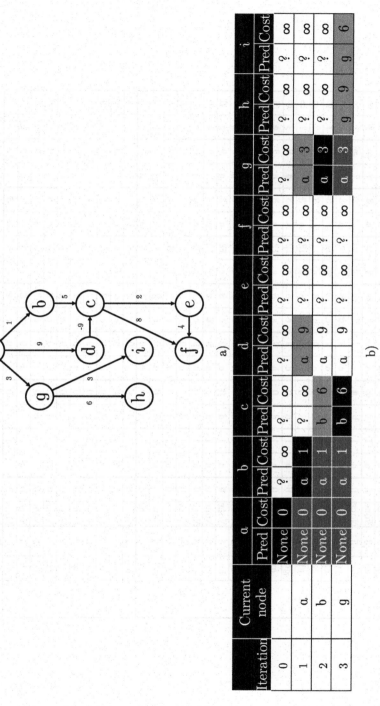

Fig. 2.23 An example of an incorrect result by the Dijkstra's algorithm in a graph with negative edges. a) The traversed graph. The numbers indicate the edge weight. b) Order of traversal. The first three iterations of the algorithm. In the third iteration, the Dijkstra's algorithm determines an incorrect shortest path to node *c* via node *b* with cost 6. The node *c* can however be reached with cost 0 by traveling via node *d*.

Iteration	Edge	Weight	a Pred	a Cost	b Pred	b Cost	c Pred	c Cost	d Pred	d Cost	e Pred	e Cost	f Pred	f Cost	g Pred	g Cost	h Pred	h Cost	i Pred	i Cost
0			None	0	?	∞	?	∞	?	∞	?	∞	?	∞	?	∞	?	∞	?	∞
1	ab	1	None	0	a	1	?	∞	?	∞	?	∞	?	∞	?	∞	?	∞	?	∞
	ad	9	None	0	a	1	?	∞	a	9	?	∞	?	∞	?	∞	?	∞	?	∞
	ag	3	None	0	a	1	?	∞	a	9	?	∞	?	∞	a	3	?	∞	?	∞
	bc	5	None	0	a	1	b	6	a	9	?	∞	?	∞	a	3	?	∞	?	∞
	ce	2	None	0	a	1	b	6	a	9	c	8	?	∞	a	3	?	∞	?	∞
	cf	8	None	0	a	1	b	6	a	9	c	8	c	14	a	3	?	∞	?	∞
	dc	-9	None	0	a	1	d	0	a	9	c	8	c	14	a	3	?	∞	?	∞
	ef	4	None	0	a	1	d	0	a	9	c	8	e	12	a	3	?	∞	?	∞
	gh	6	None	0	a	1	d	0	a	9	c	8	e	12	a	3	g	9	?	∞
	gi	3	None	0	a	1	d	0	a	9	c	8	e	12	a	3	g	9	g	6
2	ab	1	None	0	a	1	d	0	a	9	c	8	e	12	a	3	g	9	g	6
	ad	9	None	0	a	1	d	0	a	9	c	8	e	12	a	3	g	9	g	6
	ag	3	None	0	a	1	d	0	a	9	c	8	e	12	a	3	g	9	g	6
	bc	5	None	0	a	1	d	0	a	9	c	8	e	12	a	3	g	9	g	6
	ce	2	None	0	a	1	d	0	a	9	c	2	e	12	a	3	g	9	g	6
	cf	8	None	0	a	1	d	0	a	9	c	2	c	8	a	3	g	9	g	6
	dc	-9	None	0	a	1	d	0	a	9	c	2	c	8	a	3	g	9	g	6
	ef	4	None	0	a	1	d	0	a	9	c	2	e	6	a	3	g	9	g	6
	gh	6	None	0	a	1	d	0	a	9	c	2	e	6	a	3	g	9	g	6
	gi	3	None	0	a	1	d	0	a	9	c	2	e	6	a	3	g	9	g	6
3	ab	1	None	0	a	1	d	0	a	9	c	2	e	6	a	3	g	9	g	6
	ad	9	None	0	a	1	d	0	a	9	c	2	e	6	a	3	g	9	g	6
	ag	3	None	0	a	1	d	0	a	9	c	2	e	6	a	3	g	9	g	6
	bc	5	None	0	a	1	d	0	a	9	c	2	e	6	a	3	g	9	g	6
	ce	2	None	0	a	1	d	0	a	9	c	2	e	6	a	3	g	9	g	6
	cf	8	None	0	a	1	d	0	a	9	c	2	e	6	a	3	g	9	g	6
	dc	-9	None	0	a	1	d	0	a	9	c	2	e	6	a	3	g	9	g	6
	ef	4	None	0	a	1	d	0	a	9	c	2	e	6	a	3	g	9	g	6
	gh	6	None	0	a	1	d	0	a	9	c	2	e	6	a	3	g	9	g	6
	gi	3	None	0	a	1	d	0	a	9	c	2	e	6	a	3	g	9	g	6

Fig. 2.24 An example of the Bellman-Ford algorithm applied to the graph depicted in Fig. 2.23. During each iteration, each edge is evaluated to update the

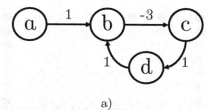

a)

Iteration	Edge	Weight	a		b		c		d	
			Pred	Cost	Pred	Cost	Pred	Cost	Pred	Cost
0			None	0	?	∞	?	∞	?	∞
1	ab	1	None	0	a	1	?	∞	?	∞
	bc	-3	None	0	a	1	b	-2	?	∞
	cd	1	None	0	a	1	b	-2	c	-1
	db	1	None	0	d	0	b	-2	c	-1
2	ab	1	None	0	d	0	b	-2	c	-1
	bc	-3	None	0	d	0	b	-3	c	-1
	cd	1	None	0	d	0	b	-3	c	-2
	db	1	None	0	d	-1	b	-3	c	-2
3	ab	1	None	0	d	-1	b	-3	c	-2
	bc	-3	None	0	d	-1	b	-4	c	-2
	cd	1	None	0	d	-1	b	-4	c	-3
	db	1	None	0	d	-2	b	-4	c	-3
4	ab	1	None	0	d	-2	b	-4	c	-3
	bc	-3	None	0	d	-2	b	-5	c	-3

b)

Fig. 2.25 The Bellman-Ford algorithm applied to a graph with a negative cycle. a) A graph with a negative cycle. The sum of weights along the path $[b, c, d]$ is -1. b) The BF algorithm. The maximum expected number of iterations is $|V| - 1 = 3$. An update of the cost during the fourth iteration indicates the presence of a negative cycle which triggers the termination of an algorithm.

2.7.1.5   A* (A-star) algorithm

The Dijkstra's and Bellman-Ford algorithms exclusively rely on weight and connectivity information. In practical graphs, additional information is often available that can assist in finding the shortest path. Consider a routing problem within a two-dimensional space, as illustrated in Fig. 2.26a. The grid graph is used to model the layout space. If the path is determined using the Dijkstra's algorithm, more than 95% of the nodes are traversed, as shown in Fig. 2.26b. By incorporating location

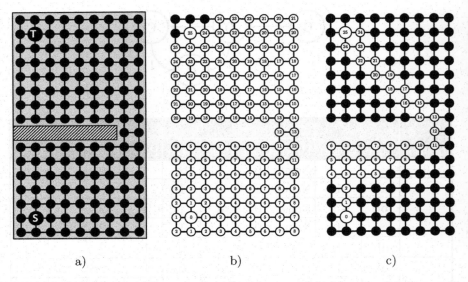

a)   b)   c)

Fig. 2.26 Finding a path within a two-dimensional layout. a) Initial layout modeled as a grid graph. b) Path finding using the Dijkstra's algorithm. The hollow nodes denote traversed nodes. The numbers indicate the distance from the source. c) Path finding using the A* algorithm. The Euclidean distance from the target is used to determine the direction for traversal. Significantly fewer nodes are therefore traversed using the A* algorithm.

information, the path between the source and the target nodes can be more efficiently determined.

Best-first search (also known as informed search [118]) is the family of algorithms that complement graph information with *heuristics* that assist the algorithms in determining the most promising direction of traversal. The A* algorithm is considered an extension of the Dijkstra's algorithm. In the Dijkstra's algorithm, those nodes that can be reached with the least cost are expanded. Node u with the smallest distance from source c_u is used as the next node. In the A* algorithm, an additional guiding heuristic h_u is incorporated into the analysis process. The next node for traversal is selected based on the smallest combined score $c_u + h_u$. Consider the traversal shown in Fig. 2.26c. The Euclidean distance from the target is used as a heuristic. Those nodes closer to the target are more likely to be explored, finding the shortest path faster while exploring fewer nodes.

2.7.2 Spanning tree

A *spanning tree* of a simple graph $G = (V_G, E_G)$ is a subgraph $T = (V_T = V_G, E_T \subseteq E_G)$, containing all nodes of G while containing no cycles. Many spanning trees can be generated for the same graph. 16 spanning trees can, for

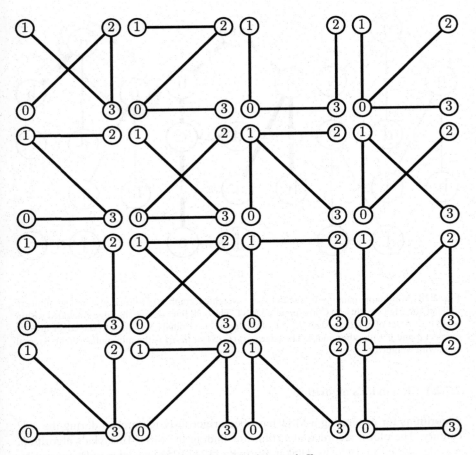

Fig. 2.27 All 16 possible spanning trees for a complete graph K_4.

example, be generated for a complete graph with four vertices K_4, as shown in Fig. 2.27. The *minimum spanning tree* (MST) is the spanning tree whose sum of edge weights is minimum. An example of a MST T_m is illustrated in Fig. 2.28. Observe that the sum of edge weights in T is larger than in T_m. MST are found in a wide range of modern engineering problems, including wireless communications networks [119, 120], image classification [121], object recognition [122], and VLSI routing [123]. Efficient algorithms have been developed for determining a MST. Three classic spanning tree algorithms are discussed in this subsection.

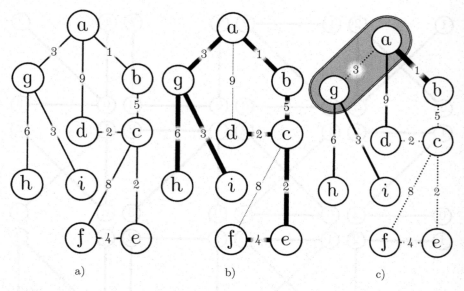

Fig. 2.28 Minimum spanning tree (MST) of a weighted graph. a) Original graph. The numbers indicate the edge weights. b) Corresponding MST. The bold lines denote the edges included within a MST. c) A set of external edges E_s^{ext} (solid edges) for terminals $S = \{a, g\}$. The edges in E_s^{ext} connect a and g with other nodes. The edge $e_{min}(G, S) = \{a, b\}$ is the minimum-weight external edge with weight 1.

2.7.2.1 Borůvka's algorithm

Algorithms for generating a MST have been rigorously researched during the 20th century. The oldest recorded algorithm for finding the MST, Borůvka's algorithm, was developed in 1926 by Otakar Borůvka [124, 125] and later rediscovered by Choquet in 1938 [126], Florek *et al.* in 1951 [127], and Sollin in 1965 [128]. Suppose set $S \subset V_G$ is a proper subset of node set of a simple graph G. Define the set of external edges E_s^{ext} connecting the nodes in S with the nodes outside S, *i.e.*,

$$E_s^{ext} = \{\{u, v\}|u \in S, v \notin S, \{u, v\} \in E_G\}. \tag{2.15}$$

The minimum-weight external edge $e_{min}(G, S)$ is

$$e_{min}(G, S) = \{u, v\}|\{u, v\} \in E_s^{ext}, w(\{u, v\}) \le w(\{w, z\}) \forall \{w, z\} E_s^{ext}, \tag{2.16}$$

where $w(\{u, v\})$ is the weight of an edge $\{u, v\}$. A set S and minimum-weight external edge are illustrated in Fig. 2.28c. The primary principle behind the Borůvka's algorithm is the observation that for each subset of nodes $S \subset V_G$, a minimum-weight external edge e is contained in a MST T, *i.e.*, $e \in E_T$. Suppose that the

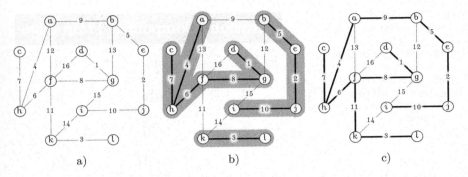

Fig. 2.29 The Borůvka's algorithm for finding a MST. a) Initial graph. Each node is considered a component. b) Graph components after the first iteration. Three components are determined. c) Final MST.

contrary is true and a MST T_1 does not contain edge e but contains a different edge $e_1 \in E_T$. By adding edge e to the MST, a cycle containing both e and e_1 is created within the MST. By deleting edge e_1, a new tree T_2 is obtained whose sum of edge weights is smaller than the sum of edge weights in T_1. T_1 is therefore not a MST, leading to a contradiction. A MST therefore always contains a minimum-weight external edge for each subset of a node set.

The algorithmic procedure is illustrated in Fig. 2.29. An edgeless forest $F = (V_F = V_G, E_F = \varnothing)$ is created from the node set of graph $G = (V_G, E_G)$. A set of connected components within F is initially

$$C = \{\{v\} | v \in V_G\}. \tag{2.17}$$

The minimum-weight external edge is determined for each connected component, producing a set of edges $A = \{e_{min}(S) | S \in C, \}$ that is added to the edge set E_F. The set of connected components within forest F is updated, and the process is repeated until F is connected. The Borůvka's algorithm requires $\log |V|$ iterations to complete, since the number of connected components is at least halved during each iteration. Determining the minimum-weight external edge can be achieved in linear time. The Borůvka's algorithm therefore exhibits a worst case time complexity of $O(|E| \log |V|)$.

2.7.2.2 Prim's algorithm

The second oldest MST algorithm was discovered by Jarnik [129] in 1929 and later rediscovered in the 1950's by Kruskal, Prim, Loberman, and Weinberger, and Dijkstra [125]. Similar to the Borůvka's algorithm, the Prim's algorithm relies on finding the minimum-weight external edge. The algorithm starts by creating a graph $T = (V_T = \{u\}, E_T = \varnothing)$ containing an arbitrary node $u \in V_G$ and no edges.

Fig. 2.30 Progress of the Prim's algorithm applied to the graph shown in Fig. 2.29a. Node a is used as the initial component. The minimum-weight external edge is used to determine which node is added to the component.

Iteration	Component	Min edge
1	a	ah
2	ah	fh
3	ahf	ch
4	achf	fg
5	achfg	dg
6	acdhfg	ab
7	abcdhfg	be
8	abcdehfg	ej
9	abcdehfgj	ij
10	abcdehfgij	fk
11	abcdehfgijk	kl
Result	abcdehfgijkl	

During each iteration, the minimum-weight external edge $e = e_{min}(G, V_T)$ of V_T within graph G is determined. Edge e and node $v \notin V_T$ adjacent to e are added, respectively, to E_T and V_T. This procedure is repeated until $V_T = V_G$, indicating completion of the MST. The progress of the Prim's algorithm applied to the graph shown in Fig. 2.29a is shown in Fig. 2.30. The runtime of the Prim's algorithm depends upon the implementation and graph characteristics. By using an adjacency matrix, the runtime is $O(|V|^2)$. For sparse graphs where the size of the graph is proportional to the order, the computational complexity is reduced by applying a binary heap data structure, yielding a runtime of $O(|E| \log |V|)$ [130].

2.7.2.3 Kruskal's algorithm

The existence of the minimum-weight external edge for any subset of nodes within a MST implies that an edge with the smallest weight is within a MST. A MST can

Fig. 2.31 Progress of the Kruskal's algorithm applied to the graph shown in Fig. 2.29a. The MST is constructed by iteratively adding edges with minimum weight while avoiding cycles. The column *Min edge* lists the edges with minimum weight added to the MST. The edges that could not be added to the MST so as not to create cycles are listed in column *Skipped*.

Iteration	V_F	Min edge	Skipped
1	abcdefghijkl	ac	
2	a c h	ah	
3	abc h	ab	
4	abc h k	hk	ch
5	abc fgh k	fk	
6	abc fgh k	fg	
7	abc fghi k	gi	
8	abcd fghi k	df	ik
9	abcd fghijk	ij	fh, dg, bg
10	abcdefghijk	ej	
11	abcdefghijkl	kl	bj, be

therefore be constructed by iteratively adding edges with the smallest weight while avoiding cycles. This process is the essence of the Kruskal's algorithm developed by Kruskal in 1956 [131]. Application of the Kruskal's algorithm to the graph shown in Fig. 2.29a is illustrated in Fig. 2.31. The edge set E_G is initially sorted from the smallest weight to the largest weight, producing an ordered sequence P. Similar to the Borůvka's algorithm, an empty forest graph $F = (V_F = V_G, E_F = \varnothing)$ is created. During each iteration, an edge $e \in P$ with the smallest weight is considered. If adding e to E_F does not create a cycle, an edge is added to the edge set E_F. Edge e is removed from P and the process repeats until forest F is connected. The runtime of the Kruskal's algorithm is dominated by the edge sorting process that is typically completed in $O(|E| \log |E|)$ time.

2.7.2.4 Advanced MST Algorithms

Borivka's, Prim's, and Kruskal's algorithms belong to the class of greedy algorithms, where a locally optimal decision is made during each iteration [132]. Unlike $\mathcal{NP}$-hard problems, an optimal MST can be generated using a greedy approach [132]. Further development of the MST theory has produced algorithms that run in nearly linear time. In 1987, Fredman and Tarjan augmented Prim's algorithm by limiting the size of a tree generated by the Prim's algorithm [116]. A subset of nodes $S \subset V_G$ is initially selected. The Prim's algorithm is run from each node $n \in S$ until the size of a subtree exceeds a threshold k or the tree joins another subtree. Each subtree is contracted into a single node, and the process repeats until all of the subtrees are connected. The contracted subtrees are expanded, yielding the MST. Fredman and Tarjan showed that each iteration runs in $O(|E| + |V| \log k)$ time. By judiciously choosing the threshold k, the number of iterations can be minimized to $O(\log^* |V|)$, where $\log^* |V|$ is an *iterated logarithm*, the number of times a logarithm function should be applied to produce a result less than or equal to 1. The iterated logarithm is an extremely slowly increasing function recursively defined as

$$\log^*(x) \equiv \begin{cases} 0, & \text{if } x \leq 1, & \text{(2.18a)} \\ 1 + \log^*(\log x), & \text{otherwise,} & \text{(2.18b)} \end{cases}$$

where x is an arbitrary real positive number. For example, $\log^*(x) = 2$ for $x \in [16, 3, 814, 279]$ ($x \in [\lceil e^e \rceil, \lfloor e^{e^e} \rfloor]$), while $\log^*(x) \leq 4$ for $x \leq \lfloor e^{e^{e^e}} \rfloor \approx 2.33 \times 10^{1,656,520}$.

Further developments in subgraph contraction has yielded an even faster algorithm, proposed by Chazelle [133]. In this algorithm, the graph is initially decomposed into a disjoint set of *contractible* subgraphs, *i.e.*, those subgraphs whose intersection with the MST is a connected tree, as illustrated in Fig. 2.32. A MST is found for each contractible subgraph, and the graph is reduced by converting each subgraph into a single node. This recursive procedure completes in $O(|E|\alpha(|E|, |V|))$ time, where $\alpha(m, n)$ is the inverse Ackermann function [134], increasing at a slower rate than the iterated logarithm. Using randomized methods, an expected linear time algorithm was developed in 1995 by Karger *et al.* [135], approaching the theoretical lower limit $O(|E|)$ for finding a MST.

2.7.2.5   Steiner tree

A MST connects the entire node set V_G of graph G. Many practical applications, however, connect only a subset of nodes, called *terminals* $S \subseteq V_G$. A *Steiner Minimum Tree* (SMT) is a connected subgraph tree $T = (V_T, E_T \subseteq E_G)$ with minimum weight whose node set contains all terminals, *i.e.*, $S \subseteq V_T \subseteq V_G$. An example of SMT is illustrated in Fig. 2.33. Observe that in addition to the terminals, a SMT can contain additional nodes to minimize the total weight of the edges. The non-terminal nodes within a SMT are commonly called *Steiner nodes*. Despite the similarity between the MST and SMT problems, the complexity of MST and SMT is drastically different. The MST can be determined in nearly linear time [133, 135]. No polynomial time algorithm exists for finding a SMT within a target graph unless $\mathcal{P} = \mathcal{NP}$ [99].

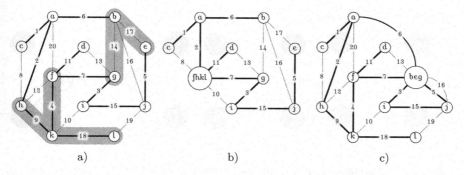

Fig. 2.32 Example of contractible subgraph. a) Initial graph. Edges belonging to MST are depicted with solid lines, and the remaining edges are depicted with dotted lines. Two subgraphs, G_1 and G_2, are considered in this example with node sets, respectively, $V_1 = f, h, k, l$ and $V_2 = b, e, g$. b) Contraction of subgraph G_1. Those edges whose endpoints are both in V_1 are discarded. The nodes in V_1 are combined into a single node. Edges incident to nodes in V_1 are incident to the combined node $fhkl$ after contraction. The tree structure is retained after contraction. Subgraph G_1 is therefore contractible. c) Contraction of subgraph G_2. Cycles within a MST are produced during the contraction (e.g., (beg, i, j)). Subgraph G_2 is therefore not contractible.

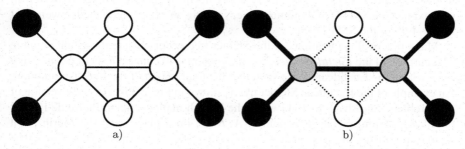

Fig. 2.33 Example of a Steiner minimum tree (SMT). a) Initial graph. All edges have equal weight. The solid circles denote the terminal nodes. b) The SMT utilizes two Steiner nodes. The bold lines denote the edges belonging to a SMT.

A SMT can however be approximated using a MST. Consider a complete graph $G_K = (V_G, E_K)$, where weight $w(e)$ of edge $e \in E_K$ denotes the shortest path between the endpoints of e within graph G. Graph G_K is called the *metric closure* of G [136]. $G_K[S]$ is a subgraph of G_K induced by the set of terminals S, and is illustrated in Fig. 2.34. A MST of $G_c[V_T]$ can be converted into a Steiner tree T_{apx} of G. The sum of weights of this Steiner tree $T^{apx} = (V_T^{apx}, E_T^{apx})$ is shown to be no greater than [117]

$$\sum_{e \in E_T^{apx}} w(e) = 2\left(1 - \frac{1}{|S|}\right) \sum_{e \in E_T} w(e), \tag{2.19}$$

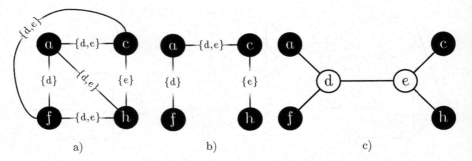

Fig. 2.34 Construction of a Steiner Minimum tree (SMT) by metric closure of the graph. a) Subgraph of the metric closure of the graph shown in Fig. 2.33a induced by the set of terminals $S = \{a, c, f, h\}$. The nodes along a shortest path are shown in curly brackets. b) The MST of the metric closure. c) SMT constructed from the MST. Note that the algorithm is an approximation of the SMT and does not guarantee optimality.

i.e., the total weight of edges within T^{apx} is at most two times greater than the sum of the edge weights of a SMT. This upper bound was improved to $(1 + \frac{\ln 3}{2}) \approx 1.55$ by Robins and Zelikovsky [137]. By applying linear programming, Byrka *et al.* approximated a SMT in polynomial time, yielding an expected total edge weight of $\ln 4 \approx 1.39$ of the SMT [138].

Many practical applications of the Steiner tree occur outside the graph domain. The purpose of the *Euclidean Steiner tree* problem is to connect a set of points using lines within the Euclidean space such that the total length of the lines is minimum. An example of the Euclidean Steiner tree is shown in Fig. 2.35b. An early version of this problem dates back to 1643 when the French mathematician Pierre de Fermat posed a question: given three points $\{A, B, C\}$ on a plane, find the fourth point D that minimizes the sum of distances from D to A, B, and C [139, 140]. The earliest published solution to this problem is attributed to Evangelista Torricelli in 1644 [141]. Jarnik and Kossler [139] are regarded as the first mathematicians who formulated the modern version of the Euclidean Steiner tree problem in 1934: *find the shortest network connecting n points in a plane*. Melzak is regarded as the author of the first algorithm for constructing a Euclidean Steiner tree [142]. In the Melzak's algorithm, a pair of points is iteratively replaced with an equivalent single point, thereby reducing the n-point problem to $n - 1$ points. The complexity of the algorithm is however exponential, making the Melzak's algorithm impractical for large networks. Garey, Graham, and Johnson demonstrated in 1977 that the problem belongs to the class of $\mathcal{NP}$-hard problems [143]. In 1966, Hanan studied a *Rectilinear Steiner Minimum Tree* (RSMT) problem, where the set of points is connected using orthogonal lines [144]. Hanan showed that the optimal solution is contained within the grid created by drawing the horizontal and vertical lines through the target points, subsequently called a Hanan grid, as illustrated in Fig. 2.35c. The RSMT problem is of particular interest in VLSI routing, where rectilinear interconnects are typically used [145].

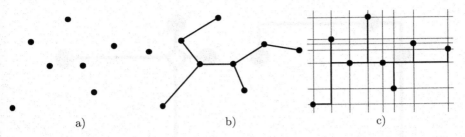

Fig. 2.35 Construction of a SMT within a planar space. a) Initial arrangement of points. b) Euclidean SMT. c) Manhattan SMT constructed using a Hanan grid.

Interest in the RSMT problem at the end of the 20th century was driven in no small part by the significant focus placed on VLSI routing automation [146]. One of the oldest methods for approximating a RSMT is based on constructing a MST within a Hanan grid, followed by improvements using heuristics [147]. A greedy approach for improving a spanning tree is 1-Steiner point insertion [148]. A 1-Steiner point is defined as a point whose addition to the node set reduces the length of the MST. Insertion of a 1-Steiner point is illustrated in Fig. 2.36. Observe that by adding three 1-Steiner points, the total length of the MST is significantly reduced. An iterative 1-Steiner, proposed by Kahng and Robins [149], achieved 11% improvement in wire length as compared to the MST in $O(n^3)$ time. An edge-based heuristic, proposed by Borah, Owens, and Irwin [150], achieves a similar improvement of a MST length in $O(n^2)$ time.

Further developments in Steiner trees include adaptation of Steiner trees to practical problems. Two broad classes of Steiner tree algorithms in VLSI include the Length-Restricted Steiner Minimum Tree (LRSMT) [151] and the Obstacle-Avoiding Steiner Minimum Tree (OASMT) [152]. In [123], for example, the Bounded Radius Spanning Tree is proposed to limit the parasitic impedance and Elmore delay [153] of the corresponding wire. The primary motivation for the LRSMT approximation algorithms is to limit the parasitic impedance of the resulting wires. The length of the resulting tree can often be larger than the optimal Steiner tree, as illustrated in Fig. 2.37 [123]. In OASMT, a practical layout is considered where, due to congestion, parts of the layout are unavailable for routing [154, 155]. Extensions to non-rectilinear Steiner trees have been presented to further reduce the total wirelength [156, 157]. Extension to three-dimensional routing is currently being explored [158–160].

2.7.3 Graph coloring

Coloring is one of the fundamental problems in graph theory. The problem originates from the classic Four Color theorem, first posed by Francis Guthrie in

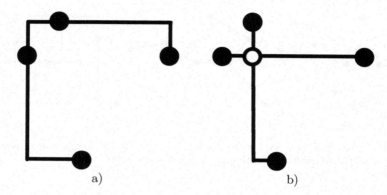

Fig. 2.36 Insertion of a 1-Steiner point. a) Minimum spanning tree. Solid circles denote terminals. b) Steiner tree after addition of a 1-Steiner point (hollow circle). The total length of a tree is reduced by 17.3%.

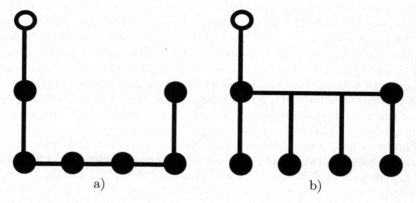

Fig. 2.37 Length-Restricted Steiner Minimum Tree (LRSMT) construction. a) Minimum length tree. The distance from the main terminal node (hollow circle) to other terminals is unbalanced. b) LRSMT. The difference in distance from the main terminal node is reduced.

1852 [161] who noticed that only four colors are sufficient for coloring a map of English counties (see Fig. 2.38.):

> "if a figure be anyhow divided, and the compartments differently coloured, so that figures with any portion of common boundary line are differently coloured–four colours may be wanted, but not more [162]."

In 1852, this theorem was brought to the attention of Augustus De Morgan [162], who recognized the complexity of the problem despite the simplicity of the formulation. Widespread attention to the theorem occurred in 1878 when Arthur Cayley made a query to the London Mathematical Society and the Royal Geographical Society about this problem [162]. The graph-theoretic equivalent of the four color theorem is attributed to Tait [163], who, in 1880, suggested replacing

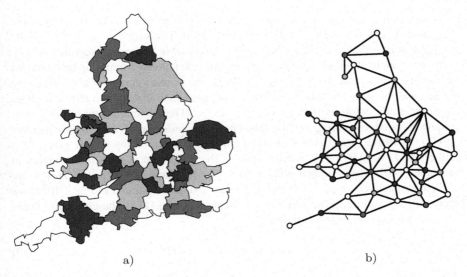

Fig. 2.38 Four color theorem originated in the middle of the 19th century when Francis Guthrie noted that only four colors are sufficient when coloring the map of England. a) Map of contiguous English counties colored using four colors, and b) an equivalent graph. The nodes represent the counties, and the edges connect adjacent regions.

districts with points, and connecting the points whose corresponding districts share a boundary. In 1890, Heawood proved a five color theorem, a weaker version of the problem [164]. The proof of the original theorem however required almost a century from Cayley's original query. In 1977, Kenneth Appel and Wolfgang Haken published the proof of the four color theorem where the theorem is reduced to 1,834 configurations that were verified by a computer [165, 166].

The four coloring theorem sparked the field of *graph coloring*. In the process of *node coloring*, the nodes of a graph are assigned labels, such that no two nodes incident to the same edge share the same color. Formally, graph coloring is a map,

$$A : V \rightarrow C, \tag{2.20}$$

such that

$$A(u) \neq A(v), \iff \{u, v\} \in E_G, \tag{2.21}$$

where $C = \{c_1, c_2, ..., c_k\}$ is a set of colors. A *chromatic number* $\chi(G)$ is the minimum number of colors $|C|$ required to color graph G. A graph whose chromatic number is $\chi(G) = k$ is often called k-chromatic, and k-colorable if $k \geq \chi(G)$.

Different variations of graph coloring problems exist that find applications in engineering. The purpose of *equitable* graph coloring is the assignment of colors to nodes $[c_1, c_2, ..., c_k]$ of a graph, such that for any pair of colors $\{c_i, c_j\}$, the number

of nodes colored with color c_i and c_j differs by at most one [167]. An example
of equitable graph coloring is shown in Fig. 2.39b. The smallest number of colors
required for equitable coloring is called the *equitable chromatic number* $\chi_=(G)$.
Important applications of equitable coloring include parallel computing and wireless
sensor networks [168, 169]. In *edge coloring*, the primary object of coloring is the
edges, and the goal is to assign colors to the edges such that no two adjacent edges
have the same color, as illustrated in Fig. 2.39c. The minimum number of colors
required for edge coloring is called the *chromatic index* or *edge chromatic number*
$\chi'(G)$. In 1964, Vizing proved that the chromatic index of any simple graph G is
either $\Delta(G)$ or $\Delta(G) + 1$, where $\Delta(G)$ is the maximum degree of any vertex in
a graph [170]. These graphs with $\chi'(G) = \Delta(G)$ and $\chi'(G) = \Delta(G) + 1$ are
called, respectively, type 1 and type 2 graphs and are illustrated in Fig. 2.40. More
generally, according to the generalized Vizing theorem [171], the chromatic index
of a connected multigraph is bound by

$$\chi'(G) \leq \min(\Delta(G) + \mu(G)), \tag{2.22}$$

where $\mu(G)$ is the maximum multiplicity within the graph. Edge coloring is found
in error correction [172], link scheduling in sensor networks [173], and scheduling
of communications [174]. In fractional coloring, the nodes of a graph are assigned
sets of colors. The adjacent nodes are required to have no colors in common, as
depicted in Fig. 2.39d. Fractional coloring can be found in resource allocation and
deadlock resolution in distributed systems [175].

2.7.4   Topological sorting

Many applications of a DAG require finding a topological ordering of a graph. Two
classical algorithms for topological sorting are the Kahn's algorithm [176], and DFS
sorting [177]. Those nodes with zero indegree are placed into a list L, as a queue or
stack. Depending upon the data structure, the topological sorting may differ. Both
structures, however, produce a valid topological sorting of a DAG. During each
iteration, node u is removed from L and placed into the final order. The indegree
of the successors of node u is decremented (reduced by 1). Successors of u whose
indegree is decremented to zero are placed in L. The process repeats until the list is
empty.

Consider the example DAG shown in Fig. 2.41a. The topological sorting process
using a stack-based version of the Kahn's algorithm is shown in Fig. 2.41b, and the
result is illustrated in Fig. 2.41c. Note that all of the edges in Fig. 2.41c are directed
rightward, indicating the correctness of the ordering. The process of the queue-based
Kahn's algorithm and the resulting ordering are shown, respectively, in Figs. 2.41d
and 2.41e. Observe that a queue and stack produce different orderings. Both of the
orderings are valid and satisfy (2.13). The total number of iterations of the Kahn's
algorithm is $|V_G|$, since every node is processed. A total of $|E_G|$ indegree decrement

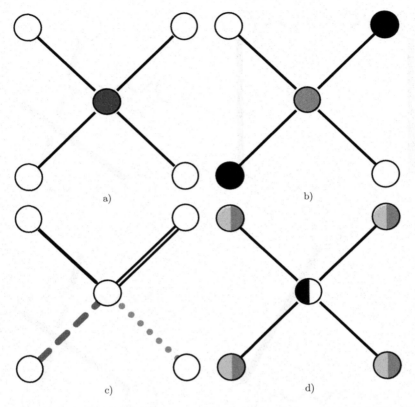

Fig. 2.39 Coloring types. a) Regular coloring, b) equitable coloring, c) edge coloring, and d) fractional coloring.

operations are performed during the algorithm. The complexity of the algorithm is therefore linear, $O(|V_G| + |E_G|)$. In a DAG, all of the nodes within the node set are processed before the list is empty. In the presence of cycles, however, not all nodes are processed. Consider the example depicted in Figs. 2.42a and 2.42b. After the first iteration, a is removed from the list. The list is empty, but none of the unprocessed nodes can be enqueued. The algorithm therefore terminates prematurely, indicating the presence of a cycle.

An alternative method for topological sorting is DFS traversal, as described in Subsection 2.7.1. Recall that DFS traversal utilizes a stack. Topological sorting using DFS is illustrated in Figs. 2.43. Observe that if u is the ancestor of v, node u is placed into the stack before node v. Since the stack is a LIFO data structure, node v is removed from the stack before node u. By recording the order of the node removal process, a reverse topological sorting is obtained. Starting DFS from any node produces a valid topological sort. Suppose an arbitrary node u is selected as the source. All of the nodes reachable from u are traversed during the DFS until the

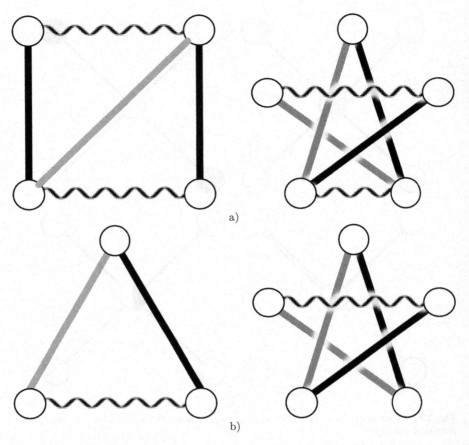

Fig. 2.40 Edge coloring classes. a) Class one graphs. The chromatic index of these graphs is equal to Δ, the maximum degree of any node within a graph. b) Class two graphs. The chromatic index of these graphs is $\Delta + 1$.

stack is empty. DFS is repeated from another unvisited node v until all of the nodes are marked as visited. In Fig. 2.43d, for example, the first DFS traversal from node d leaves nodes a, b, c, e, and h unmarked. None of these nodes is a descendant of d. A valid ordering can therefore be produced by repeating DFS from any of these nodes. The complexity of the DFS algorithm is $O(|V_G| + |E_G|)$, similar to the Kahn's algorithm. Unlike the Kahn's algorithm, however, cycle detection is not inherent to DFS and requires keeping track of the nodes within the stack. For example, the nodes can be marked during pushing into the stack and unmarked during popping from the stack. The cycle can be detected if a marked node (*i.e.*, a node already stored in the stack) is encountered. Consider the example shown in Fig. 2.42. During the fifth iteration, node b should be added to the stack. Node b is, however, already stored within the stack, indicating the presence of a cycle containing b and d.

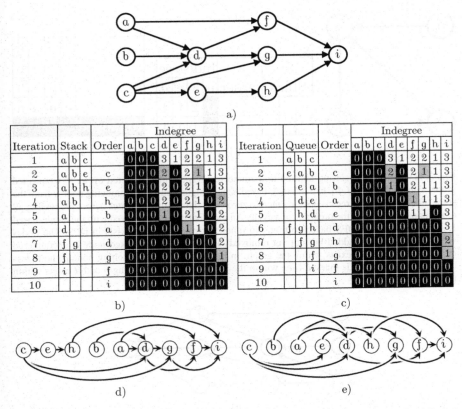

a)

b)

Iteration	Stack	Order	a	b	c	d	e	f	g	h	i
						Indegree					
1	a b c		0	0	0	3	1	2	2	1	3
2	a b e	c	0	0	0	2	0	2	1	1	3
3	a b h	e	0	0	0	2	0	2	1	0	3
4	a b	h	0	0	0	2	0	2	1	0	2
5	a	b	0	0	0	1	0	2	1	0	2
6	d	a	0	0	0	0	0	1	1	0	2
7	f g	d	0	0	0	0	0	0	0	0	2
8	f	g	0	0	0	0	0	0	0	0	1
9	i	f	0	0	0	0	0	0	0	0	0
10		i	0	0	0	0	0	0	0	0	0

c)

Iteration	Queue	Order	a	b	c	d	e	f	g	h	i
						Indegree					
1	a b c		0	0	0	3	1	2	2	1	3
2	e a b	c	0	0	0	2	0	2	1	1	3
3	e a	b	0	0	0	1	0	2	1	1	3
4	d e	a	0	0	0	0	0	1	1	1	3
5	h d	e	0	0	0	0	0	1	1	0	3
6	f g h	d	0	0	0	0	0	0	0	0	3
7	f g	h	0	0	0	0	0	0	0	0	2
8	f	g	0	0	0	0	0	0	0	0	1
9	i	f	0	0	0	0	0	0	0	0	0
10		i	0	0	0	0	0	0	0	0	0

d)

e)

Fig. 2.41 Topological ordering using the Kahn's algorithm. a) An example DAG. b) Stack-based Kahn's algorithm. During each iteration, nodes with zero indegree are placed into the stack. The top node u is removed from the stack, and the indegree of the successors is decremented. The process repeats until the stack is empty. c) Queue-based Kahn's algorithm. The process is identical to the stack-based Kahn's algorithm except the order of processing the zero-degree nodes. d) Result of the stack-based Kahn's algorithm, and e) result of the queue-based Kahn's algorithm.

Checking the membership of an element requires a more advanced data structure as compared to a stack, potentially degrading the computational performance of the algorithm.

2.8   Summary

Basic graph terminology is revisited in this chapter. Based on the existence or absence of parallel edges and loops, a graph belongs to a class of pseudographs, multigraphs, graphs with loops, or simple graphs. Based on the edge orientation, a graph is classified as directed or undirected. Additional information can be

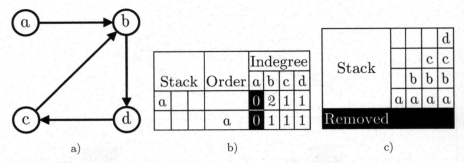

Fig. 2.42 Topological sorting applied to a connected graph with a directed cycle. a) Connected graph with cycle $[b, c, d, b]$. b) The stack-based Kahn's algorithm. After removing node a from the stack, none of the indegrees is decremented to zero. The algorithm terminates before processing all of the nodes, indicating the presence of a cycle. An identical result is achieved with the queue-based Kahn's algorithm. c) DFS based sorting. No inherent cycle detection exists in DFS. The nodes within the stack (*i.e.*, added but not yet removed) are marked. If a marked node is encountered during DFS, the cycle exists within the graph. In this example, upon reaching node d, node b is detected. Since node b is marked (*i.e.*, within the stack), a cycle containing b and d exists within the graph.

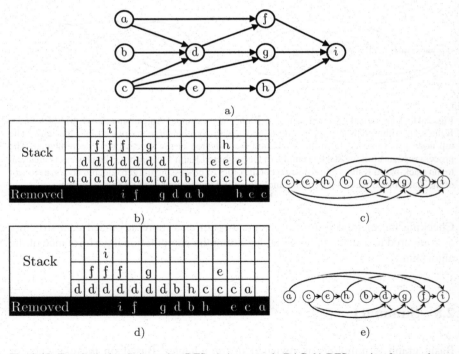

Fig. 2.43 Topological ordering using DFS. a) An example DAG. b) DFS starting from nodes a, b, and c, and c) the resulting ordering. The ordering is obtained by reversing the order of removal from the stack. This ordering is coincidentally identical to the ordering obtained using the queue-based Kahn's algorithm. d) DFS starting from nodes d, b, h, c, and a, and e) the resulting ordering. Note that a valid ordering is obtained despite starting from nodes with nonzero indegree.

embedded into the nodes and edges, such as edge weights and node attributes. Trees, bipartite graphs, and directed acyclic graphs are frequently encountered in practical applications and are each described in this chapter. Classical graph-based problems are presented, including pathfinding, spanning and Steiner tree construction, graph coloring, and topological sorting. The algorithms discussed in this section provide a rigorous framework for the design and analysis of a large variety of practical systems. VLSI is an important application of graph theory. The application of graph theory to VLSI circuits and systems is discussed in the following chapter.

Chapter 3
Graphs in VLSI circuits and systems

The history of engineering is characterized by the gradual increase in the complexity of systems. The birth and development of Very Large Scale Integration (VLSI) has followed a similar path. Early computing systems, while containing thousands of elements, were relatively simple in complexity, permitting *ad hoc*, often manual, design practices which required only a small group of people. For example, Z1, the first relay computer, was built in 1938 by Konrad Zuse and several of his fellow students in a living room of an apartment [17]. In contrast, modern VLSI systems consist of many billions of devices, employ a rigorous approach to the design process, and require collaboration of many hundreds to thousands of people with expertise ranging from material physics to software engineering.

These complex systems cannot be efficiently designed or even fully comprehended by a single human individual. *Abstraction* is a powerful tool for managing the complexity of sophisticated systems, where the fine details of a structure are omitted to enable the design process at higher levels of abstraction [75, 178]. From a cognitive perspective, abstraction is a process of compressing information [179]. Complicated objects and phenomena are reduced into a more manageable size, facilitating design and analysis at a higher level. In developing complex systems, abstraction is repeatedly applied, separating the design process into multiple abstraction layers. Systems employing layered abstraction are not limited to engineering and are often encountered in all types of endeavors requiring large scale collaboration. A government, for example, is a multilayer system [180]. Issues managed nationally, such as currency, military, and foreign affairs, influence an entire country. Information is typically processed in an aggregate form, focusing on trends rather than details. Policies at the national layer constitute a framework for governments at the lower layers. While a significant overlap often exists between the functions of national and regional governments, such as taxation and justice, the focus of regional governments is relatively narrow. Decisions and policies are however more nuanced, since a more precise understanding is possible at the regional layer. For example, while the U.S. constitution contains approximately

R. Bairamkulov, E. G. Friedman, *Graphs in VLSI*,
https://doi.org/10.1007/978-3-031-11047-4_3

4,400 words, constitutions in the 50 U.S. States are, on average, 34,000 words long [181]. Lower layer governing structures (e.g., municipal or county governments) often exist to oversee local affairs such as public facilities, housing, school systems, and emergency services.

In engineering, layered abstraction is utilized, for example, in software engineering [182], Internet Protocol Suite [183], artificial intelligence [184], and VLSI [185]. Dividing a design problem into multiple separate levels brings three major advantages to the development process.

1. **Focus.** Each abstraction layer is concentrated on a clearly defined set of design objectives. The characteristics of the other abstraction layers are assumed reliable and immutable. The design process therefore assumes correct functionality within the other abstraction layers.
2. **Simplification.** Complex systems contain an excessive number of parameters that complicate the design and analysis process. By applying layers of abstraction, redundant information is compressed or discarded. Only the most relevant parameters are retained, greatly accelerating the system development process.
3. **Generalization.** Solutions within a particular layer do not typically rely on specific characteristics of the other layers. These solutions can therefore be generalized and applied to a wide range of systems.

The process of developing VLSI circuits and systems is largely hierarchical, as illustrated in Fig. 3.1. Four abstraction layers are identified, namely, register transfer, logic, circuit, and layout. Additional layers beyond the scope of VLSI exist that encompass software engineering and semiconductor device and materials development. In this context, VLSI can be viewed as a link connecting materials and systems. A product development flow of a general VLSI system is shown in Fig. 3.2 [186]. The integrated circuit design process is essentially a series of transformations from the highest abstraction layer (behavioral description) to the lowest abstraction layer (physical layout).

Graph theory plays an important role in facilitating these transformations. By applying a graph representation, a system is significantly simplified while retaining essential information. The importance of graph theory as a method for abstracting the VLSI design process is discussed in Section 3.1. Four layers of the VLSI design process are identified. At the register transfer layer, graphs facilitate the analysis of data flow within an IC, as described in Section 3.2. A graph-based analysis at the gate layer is introduced in Section 3.3. In Section 3.4, application of graph theory to circuit analysis is presented. Design issues at the physical layer are primarily resolved using graph-based methods, as described in Section 3.5.

3.1 Graphs as a VLSI abstraction tool

From the most general perspective, an integrated circuit is a network of several on-chip systems, multiple power grids, thousands of functional modules, billions

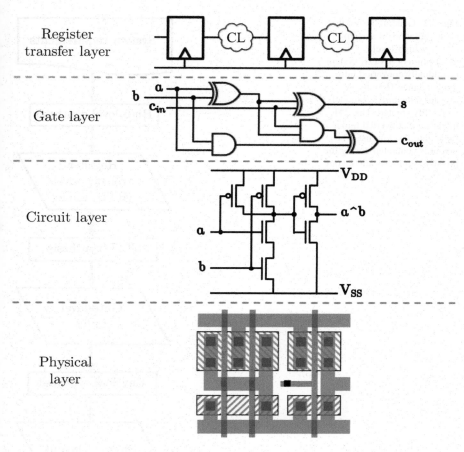

Register
transfer layer

Gate layer

Circuit layer

Physical
layer

Fig. 3.1 Design hierarchy in VLSI. At the register transfer layer, these macroblocks are transformed into a network of memory blocks connected by wires and combinatorial logic. At the logic layer, a gate-level representation of the system is the primary focus. The transistors within the logic gates are the focus of the circuit layer. At the physical layer, the circuits are transformed into a physical layout.

of registers, and many tens of billions of transistors. Graphs are highly effective in managing the hierarchical design of these complex VLSI circuits and systems. A graph representation of a system can be adjusted to suit the requirements of a particular abstraction layer.

Early electronic systems before the 1970's, composed of hundred of transistors, were designed at the gate and physical layers [187]. The relative simplicity of these early electronic systems supported an *ad hoc* design process, permitted a lack of standardization, and allowed the design process to focus at lower levels of abstraction. The increase in the complexity of microelectronic systems has, however, significantly increased the workload. Manual drawing of IC layouts, for

Fig. 3.2 General design flow for a digital VLSI system [186]. A high-level description of a VLSI system is gradually converted into more detailed formats. A register transfer level model is initially created. The RTL models are converted into a logic gate-level netlist. The layout is generated during the physical synthesis and layout processes.

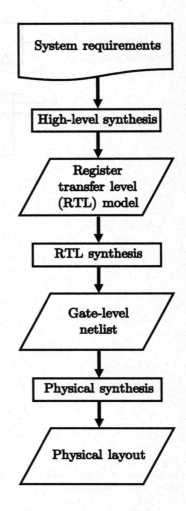

example, demanded a considerable amount of effort. In one estimate, 1,100 labor hours were required to complete a mask for a 'truly large' array of the time (800 to 1,000 elements) [39]. Furthermore, verification of a logic unit with 6,000 active elements was reported as manually intractable, exhibiting an unacceptably high 1% error rate [39].

The complexity of a manual IC design effort motivated the adoption of higher abstraction layers into the design process. For example, a methodology based on standard cells was presented in 1968 to accelerate the design process and improve reliability [39]. A 77% gain in labor productivity was reported (from 1,100 to 250 labor hours) at the cost of 10% to 20% larger on-chip area. By the early 1970's, the standard cell-based design process was widely adopted in the large scale integration (LSI) industry. Design with standard cells allowed the application of abstract graph theoretic techniques to IC design problems. Planar routing is one of the earliest applications of graph theory in automating the microelectronic system design process. IC wire routing algorithms, such as channel routing [188] and intercellular wiring [189–191], incorporated graph-based algorithms. With the advent of design methodologies based on standard cells and macroblocks, circuit partitioning algorithms were developed. Heuristic graph cut algorithms, such as the classic Kernighan-Lin algorithm [40], were integrated into the automated layout process. Other notable early applications of graph theory in LSI/VLSI include delay testing [192], system-level verification [193], and task scheduling [194]. In the upcoming sections, applications of graph theory to VLSI are reviewed. In Section 3.2, register allocation, task scheduling, and synchronization at the register transfer layer are presented. Logic optimization at the gate layer is reviewed in Section 3.3. Several applications of graph theory at the circuit and physical layers are presented, respectively, in Sections 3.4 and 3.5.

3.2   Register transfer level

At the register transfer level (RTL), a VLSI circuit is expressed as a network of interconnected blocks, as illustrated in Fig. 3.3. These blocks consist of many primitive blocks that perform a particular function. At the RTL, the functional behavior of a block is the primary focus, while the internal structure of the functional block is rarely considered. The integrated system development process is drastically

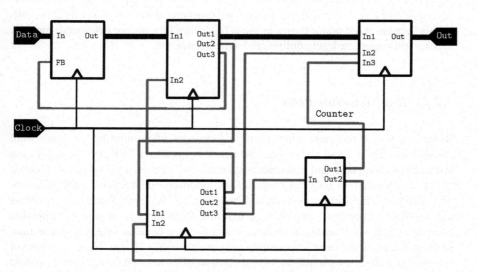

Fig. 3.3 Example of a VLSI system from an RTL perspective. The system consists of multiple interconnected functional blocks. The data flow within the system is synchronized by a common clock signal.

accelerated by utilizing RTL techniques [185, 187]. RTL design is therefore an integral part of any VLSI system development effort [195]. Adoption of the RTL design paradigm was, however, a gradual process. An early version of a register transfer language for describing the high-level structure of a hardware system was first presented in 1962 [196].

A higher level design paradigm was further advanced with the advent of modular design, as proposed in the seminal paper by W. A. Clark and colleagues in 1967 [197–199]. Compound devices, such as adders, registers, control devices, and memory units, were merged into standard 'macromodules.' In Clark's vision, an 'electronically-naive' designer could create an arbitrarily complex computer from these macromodules. Many of the features of modern RTL design processes were described. For example, two groups of macromodules were identified. The 'processing network' provides transfer, storage, and transformation of data, while the sequencing network ensures the correct flow of data. This prescient vision gained significant support in both the academic and industrial communities. In [200], for example, a 500 fold reduction in hardware due to macromodular systems was estimated. Similar systems, such as Register Transfer Modules [201] and Computer Modules [202], have been proposed. Register Transfer Modules were used in the design of the PDP-16 minicomputer by Digital Equipment Corporation [203].

By the early 1980's, VLSI systems were primarily designed at the RTL [204]. Verification gradually transitioned to RTL, replacing logic level simulation [205]. Hardware description languages, such as the Very High Speed Integrated Circuit (VHSIC) Hardware Description Language (VHDL) and Verilog Hardware Description Language (Verilog HDL), were quickly adopted in the 1980's for describing and verifying VLSI circuits and systems [206, 207]. Methodologies for creating an RTL description of a system based on a behavioral description were developed [208, 209]. In this section, three major topics in the RTL design process are discussed; namely, register allocation, task scheduling, and synchronization.

3.2.1   Register allocation

Before the rise of programmable computing systems, all computing machines were capable of only a small set of predetermined operations [210]. For example, the arithmometer patented and manufactured in the 19th century by Thomas de Colmar [211] was limited to only four operations, addition, subtraction, multiplication, and division. The arithmometer could however not be reprogrammed to process text input or to perform symbolic calculations. With the advent of programmable computers, the von Neumann computer architecture became highly popular (and remains today as the standard computer architecture) [212]. The basic structure of this architecture is shown in Fig. 3.4 and consists of three main components; random access memory (RAM), a central processing unit (CPU), and input/output (I/O) interfaces. The computer program and data are stored in the memory. The control unit of the CPU determines the sequence of operations and necessary operands and

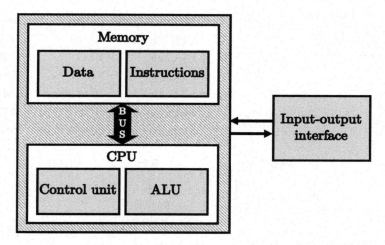

Fig. 3.4 The Von Neumann architecture is a reprogrammable architecture initially developed for early programmable computers. The architecture consists of a CPU, memory, and input-output interface. The memory stores data and instructions. The shared data and instruction bus provides communication between the CPU and memory.

fetches the operands from the memory. The ALU performs the arithmetic and logic operations as instructed by the control unit of the CPU. The output of the ALU is sent to the memory or to an output interface, such as a monitor.

The reduced instruction set computer (RISC) architecture, prevalent in modern computing systems, is largely based on the original von Neumann architecture [213]. A significant modification has however been made to the communication between the RAM and CPU. The delay of an ALU in early computing systems, such as the EDVAC, which was completed in 1949, was comparable to the latency of the memory access time [213]. With the development of faster ALUs, memory access time has become the primary bottleneck, severely limiting performance. To reduce the memory access time, a hierarchical memory structure was developed which remains in use today in modern computing systems [214]. The size and latency of the memory hierarchy in a typical desktop computer are listed in Table 3.1. The fastest memory type is the on-chip registers that require negligible access time. These registers are located within the CPU, in close proximity to the ALU and control units, to ensure minimal latency. Due to space constraints, however, the number of these registers is limited. A modern CPU typically contains between 32 and 64 registers [215, 216].

In computer engineering, a variable is a reference to a specific value – a datum stored within the memory. During a variable definition, a certain value is linked to a symbolic name. An example of a variable definition is illustrated in Fig. 3.5a. The symbolic name x is associated with the value 2016. The primary purpose of the registers is the temporary storage of the variables in proximity of a CPU. In Fig. 3.5b, for example, the variable x is stored in register R1. If, during execution of

Table 3.1 Typical capacity and latency (in 2019) at different levels of the memory hierarchy [217].

Memory type	Latency	Size
CPU Registers	300 ps	2 KB
L1 Cache	1 ns	64 KB
L2 Cache	3 to 10 ns	256 KB
L3 Cache	10 to 20 ns	8 MB
Memory	50 to 100 ns	32 GB
Storage	50 to 100 μs	256 GB

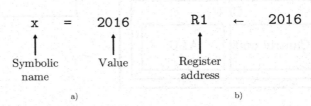

a)   b)

Fig. 3.5 Variables in computer engineering. a) Variable definition. The value 2016 is linked to the symbolic name x. b) The value 2016 is stored in register R1 after register allocation. Operations involving variable x will access R1.

a program, all of the registers are occupied, additional variables cannot be stored within a register without displacing existing variables. This situation is called a register spill [218] and results in a variable being stored in a lower tier memory. Memory at the lower tiers is located at a greater distance from the CPU, producing greater latency during reading and writing. Avoiding a register spill significantly enhances performance by reducing the memory latency.

Register allocation is an important process assigning program variables to the registers [217]. Consider a set of instructions, as shown in Fig. 3.6a. The operations on variable x are completed before the operations on variable y commence. The register occupied by x should therefore be vacated for use by variable y. A variable is called *live* during the period between the variable definition until the last use of the variable. The range of program lines during which a variable is live is called a *live range*. The live ranges of variables x and y, as depicted in Fig. 3.6a, do not overlap. The program shown in Fig. 3.6a therefore requires only one register despite operating with two variables.

Consider the program shown in Fig. 3.6b. The live range of the variable a overlaps with the live ranges of b and c. Different registers are therefore necessary to store variables a and b since the live ranges of these variables overlap. The conflict between the live ranges can be modeled as an *interference graph*. The interference graph for the program shown in Fig. 3.6b is shown in Fig. 3.6c, where the nodes represent variables, and the edges represent interference between the respective live ranges. Any two variables connected with an edge in an interference graph cannot use the same register during the course of a program. Allocation of registers to variables can be represented as a *graph coloring* problem [219, 220].

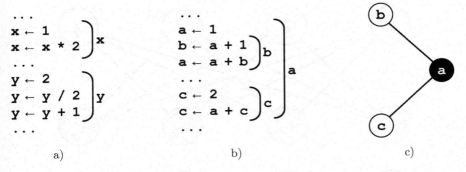

a)   b)   c)

Fig. 3.6 Live ranges used for register allocation within a computer program. a) Non-overlapping live ranges. Variables x and y can share the same register. b) Overlapping live ranges. Variable c can be assigned to the register of variable b, but not a, since the live ranges of a and c overlap. c) Coloring of the interference graph for case (b). Register allocation can be viewed as a graph coloring problem [219, 220], where the live ranges are represented by nodes and the interference is represented by edges.

Recall from Subsection 2.7.3 that the purpose of graph coloring is to label (color) each node to ensure no adjacent nodes share the same label. In the context of register allocation, the nodes represent variables and the color of the nodes represents the registers.

The problem of graph coloring is $\mathcal{NP}$-hard [10] and therefore requires heuristic methods to approach a near optimal solution. Several classes of graphs can however be colored in subexponential time. Bipartite graphs, for example, can be colored in linear time using two colors (see Fig. 3.7). Many heuristic algorithms have been proposed to determine a near optimal coloring of a general graph in polynomial time. A simple approach to graph coloring is a greedy algorithm completed in linear time [221]. Suppose the colors for coloring are $[c_1, c_2, ...]$, and the graph is traversed in an arbitrary order. A minimum available color is assigned to a node, i.e., if the node can be colored with c_i or c_j, the color c_i is chosen if $i < j$. Node ordering is crucial for reducing the number of colors used by an algorithm. For example, greedy coloring in the order shown in Fig. 3.7a requires only two colors, while the order shown in Fig. 3.7b requires $|V|/2$ colors.

Chaitin's coloring algorithm is the first coloring algorithm applied to the register allocation problem [219]. The algorithm modifies the greedy coloring by providing a coloring order. The number of colors k is initially chosen for coloring. Those nodes with degree less than k are removed from the graph, and the order of removal is recorded. If no node with degree less than k exists, the corresponding variable can be spilled, i.e., moved to the cache or memory. After all of the nodes are removed, the graph is colored in the reverse order of removal, i.e., those nodes removed first are colored last.

Register allocation has undergone significant advances since 1981. Register allocation in modern computing systems utilizes advanced coloring algorithms combined with non-graph theoretic algorithms, such as linear scan [222]. The

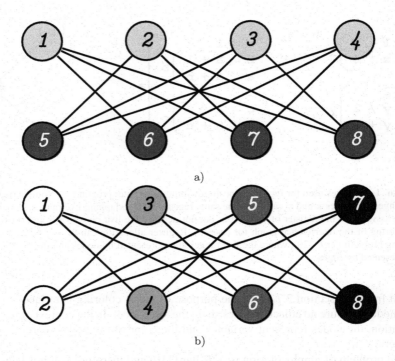

Fig. 3.7 Greedy coloring of a bipartite graph. a) Optimal coloring order requiring only two colors. b) Suboptimal coloring requiring four colors

complexity and performance of coloring algorithms have also been improved. Advanced coloring algorithms perform coloring in $O(log(n))$ time or faster [223–226], producing fast and efficient register allocation.

3.2.2   Task scheduling

Many processes inside and outside engineering require a strict order of operations. Consider, for example, the process of preparing a salad. Vegetables first need to be washed. Some of the vegetables require peeling, followed by slicing. After the vegetables are mixed in a bowl, a dressing is poured over the salad. Observe that a strict order of precedence exists throughout this process, e.g., mixing occurs only after slicing.

In VLSI systems, most operations depend upon the result of previous operations. Similar to the salad preparation process, certain operations cannot be started before the preceding operation is completed. *Task scheduling* is the process of determining an order of execution such that a target metric, such as latency or throughput, is minimized. The set of tasks within a process can be represented in graphical form.

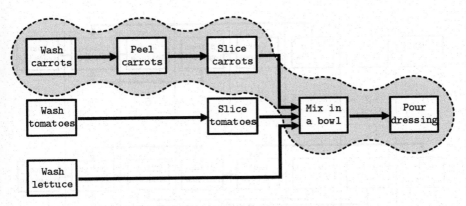

Fig. 3.8 Task graph for salad preparation. Directed edges establish a task precedence. The shaded region represents a critical path, the longest chain of dependent tasks.

Fig. 3.9 Example of a circular task dependence. Task B awaits completion of task A, Task C awaits completion of task B, and task A awaits completion of task C. None of the tasks can therefore start, producing a deadlock.

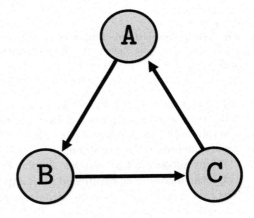

A directed graph describing the salad preparation process is shown in Fig. 3.8. The nodes in the graph represent the tasks, and the edge directions establish the task precedence. If nodes i and j within a task graph are connected with edge $i \to j$, task j cannot be started before task i is completed. Observe that no directed cycles exist within a task graph. To illustrate the inadmissibility of cycles within a task graph, consider a task schedule that contains a cycle, as illustrated in Fig. 3.9. In cycle ABC, task B waits for the result of task A, task C waits for the result of task B, and task A waits for the result of task C. No task can commence, since a cyclic dependence exists, producing a *deadlock*. A functional task graph is a directed acyclic graph (DAG), previously introduced in Section 2.5. The task graph establishes a strict precedence between tasks, prohibiting cyclic dependencies.

The objective of the task scheduling process is to establish the order of execution while respecting any precedence constraints. This task is equivalent to a *topological sorting* of a task graph $G(V, E)$ – a process of finding an ordered sequence of vertices of G such that node $v \in V$ appears only after the appearance of all of

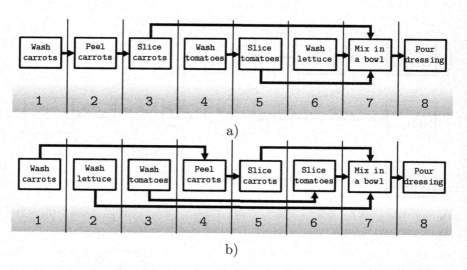

Fig. 3.10 Topological sort of the task graph shown in Fig. 3.9. Both of the schedules in (a) and (b) are valid, since the sequence number of each task is greater than the sequence number of the preceding tasks.

the predecessors [227]. Recall from Subsection 2.7.4 that many valid topological orders may exist for a particular DAG. For example, the salad preparation process can follow either sequence shown in Figs. 3.10a and 3.10b. Both of these sequences satisfy the precedence constraints.

Many modern VLSI systems support parallel execution of processes. A task graph can therefore be partitioned to split the workload among multiple processors. The gain in performance due to operating multiple processors depends upon the task schedule [97]. An inefficient task schedule underutilizes the available processing resources. In the salad preparation example, parallel processing is analogous to multiple chefs preparing a single dish. Suppose two chefs are preparing a salad, and each task requires one time unit. An example task schedule for two chefs is depicted in Fig. 3.11a. A 25% speedup is achieved by parallel execution, requiring six time units. The tasks are however partitioned suboptimally. Consider the partition shown in Fig. 3.11b. The execution time is reduced to five time units by adjusting the task schedule. Observe that the process execution time cannot be further reduced by adding another chef to the process. A lower limit on process execution time exists that prevents a process from being completed even if the number of processors is unlimited. The longest sequence of tasks is called a *critical path* and is shaded in Fig. 3.8. The length of the critical path determines the minimum time required to complete the process.

Different variations of the scheduling problem exist. In the simplest case, the tasks require the same time on each of the processors. Practically, the time required to complete a task may vary. Furthermore, many practical systems are heterogeneous, where a specific processor performs faster or slower on a particular

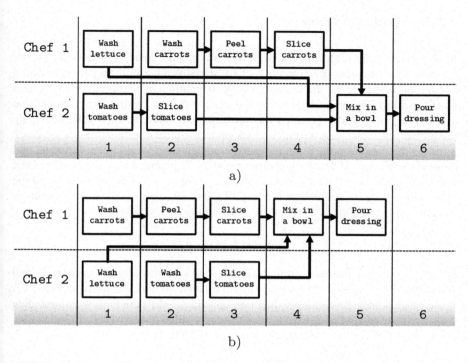

Fig. 3.11 Task scheduling with two chefs. Each task requires one time unit. a) Suboptimal task scheduling. Chef 2 is idle for two time units. b) Optimal task scheduling, requiring one fewer time unit. The execution time cannot be further optimized, since the execution time is equal to the length of the critical path.

task. Communication costs further complicate the scheduling process. If an isolated sequence of tasks is performed on a single processor A, no communication between the processors is needed. If however a preceding task is performed by a different processor B, the result of the computation must be delivered to processor A. The communication time is typically not negligible and can significantly degrade the performance. The problem of parallel task scheduling is $\mathcal{NP}$-hard. Different optimization algorithms have been proposed in the literature to approximate optimal task allocation among several processors [228]. Two classic algorithms for task scheduling in heterogeneous systems are Heterogeneous-Earliest-Finish-Time (HEFT) and Critical-Path-on-a-Processor (CPOP) [229]. In HEFT, the task graph is traversed in reverse order, inserting the tasks into a schedule to minimize the total execution time. In CPOP, the task graph is traversed in forward order. The tasks along a critical path are assigned to a single processor to minimize communication costs, while other tasks are scheduled to minimize the task completion time. Both of these algorithms exhibit complexity $O(|E||V|)$, where $|E|$ is the number of edges in a task graph and $|V|$ is the number of processors [97].

3.2.3   Synchronization

Most modern high performance VLSI systems require synchronization to ensure the data flow is temporally correct. A clock signal is a periodic signal establishing a temporal reference for the memory elements within a synchronous VLSI system (such as a flip flop or latch). During the design of a sequential logic system, the clock signal is typically assumed to simultaneously arrive at each gate, providing the global time reference. Within a clock period, a local combinatorial logic block retrieves the data from the input registers, completes the local logical function, and delivers the processed data to the output registers. In practical systems, however, the clock signal travels over long distances (e.g., possibly across the entire IC) and the propagation speed is finite. The delivery of the clock signal to all memory units is therefore not simultaneous, producing *clock skew* s_{if} [230, 231],

$$s_{if} = t_i - t_f, \tag{3.1}$$

where t_i and t_f are the delay from the clock source to, respectively, register R_i and R_f [230]. An example of a system exhibiting clock skew is illustrated in Fig. 3.12. If the clock signal travels in the direction opposite to the data flow within the data path, as illustrated in Fig. 3.13a, the clock skew is called positive [230, 232]. Positive clock skew reduces the effective clock period of a data path,

$$T_{CP}^{eff} = T_{CP} - s_{if}, \tag{3.2}$$

where T_{CP} and T_{CP}^{eff} denote, respectively, the actual and effective clock periods. The datum should therefore be processed faster to be delivered to register R_f before the clock signal of the next period arrives at R_f. If the datum is not delivered, register R_f captures incomplete or an incorrect datum, producing a clock period violation (zero clocking). Conversely, if the clock signal travels in the direction of a data path, as illustrated in Fig. 3.13b, the clock skew is called negative [230, 232]. Negative clock skew increases the effective clock period of a data path by providing additional time for the datum to be processed by the combinatorial logic. A different condition may however occur. If the combinatorial logic completes the function before the clock signal of the same period arrives at the next register, the datum stored in the register can be destroyed, while the incorrect datum propagates through the system. This issue is called a race condition or double clocking [230, 232, 233].

A significant design focus has been to minimize clock skew [178, 186, 230, 234, 235]. Different clock distribution topologies have been proposed in the literature [232]. An H-tree, for example, is an example of a balanced clock tree that equalizes the delay between the source and all of the endpoints within a network [236]. A geometrically balanced clock tree synthesis methodology is proposed in [235]. Mesh and grid clock distribution networks provide a low impedance path between the leaves of the clock tree, reducing the difference between the clock arrival time at the registers [232]. Clock skew minimization strategies, however, require

Fig. 3.12 A sequential
system exhibiting clock skew.
The clock signal is generated
at the clock source and travels
through the clock tree toward
registers R_i and R_f. The
delay from the clock source
to registers R_i and R_f are,
respectively, t_i and t_f.

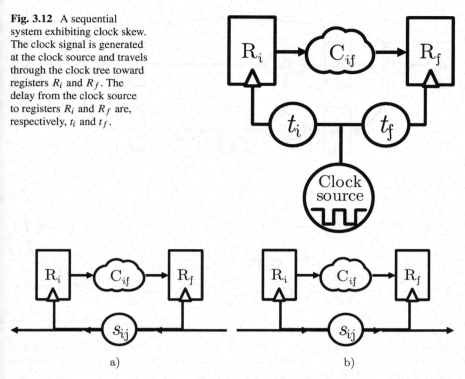

Fig. 3.13 Data and clock flow for positive and negative skew. a) Positive clock skew. The clock signal travels in the direction opposite to the flow of data. The effective clock period is reduced. This configuration is immune to race conditions. b) Negative clock skew. The clock signal travels in the same direction as the datum. The effective clock period is increased. Race conditions may occur if the data signal arrives at R_f before arrival of the clock signal within the same clock period.

significant overhead and cannot completely suppress clock skew [66]. Furthermore, minimization of clock skew does not ensure correct functionality of a synchronous system [231, 237].

In the 1990's, an alternative perspective towards clock skew optimization was developed [238–241]. Clock skew scheduling was presented as an alternative to zero skew design techniques. Rather than eliminating clock skew, the arrival time of a clock signal is deliberately controlled to ensure correct functionality while improving the performance of a synchronous system. The primary tool in clock skew scheduling is a *timing graph*, as illustrated in Figs. 3.14. The nodes of a timing graph represent synchronous elements, such as flip flops or clocked logic gates. The edges represent data paths connecting sequentially-adjacent registers [230] which may also contain combinatorial logic. Each edge is assigned two attributes which indicate the maximum and minimum propagation delay of a data signal through the corresponding data path.

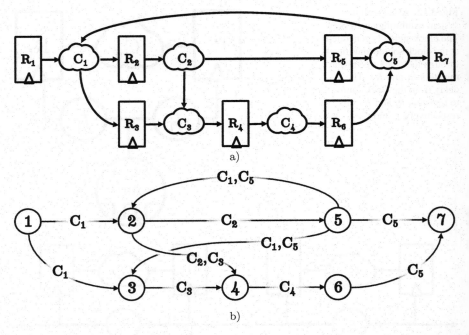

Fig. 3.14 Graphical model of a synchronous system. a) Register transfer level representation of a synchronous system. b) Timing graph. The nodes represent sequential logic elements, and the edges represent combinatorial logic.

A *permissible range* of clock skew [237, 239] is the minimum and maximum clock skew between sequentially-adjacent registers, defined as

$$PR_{if} \equiv [l_{if}, u_{if}], \tag{3.3}$$

where l_{if} and u_{if} denote, respectively, the lower and upper bounds on the clock skew. To ensure correct functionality of a synchronous system, the clock skew of each data path should be maintained within the permissible range. Linear and quadratic programming algorithms have been proposed to achieve an acceptable clock skew schedule [231, 233, 239]. A more complete description of clock skew scheduling is provided in Chapter 4.

3.3 Gate layer

A gate layer representation of a digital IC consists of interconnected combinatorial logic gates, performing Boolean operations, as illustrated in Fig. 3.15a. A logic circuit can be described by a directed graph where the nodes and edges represent,

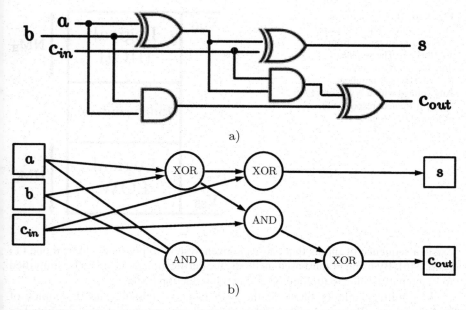

a)

b)

Fig. 3.15 Example of a half adder. a) Logic circuit. b) Equivalent directed graph. The nodes denote the gates, and the edges denote the connection between the gates. Observe that the fan-in and fan-out of the gates are equal to, respectively, the indegree and outdegree of the corresponding node in a directed graph.

respectively, the gates and connections between the gates, as depicted in Fig. 3.15b. The fan-out of a gate within a graph is equal to the outdegree of the corresponding node within the directed graph. Similarly, the fan-in of a gate is equal to the indegree of the corresponding node. Signals at the logic layer exhibit only two binary values; namely, low (typically, logical 0 or `false`) and high (typically, logical 1 or `true`). The binary signals exhibit a high tolerance to signal uncertainty by employing wide noise margins, as shown in Fig. 3.16 [242]. Communication and storage of the digital data are therefore significantly simplified. By abstracting from continuous analog levels to binary logical signals, Boolean algebra can be applied to the design of digital systems.

Verification is an important step in the development of logic circuits that ensures the correct functionality of a digital system. Fundamental issues in formal verification at the gate layer include model checking, equivalence checking, and Boolean satisfiability. *Model checking* is an important issue in system design. The state space of a logic system is exhaustively searched to verify the correspondence of a particular function to a target specification [243, 244]. A primary issue in VLSI logic design is to verify whether a particular design satisfies the target requirements [245]. Formal *equivalence checking* is a form of model checking with the objective to verify whether two logic systems produce the same logic function [246]. The objective of the *Boolean satisfiability (SAT)* problem is to determine whether there

Fig. 3.16 Noise margins in digital circuits. The voltage in a digital circuit ranges from V_{SS} to V_{DD}. The signals in digital circuits are often treated as binary numbers. Any voltage within NM_L or NM_H is treated, respectively, as logical 0 or logical 1.

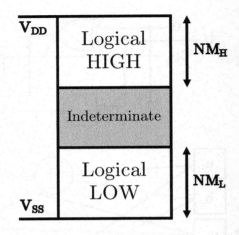

exists a sequence of inputs that causes a given system to produce a logical one (1) [208, 246]. By using graph-based methods, the verification process can be simplified to graph partitioning, path finding, and graph reduction [208].

A primitive building block at the gate layer is a logic gate. Examples of logic gates are illustrated in Fig. 3.17a. Multiple logic gates can be combined to form a logic circuit performing a particular Boolean function, as exemplified in Fig. 3.17b. Modern integrated circuits contain millions to many billions of logic gates, producing highly complex logical networks. The complexity of logic circuits in modern digital systems requires advanced techniques not only for verifying but also to represent a Boolean function. The most basic representation is the truth table, as illustrated in Fig. 3.18. Each row represents an output corresponding to a sequence of inputs. The truth table is a *canonical* description of a Boolean function. A canonical representation uniquely characterizes a Boolean expression. For example, expressions $(a + b)c$ and $ac + bc$ are described differently but the truth tables are identical, indicating that these functions are equivalent. A tabular format is only suitable for a small number of input variables since the number of rows doubles with each new variable (grows exponentially with the number of input variables). More efficient graph-based Boolean function representations exist. In this section, two of the most widely used methods are reviewed, namely, Ordered Binary Decision Diagrams and And-Inverter Graphs.

3.3.1   Ordered binary decision diagram

An *Ordered Binary Decision Diagram (OBDD)* is a directed acyclic graph (DAG) representing a Boolean function [95]. The nodes are divided into two groups. Non-terminal nodes represent the variables of a Boolean expression. Each non-terminal node x has two children, $high(x)$ and $low(x)$. Terminal (or leaf) nodes

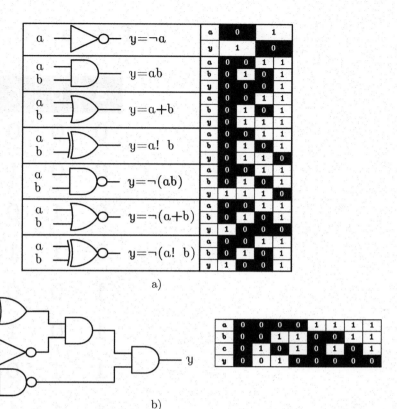

Fig. 3.17 Examples of logic gates. a) A list of primitive one and two input logic gates and truth tables. b) An example of a logic circuit composed of multiple logic gates and the associated truth table.

have a value of 0 or 1, representing the result of a logical function. To evaluate a Boolean function, an OBDD is traversed starting from the root. The traversed path is determined by the operands of the function. If a Boolean operand x has value 1, the traversal continues along the edge $high(x)$, otherwise $low(x)$ is traversed. For example, suppose an expression $a(b + c)$ is evaluated for $[a, b, c] = [0, 1, 0]$ using the OBDD shown in Fig. 3.19a. Due to the chosen variable ordering, the root vertex is a. Since $a = 0$, the edge $low(a)$ is traversed. Next, $b = 1$, and the traversal continues along the path $high(b)$. Finally, $c = 0$, and the path continues with $low(c)$, terminating at 0.

The size of an OBDD can be reduced by applying a set of reduction techniques. Formally, two rules govern the reduction process.

- If both of the outgoing edges of node u are directed towards the same node v, node u can be removed. The incoming edges of node u can point directly to node v.

Fig. 3.18 Truth table for
function $f = a(b + c)$.

$$f = a\ (b + c)$$

a	b	c	f
0	0	0	0
0	0	1	0
0	1	0	0
0	1	1	0
1	0	0	0
1	0	1	1
1	1	0	1
1	1	1	1

- If nodes u and v correspond to the same variable and the subtrees are identical, node v can be removed. The incoming edges of node v are redirected to u.

Using these two rules, an OBDD can be transformed into a Reduced OBDD (ROBDD). Observe, for example, that if $a = 0$, the value of b and c does not affect the result of the expression. Edge $low(a)$ is therefore connected directly to node 0, as illustrated in Fig. 3.19b. Similarly, if $b = 1$, the value of node c is irrelevant. Edge $high(b)$ is therefore connected to node 1. Finally, identical nodes are merged, yielding the OBDD depicted in Fig. 3.19c.

Using OBDD, the satisfiability problem is reduced to finding a path terminating at node 1 [95]. An OBDD is not a canonical representation of a Boolean function, necessitating additional processing for equivalence checking. An ROBDD, however, is a canonical representation, since the ROBDD is unique for a given function and variable order. The major issue pertaining to ROBDDs is the dependence on the variable ordering. Compare the two ROBDDs for function $a(b + c)$ shown in Figs. 3.19c and 3.19d. Changing the variable order reduces the number of edges by 50% and the number of non-terminal nodes by 25%. Finding an appropriate variable

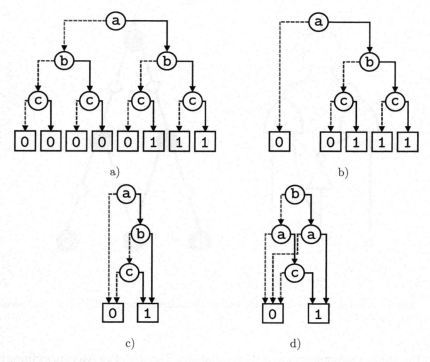

Fig. 3.19 Ordered Binary Decision Diagrams (OBDD) for a Boolean function $f = a(b + c)$ and variable order (a, b, c). a) Original OBDD. The solid and dashed edges leaving edge x represent, respectively, paths $high(x)$ and $low(x)$. b) OBDD after eliminating nodes b and c for the case $a = 0$. c) Reduced OBDD (ROBDD). d) Inefficient ROBDD due to suboptimal variable order (b, a, c).

order is critical since the worst case size of a ROBDD grows exponentially with the number of variables [208].

3.3.2 And-inverter graph

Another graph-based representation of a Boolean function is the *And-Inverter Graph* (AIG). There are three types of nodes in an AIG, namely, primary inputs, AND nodes, and primary outputs. Primary inputs represent the operands of a Boolean function and have zero indegree. Similarly, primary outputs denote the result of a Boolean function, and therefore have zero outdegree. The inner nodes within an AIG represent a two input AND gate. Edges within an AIG can be non-inverting or inverting. A signal traveling through an edge is unaltered in the former case and inverted in the latter case. Examples of Boolean functions modeled as an AIG are shown in Fig. 3.20.

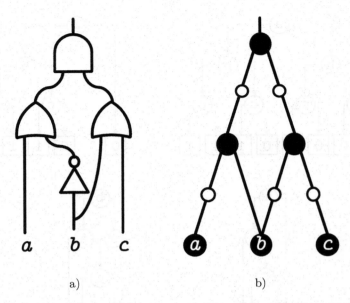

a)   b)

Fig. 3.20 Conversion of a Boolean circuit into an And-Inverter graph (AIG). a) Original circuit. b) Equivalent AIG.

The primary advantage of an AIG is scalability. The size of an AIG grows linearly with the size of a circuit as compared to a ROBDD whose worst case growth rate is exponential [247]. An AIG, however, is not a canonical format for representing a Boolean function. Both of the AIGs shown in Fig. 3.21, for example, represent the same Boolean function. To mitigate this issue, a structural hashing technique is presented in [248–250]. A functionally reduced AIG (FRAIG), proposed in [247], applies structural hashing to produce a 'semi-canonical' AIG where no two nodes execute the same function. A FRAIG is however not canonical, since two different FRAIGs can execute the same function.

Despite the non-canonicity complicating the structural analysis, the efficiency of an AIG makes this structure preferable for many applications. In tree balancing, for example, the maximum number of levels within an AIG is reduced, thereby decreasing the delay of a critical path [251]. An AIG-based path balancing methodology is proposed in [252] for deep pipelines, yielding a considerable reduction in area and static power. Despite non-canonicity, the AIGs interact well with simulation-based techniques, producing highly efficient solutions for SAT problems [253]. Using the miter technique [254], the equivalence check problem can be presented as a SAT problem, as illustrated in Fig. 3.22. To verify the equivalence of two logic circuits, the output of the two circuits is connected to the inputs of an XOR gate. If no combination of inputs produces a logical 1 in the miter circuit, the circuits are equivalent.

Fig. 3.21 Identical Boolean functions can be represented differently using an AIG. a) $ab + bc$, and b) $b(a + c)$. Both (a) and (b) describe the same function.

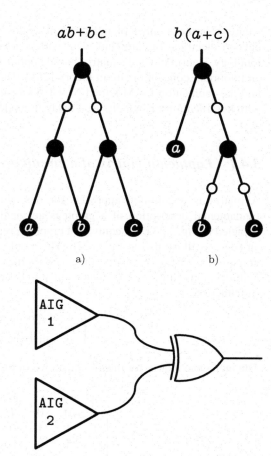

a)   b)

Fig. 3.22 Miter technique for checking the equivalence of two AIGs. The AIGs are connected to a XOR gate. The SAT problem is solved for this structure. If any input configuration produces output 1, the AIGs are not equivalent.

3.4   Circuit layer

By far the most common application of graph theory in VLSI is the analysis of electrical circuits. Graph-based analysis of electrical circuits was conceived in 1845 by Gustav Robert Kirchhoff, then a 21 year old physics student at the University of Königsberg [255]. In [256], he postulated two fundamental laws of electrical circuits, commonly known as Kirchhoff's current law (KCL) and Kirchhoff's voltage law (KVL). To determine the *cycle basis*, i.e., the number of independent cycles within a circuit, he inadvertently used the concept of a spanning tree and showed that

$$|E| = \mu + |V| - 1, \tag{3.4}$$

where μ is the number of independent cycles. Kirchhoff's laws have been used beyond electrical circuits and were widely used by Henri Poincaré in algebraic topology. Henri Poincaré's matrix-based approach introduced an incidence matrix – an important concept in circuit theory [257]. The works of Kirchhoff and Poincaré laid the foundation for Modified Nodal Analysis, the circuit analysis method that, with modifications, is widely used today in modern circuit simulation [64].

3.4.1   Laplacian matrix of a circuit graph

A matrix is a common approach for describing the properties of a graph. A fundamental description of a graph is an incidence matrix. Consider the circuit depicted in Fig. 3.23a. An equivalent graph is created by replacing each resistor with an edge, as illustrated in Fig. 3.23b. The incidence matrix Y for a loopless simple graph G is a $|V| \times |E|$ matrix that encodes the connectivity within a graph. For an undirected graph $G(V, E)$, an element $y \in Y$, corresponding to edge e and node v, is defined as

$$y_{ve} \equiv \begin{cases} 1, & \text{if } e \text{ is incident to } v, \quad (3.5a) \\ 0, & \text{otherwise.} \quad (3.5b) \end{cases}$$

The incidence matrix for the circuit shown in Fig. 3.23a is

$$Y = \begin{matrix} & \begin{matrix} e_1 & e_2 & e_3 & e_4 & e_5 \end{matrix} & \\ & \begin{bmatrix} 1 & 0 & 0 & 1 & 0 \\ 1 & 1 & 1 & 0 & 0 \\ 0 & 1 & 0 & 0 & 1 \\ 0 & 0 & 1 & 1 & 1 \end{bmatrix} & \begin{matrix} n_1 \\ n_2 \\ n_3 \\ n_4 \end{matrix} \end{matrix}. \quad (3.6)$$

Since each edge in G is connected to exactly two nodes, the number of nonzero entries in each column is two.

A directed graph $G_d(V, E_d)$ is an orientation of the underlying circuit graph G (see Subsection 2.1.5) produced by arbitrarily choosing the direction of current within a circuit. The direction of the edges in E_d corresponds to the assumed direction of current between the nodes. An example of a directed graph, corresponding to the circuit shown in Fig. 3.23a, is depicted in Fig. 3.23c. The incidence matrix Y_d for a loopless directed graph G_d is a $|V| \times |E_d|$ matrix whose element $y_d \in Y_d$, corresponding to edge e and node v, is defined as

$$y_{ve} \equiv \begin{cases} 1, & \text{if } e \text{ leaves } v, \quad (3.7a) \\ -1, & \text{if } e \text{ enters } v, \quad (3.7b) \\ 0, & \text{otherwise.} \quad (3.7c) \end{cases}$$

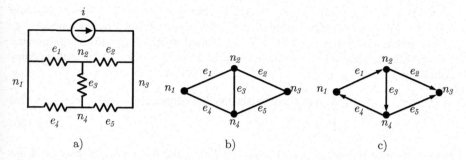

Fig. 3.23 Graph model of an electrical circuit. a) A resistive circuit with four nodes, five resistors, and a current source, b) equivalent undirected graph model, and c) equivalent directed graph model, where the direction of the edges indicates the assumed direction of current.

The incidence matrix for the circuit shown in Fig. 3.23c is

$$
Y_d = \begin{array}{c}
\begin{array}{ccccc} e_1 & e_2 & e_3 & e_4 & e_5 \end{array} \\
\begin{bmatrix}
1 & 0 & 0 & -1 & 0 \\
-1 & 1 & 1 & 0 & 0 \\
0 & -1 & 0 & 0 & -1 \\
0 & 0 & -1 & 1 & 1
\end{bmatrix}
\begin{array}{c} n_1 \\ n_2 \\ n_3 \\ n_4 \end{array}
\end{array}.
\tag{3.8}
$$

Observe that the sum of each column in Y_d is zero, since each edge leaves and enters exactly one node. A reduced incidence matrix Y_d^g is obtained by assigning one node as reference (ground), removing the corresponding row from Y_d. By grounding node n_4, the reduced incidence matrix Y_d^g becomes

$$
Y_d^g = \begin{array}{c}
\begin{array}{ccccc} e_1 & e_2 & e_3 & e_4 & e_5 \end{array} \\
\begin{bmatrix}
1 & 0 & 0 & -1 & 0 \\
-1 & 1 & 1 & 0 & 0 \\
0 & -1 & 0 & 0 & -1
\end{bmatrix}
\begin{array}{c} n_1 \\ n_2 \\ n_3 \end{array}
\end{array}.
\tag{3.9}
$$

The reduced incidence matrix of a directed graph is a basis of the classic Kirchhoff's Laws [258]. Kirchhoff's Current Law in matrix form is

$$
Y_d^g \mathbf{J} + \mathbf{Q} = \mathbf{0}_{|V|},
\tag{3.10}
$$

where $\mathbf{J} \in \mathbb{R}^{|E_d| \times 1}$ is the vector of current through each branch within a network, $\mathbf{Q} \in \mathbb{R}^{(|V|-1) \times 1}$ is the vector of external current injection, and $\mathbf{0}_n$ is the zero column vector with length n. Similarly, Kirchhoff's Voltage Law is

$$
\mathbf{W}(Y_d^g)^T = \mathbf{V}^{\mathbf{g}},
\tag{3.11}
$$

where $\mathbf{W} \in \mathbb{R}^{|E_g| \times 1}$ is the vector of voltage drops across the branches of G_d, and $\mathbf{V^g} \in \mathbb{R}^{(|V|-1) \times 1}$ is the vector of node potentials relative to the reference node.

The adjacency matrix is a different representation of a graph. For an undirected loopless simple graph G, the adjacency matrix A is a $|V| \times |V|$ matrix with an entry defined as

$$a_{ij} \equiv \begin{cases} 1, \text{ if there exists an edge connecting } i \text{ and } j & (3.12a) \\ 0, \text{ if } i = j. & (3.12b) \end{cases}$$

Note that since the graph contains no self-loops, the diagonal elements of A are all zero. In graphs corresponding to practical circuits, any node is adjacent to only a few neighbors, producing a sparse adjacency matrix. When stored in computer memory, the adjacency matrix is often represented as an adjacency list, requiring approximately $O(|V| + |E|)$ space within memory, as compared to $O(|V|^2)$ space required by a full adjacency matrix.

Practical graphs in VLSI are characterized by weights which represent the conductance of the edge. The weighted adjacency matrix is therefore generalized as A^w where

$$a_{ij}^w \equiv \begin{cases} g_{ij}, \text{ if } i \neq j & (3.13a) \\ 0, \text{ if } i = j, & (3.13b) \end{cases}$$

where g_{ij} is the conductance of the edge connecting i and j. The adjacency matrix for the circuit shown in Fig. 3.23a is

$$A^w = \begin{array}{c} \begin{array}{cccc} n_1 & n_2 & n_3 & n_4 \end{array} \\ \begin{bmatrix} 0 & g_1 & 0 & g_4 \\ g_1 & 0 & g_2 & g_3 \\ 0 & g_2 & 0 & g_5 \\ g_4 & g_3 & g_5 & 0 \end{bmatrix} \begin{array}{c} n_1 \\ n_2 \\ n_3 \\ n_4 \end{array} \end{array}. \qquad (3.14)$$

The sum of the entries along a given row of an adjacency matrix produces the degree of the corresponding node. Degree matrix D is a $|V| \times |V|$ diagonal matrix with entry d_{ij} defined as

$$d_{ij} \equiv \begin{cases} d(n_i), & \text{if } i = j & (3.15a) \\ 0, & \text{otherwise.} & (3.15b) \end{cases}$$

The degree matrix of the circuit shown in Fig. 3.23a is

$$D = \begin{array}{cccc} n_1 & n_2 & n_3 & n_4 \\ \left[\begin{array}{cccc} g_1 + g_4 & 0 & 0 & 0 \\ 0 & g_1 + g_2 + g_3 & 0 & 0 \\ 0 & 0 & g_2 + g_5 & 0 \\ 0 & 0 & 0 & g_3 + g_4 + g_5 \end{array}\right] & \begin{array}{c} n_1 \\ n_2 \\ n_3 \\ n_4 \end{array} \end{array}. \quad (3.16)$$

Observe that the sum of the entries in each row of A_w is equal to the diagonal entry in D.

Subtracting the adjacency matrix from the degree matrix produces the conductance matrix (or weighted Laplacian matrix) L [259], an important matrix in circuit theory,

$$L = D - A. \quad (3.17)$$

Entry l_{ij} within L is

$$l_{ij} = \begin{cases} d(n_i), & \text{if } i = j \\ -g_{ij}, & \text{otherwise.} \end{cases} \quad \begin{array}{l} (3.18\text{a}) \\ (3.18\text{b}) \end{array}$$

The conductance matrix can be derived from the incidence matrix,

$$L = YDY^T. \quad (3.19)$$

The Laplacian matrix of the circuit shown in Fig. 3.23a is

$$L = \begin{array}{cccc} n_1 & n_2 & n_3 & n_4 \\ \left[\begin{array}{cccc} g_1 + g_4 & -g_1 & 0 & -g_4 \\ -g_1 & g_1 + g_2 + g_3 & -g_2 & -g_3 \\ 0 & -g_2 & g_2 + g_5 & -g_5 \\ -g_4 & -g_3 & -g_5 & g_3 + g_4 + g_5 \end{array}\right] & \begin{array}{c} n_1 \\ n_2 \\ n_3 \\ n_4 \end{array} \end{array}. \quad (3.20)$$

The conductance matrix is the critical matrix in circuit graph analysis, widely used in virtually all modern circuit simulation tools. Important insights are based on the conductance matrix, such as the effective resistance [71]. Suppose a unit current is injected into node i and drawn from node j. The electric potentials within the circuit satisfy

$$L v = e_i - e_j, \quad (3.21)$$

where $v \in \mathbb{R}^{|V|}$ e_i is a vector with the ith entry equal to 1 and all other entries equal to zero. The effective resistance between the ith node and j^{th} node is

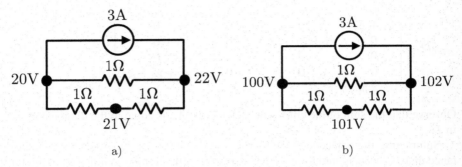

Fig. 3.24 Valid solutions for (3.21). In both (a) and (b), the potential at the nodes of the circuit satisfies (3.21).

$$R_{ij} = v_i - v_j. \tag{3.22}$$

Observe that the sum of each column and each row in L is zero. Any row in L can therefore be expressed as the linear combination of the other rows within L. The matrix L is therefore singular (non-invertible). Determining the potential at each vertex of G is therefore not possible with the conductance matrix. Infinitely many solutions satisfy (3.21). For example, both of the solutions depicted in Figs. 3.24a and 3.24b satisfy (3.21). Practical circuit analysis requires at least one reference (ground) node. By designating a particular node as a reference, the potential at this node is assumed zero. The voltage across the edges in graph G is often called a potential difference, since these voltages represent not an absolute potential but rather a difference in potential between a target node and ground. Mathematically, L_g is a grounded conductance matrix derived from L by deleting the gth row and column from L. Equation (3.21) is therefore modified as

$$L_g V_g = e_i - e_j, \tag{3.23}$$

where $g \notin \{i, j\}$ and $V_g \in \mathbb{R}^{|V|-1}$ are the vector of node voltages with the row and column for the ground node removed. In a connected graph without self-loops, L^g is invertible, since the rows in L^g are linearly independent. The voltages within the circuit are

$$V_g = L_g^{-1}(e_i - e_j). \tag{3.24}$$

This expression can be extended to determine the voltage at each node within a circuit in response to an arbitrarily injected current $Q \in \mathbb{R}^{(|V|-1)}$,

$$V^g = L_g^{-1} Q, \tag{3.25}$$

where each entry in Q is the current injected into the corresponding node. The entries in Q are negative if the current is sourced from the node.

The weighted Laplacian matrix was widely used in early computer simulation tools, such as CANCER [260] and BIAS [261]; and remains an important part of modern circuit analysis tools, particularly, SPICE [50]. The detailed description of circuit analysis process is presented in chapter 5, where a variety of techniques based on graph theory are discussed.

3.5   Physical layer

The physical layer is the lowest abstraction layer in VLSI. A circuit representation of a system is converted into a layout, as illustrated in Fig. 3.25. Procedures at the physical layer directly inform the physical nature of an integrated circuit. Many of the parameters neglected at the higher abstraction layers are important at the physical layer. The physical IC dimensions, wire pitch, and number of layers,

Fig. 3.25 Simplified flow diagram of physical design of a VLSI system. A circuit is initially partitioned into multiple blocks. The constraints on the location of the functional blocks is determined during the floorplanning stage. During the placement stage, the exact location of each block is determined. The layout of the wires connecting the blocks is synthesized during the routing stage.

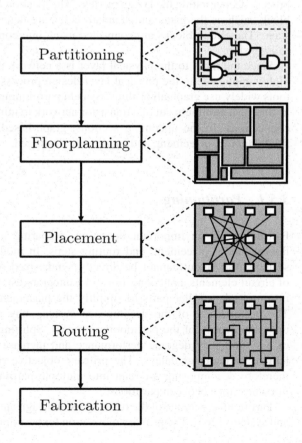

for example, are rarely considered at higher layers of abstraction, but are crucial during the physical layer design process. Similar to the gate layer, physical layer design has undergone significant advancements over the past decades. The layout of early integrated systems was manually designed, permitting local fine tuning of the physical parameters [187]. Before 1979, more than 40% of the overall labor hours was typically dedicated to the physical design process [262]. Layout synthesis became the first target of design automation [145]. According to an estimate in [262], with the rise of powerful computer-aided design tools, layout synthesis contributes only 14% to the total number of labor hours spent during the IC design process.

Automated VLSI system design at the physical layer consists of four major subproblems; namely, partitioning, floorplanning, placement, and routing [263]. System *partitioning* is the process of splitting a VLSI system into smaller parts. The separated parts can be independently processed, reducing an intractably large system design effort into several manageable parts. *Floorplanning* is the process of assigning the shape and location of each partition. The external connections of each partition are assigned to a specific location to facilitate the interconnect routing process deeper within the layout process. The location and orientation of each circuit block, such as the gates and standard cells, are determined during the *placement* stage. The circuit blocks are connected using interconnects synthesized during the *routing* step.

Since the input to the physical layer is a network of transistors, graph theory is highly applicable to the physical layer design process. Physical design automation tools widely use graph algorithms. System partitioning, for example, can be viewed as finding a minimum cut of a transistor network to minimize the number of external connections. In the upcoming sections, graph-based algorithms for partitioning, floorplanning, placement, and routing are discussed.

3.5.1   *Partitioning*

Partitioning is an important operation, preparing a system for the subsequent floorplanning, placement, and routing steps. In modern high performance VLSI systems, the layout cannot be directly synthesized due to the enormous number of circuit elements and connections. An integrated system is therefore decomposed into multiple smaller parts to simplify the placement and routing process [264]. The performance of the decomposed system may vary significantly depending upon the quality of the partitioning process. Splitting highly interconnected parts of a system may degrade performance and increase cost due to the longer wire lengths and greater delays. The primary objective of the partitioning process is therefore decomposing a circuit into multiple partitions to minimize the number of connections between partitions.

Partitioning was one of the first targets of design automation in integrated circuits and systems [265]. The primary motivation for partitioning, however, was not design simplification but rather increasing the ratio between the number of gates and pins to reduce the number of wire bonds to the off-chip discrete components [266]. One of the first examples of partitioning in the IC design process is the Large Integrated Monolithic Army Computer (LIMAC) in 1971 [267]. By proper partitioning, the number of inter-module pins was reduced by 50%. Early partitioning, however, was manual, limiting the size of the circuit being partitioned.

By representing a network of transistors as a graph, minimum cut algorithms can be applied to the partitioning process. The objective of the minimum cut of a graph $G(V, E)$ is to split vertex set V into two disjoint nonempty sets V_1 and V_2 to minimize a target metric. The minimum-cut problem is often called bisection or bipartition [268]. The edge set is split into sets of internal edges E_1 and E_2, and cut set $E_{1,2}$. The internal edges connect the nodes belonging to the same partition, while the edges in a cut set connect nodes from different partitions. A common metric in the partitioning process is the cut size $|E_{1,2}|$ or the total weight of the cut edges,

$$\sum_{e \in E_{1,2}} w(e), \tag{3.26}$$

where $w(e)$ is the weight of edge e.

Different variations of the minimum cut problem exist. In a minimum k-cut, the vertex set of a graph is divided into k disjoint sets. Certain nodes within a circuit graph may be placed within different partitions. In a minimum k-cut, the vertex set of a graph is divided into k disjoint sets. The hypergraph minimum-cut problem is an important extension of the regular minimum-cut problem, where multi-terminal nets of an integrated circuit are partitioned [269].

Efficient heuristic algorithms have been proposed to accelerate the partitioning process. The Kernighan-Lin (KL) algorithm [40] is considered the first partitioning algorithm applied to the design of integrated systems. In the KL algorithm, an unweighted graph $G(V, E)$ is bisected to equalize the size of the partitions. The algorithm starts by arbitrarily splitting node set V of a graph into two equal sets, A and B. During each iteration, a pair of nodes, $a \in A$ and $b \in B$, is chosen to ensure that swapping these nodes (i.e., placing a in B and b in A) achieves the smallest cut size. To efficiently identify the pair of nodes yielding the maximum reduction in cut size, a swapping gain metric is used. Suppose I_a is the number of neighbors of a within set A, and E_a is the number of neighbors of a within set B. Similarly, I_b and E_b denote the number of neighbors of b, respectively, in B and A. The difference between the number of external and internal connections of node a is

$$D_a = E_a + I_a, \tag{3.27}$$

which is similar for node b. The swapping gain G_{ab} is a measure of the reduction in the number of edges between A and B after swapping a and b, and is

$$G_{ab} = D_a + D_b - 2c_{ab}, \tag{3.28}$$

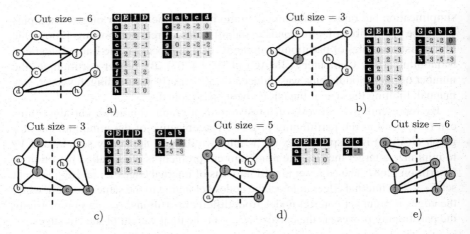

Fig. 3.26 Kernighan-Lin algorithm to partition a graph with eight nodes. a) Initial partition. The size of the partitions is equal, as required by the algorithm. In the left table, the number of external connections, number of internal connections, and the difference is calculated based on (3.27) and shown in, respectively, columns E, I, and D. In the right table, (3.28) is used to calculate the swapping gain for each pair of nodes. b) The partition after the first swap. The cut size is reduced to 3. Note that the swapped nodes are locked (shaded) and cannot be transferred to another partition. c) Partition after the second swap. The cut size is unchanged due to zero swapping gain during the swap. d) Partition after the third swap. The cut size is increased to five since the gain of swapping b and g is negative. Only two nodes are left, hence calculating the swapping gain is unnecessary but shown here for demonstrational purposes. e) Final partition.

where $c_{ab} = 1$ if an edge connecting a and b exists and 0 otherwise. With this metric, the gain due to node swapping is calculated for every pair of nodes, and the pair with the largest gain is swapped. The swapped nodes are locked and can no longer be moved during subsequent iterations. The swapping process continues until all of the nodes have been locked. The cut size after each iteration is recorded. The partition with the smallest cut set is the output of the partitioning algorithm.

An example of the KL algorithm applied to a graph with eight nodes is shown in Fig. 3.26. The initial partitioning is chosen arbitrarily and is depicted in Fig. 3.26a. Based on (3.27), the difference is initially calculated, as shown in the left table in Fig. 3.26a. The swapping gain is calculated based on (3.28) for each pair of vertices. The highest gain is $G_{ce} = 3$, indicating that by swapping nodes c and e, the cut size is reduced by 3. Nodes c and e are swapped and are therefore not considered for swapping until the end of the algorithm. The subsequent swaps are illustrated in Figs. 3.26b to 3.26e. The partitions with the smallest cut size after the first swap are determined after the first and second swaps.

Different enhancements to the KL algorithm have been proposed, such as an extension to include weighted graphs and unequally sized partitions [263]. The node swapping technique is generalized in the Fiduccia-Mattheyses (FM) algorithm to include hypergraphs and unequal partitions [270]. Unlike the KL algorithm, only a single node is transferred between partitions during each iteration. In addition, the size of the partitions is not maintained constant but rather bounded. No node can be removed or added to a partition if the size of a partition is equal, respectively, to the lower or upper bound. The FM algorithm became the basis for many subsequent partitioning algorithms that drastically improved performance. The 'foresight' method for bipartitioning was proposed by Krishnamurthy [271] to predict the cut size beyond the next iteration. In 1986, Sanchis [272] generalized this method for an arbitrary number of partitions. Gradient descent optimization is applied to the FM algorithm in [273], achieving a 40 to 50% reduction in cut size as compared to the state-of-the-art. A notable development in partitioning is the advent of multilevel clustering. In the METIS package [274], the target graph is first coarsened into a small network (on the order of hundreds of nodes). Partitioning is performed on the coarsened graph. The graph is iteratively uncoarsened, providing local corrections after each iteration. Genetic optimization has been applied to the partitioning problem in [275], improving performance and runtime. Recent developments incorporate advanced global optimization algorithms into the graph partitioning process. Examples include ant colony optimization [276] and particle swarm optimization [277].

3.5.2 Floorplanning

Floorplanning is the process of determining the shape and arrangement of the macro-blocks within a layout [263, 264]. Preliminary layout information not only aids in the subsequent placement process, but also provides valuable high-level information such as the die dimensions and location of the inputs and outputs. Each circuit partition is arranged into a rectangular block. The entire circuit constitutes a set of n rectangular blocks $M = \{m_1, \ldots, m_n\}$. The core problem in floorplanning is therefore rectangular packing (RP), the optimization problem of arranging a set of rectangles within a constrained region [278]. Two metrics are typically used to measure the quality of a floorplan [263, 279]. The first metric is the area efficiency $\eta_A(F)$ of a floorplan F,

$$\eta_A(F) = \frac{A_\square(F)}{\sum\limits_{m_i \in M} A(m_i)}, \qquad (3.29)$$

where $A_\square(F)$ is the area of the smallest rectangle enclosing a floorplan, and $A_\square(m_i)$ is the area of block m_i. The second metric is the estimated total wirelength. Precise wirelengths are not available until the routing process is completed. A connectivity matrix $C \in \mathbb{R}^{n \times n}$ is often used to estimate the total wirelength, where element c_{ij} denotes the degree of connectivity between blocks m_i and m_j [263]. The wirelength metric $L(F)$ of a floorplan F is therefore

$$L(F) = \sum_{m_i, m_j \in M} c_{ij} d_M(m_i, m_j), \tag{3.30}$$

where $d_M(m_i, m_j)$ denotes the Manhattan distance between the center point of blocks m_i and m_j, often referred to as Half-Perimeter WireLength (HWPL) [280]. A weighted sum of these metrics is often used as an objective function $Q(F)$,

$$Q(F) = w\eta_A(F) + (1 - w)\frac{L(F)}{L^*}, \tag{3.31}$$

where $w \in [0, 1]$ denotes the weight parameter which indicates the relative importance of the metrics, and L^* is a wirelength normalization parameter.

The floorplanning problem is highly complex and requires significant computational time even for a small number of blocks [278]. A graph-based floorplan representation, such as a constraint graph, is often used to simplify the representation of a floorplan [263, 281] (see Figs. 3.27a to 3.27c). Vertical and horizontal constraint graphs encode the relative position of the blocks, such as "block d should be placed to the right of block a." By considering these constraint graphs, infeasible constraints can be identified and a feasible solution can be determined.

More recent works often use a directed tree floorplan representation such as an O-tree [282] and B^*-tree [283]. Both of these tree representations can be used to unambiguously represent a floorplan. Based on the O-tree, a floorplan is constructed by performing a depth-first search (DFS) traversal. The root node represents the left boundary of a layout. Non-root nodes correspond to the circuit blocks placed tightly to the right boundary of the parent block. Child nodes of a parent are ordered from the bottom to the top. A floorplan corresponding to the O-tree can be reconstructed by using a depth-first search (DFS) traversal within a constraint graph, as illustrated in Fig. 3.28a. The B^*-tree is a binary directed tree. The root node within a B^*-tree corresponds to the left-bottom corner block within a layout. Two children of node v_i within the B^*-tree are $right(v_i)$ and $top(v_i)$, denoting the circuit blocks located, respectively, on the right and top sides of a block (see Fig. 3.28b). With these tree structures, a floorplan can be unambiguously specified using only a single graph, accelerating the floorplanning process. Recent developments apply global optimization to a tree representation of a floorplan. In [279], for example, evolutionary optimization is applied to an O-tree to minimize the combined area-wirelength metric, as described in (3.31).

3.5.3 Placement

Placement is the process of determining the precise location of the many circuit blocks within a physical layout. Similar to floorplanning, the objective of the placement process is to determine the location of the circuit blocks while complying with the physical design rules and minimizing a target metric. The placement

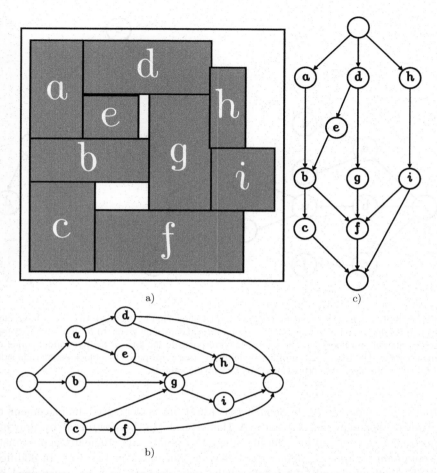

Fig. 3.27 Horizontal and vertical constraint graphs. a) Floorplan of an integrated system partitioned into nine modules. b) Horizontal constraint graph (HCG). An edge (i, j) in the HCG indicates that block i is located to the left of block j. b) Vertical constraint graph (VCG). An edge (i, j) in the VCG indicates that block i is located above block j.

process is however performed at the finer level, determining the position of smaller circuit blocks. The target metrics are also different. Total area and wirelength are typically considered during the floorplanning process. Standard metrics in placement include total wirelength, routing congestion, timing criticality, and power [263, 280].

Similar to floorplanning, a weighted sum of interconnect lengths can be used to prioritize the length of the important interconnects, such as the critical paths. A net often connects several terminals, necessitating the use of hypergraphs. A common method for estimating the length of an interconnect connecting the terminals of a hyperedge is a spanning tree, discussed in Subsection 2.7.2. A *rectilinear spanning*

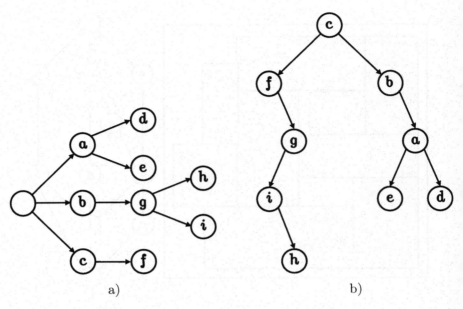

Fig. 3.28 Tree-based representation of the floorplan shown in Fig. 3.27a. a) An O-tree constructed by performing a depth-first search traversal of a vertical constraint graph. b) B^*-tree – a binary tree representation of a floorplan. The root node corresponds to the node at the left-bottom corner of the floorplan. The left successor of a node in the B^*-tree corresponds to a block on the right side, while the right successor corresponds to a block on the top side.

tree [131] and *rectilinear Steiner tree* [263] are widely used in placement to estimate the interconnect length. A Hanan grid [144], constructed by drawing horizontal and vertical lines through the target points, can efficiently approximate a minimum rectilinear Steiner tree [284], as illustrated in Fig. 3.29. In addition to the total wirelength, the length of the individual nets can be constrained. A timing driven placement procedure prioritizes the timing performance of a system. Synchronization completed at the RTL is often used to specify constraints on the length of certain wires [285, 286]. Routing congestion describes the relative ease of routing in the subsequent routing stage. Congestion maps are often used during the placement process to identify those regions where the routing is complicated and to adjust the placement to ease any congestion. Graph methods, such as Steiner tree synthesis and traversal algorithms, are frequently used to estimate the congestion [280]. For example, an A^* traversal is used in [287] to accurately and efficiently estimate the routing paths within the placement.

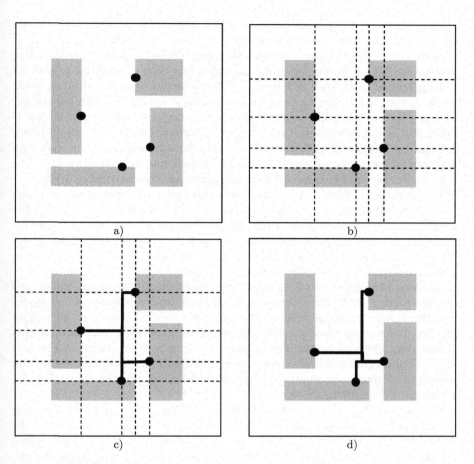

Fig. 3.29 Approximation of a minimum rectilinear Steiner tree based on a Hanan-grid. a) Initial block placement. The dots indicate the routing terminals. b) The Hanan grid is constructed by drawing lines through the terminal points. c) Based on a Hanan grid, a rectilinear minimum Steiner tree is approximated [284]. d) Final layout.

3.5.4 Routing

Interconnect routing is the final step in the physical layer design process in which the placed components are connected with wires. Similar to other physical layer procedures, the complexity of the routing process requires significant restrictions and simplifications to manage the computational complexity. Most routing problems, for example, restrict the wires to orthogonal (Manhattan) directions (also because non-perpendicular wires are difficult to manufacture) [288]. The spacing between the wires is often discretized, further reducing the solution space [289].

The history of VLSI routing can be traced to 1961, when Lee [41] proposed an algorithm for wire routing within a grid layout. The Lee's maze router is a modification of a classic breadth-first search (BFS) algorithm introduced in Subsection 2.7.1.2. Furthermore, the algorithm has been generalized to find the route in special cases, such as minimizing inter-wire crossings and avoiding bends when crossing wires. The Lee's algorithm has linear complexity with the number of nodes within a graph. The major issue in the Lee's maze router, as shown in Fig. 3.30a, is the excessively large number of traversed nodes. A^* search [290], discussed in Section 2.7.1.5, aims to minimize the search space by incorporating additional information, such as the physical location, into the routing process. The search space is therefore traversed more efficiently, as illustrated in Fig. 3.30b.

Comparison of planar routing algorithms. (a) Traversal during the Lee's maze routing algorithm. The initial and target positions are denoted with S and T. The numbers indicate the first digit of the distance from the source. Due to exhaustive traversal, the entire layout is almost traversed until the target is found. (b) In the A^* algorithm, the Euclidean distance is considered during the routing process. Fewer nodes are traversed as compared to maze routing.

Modern routing is a hierarchical process consisting of both global routing and detailed routing [263]. The coarse and detailed connections between circuit blocks are determined. In global routing, the layout is represented as a set of routing regions. The routing space is often represented in graph form, such as a channel connectivity graph or a switchbox connectivity graph, as illustrated in Fig. 3.31 [263]. Shortest path and minimum spanning tree algorithms are frequently used in routing to determine the shortest connection between two or more terminals [263] within these connectivity graphs.

In detailed routing, fine grain interconnect synthesis based on global routing is evaluated. The primary issue in detailed routing is wire congestion, since the number of nets competing for the same routing resources can be large. Detailed routing techniques are typically designed for two routing layers to permit the inter-wire crossing prohibited within a single layer. A common type of routing is channel routing [291, 292], where wires connect terminals on opposite sides of a routing channel, as illustrated in Fig. 3.32a. Channel routing is often aided by vertical and horizontal constraint graphs. The nodes in a vertical constraint graph (VCG) correspond to routing terminals. A directed edge (i, j) indicates that the terminal of node i is located directly above the terminal of node j. An example of a VCG is shown in Fig. 3.32d. Constructing a horizontal constraint graph (HCG) is similar to constructing an interference graph during the register allocation process (see Subsection 3.2.1). Horizontal ranges are determined for each of the routing terminals, as illustrated in Fig 3.32b. The nodes represent the routing nets. The edges connect two nodes within the HCG if the horizontal range of the corresponding nets overlap, as depicted in Fig. 3.32c.

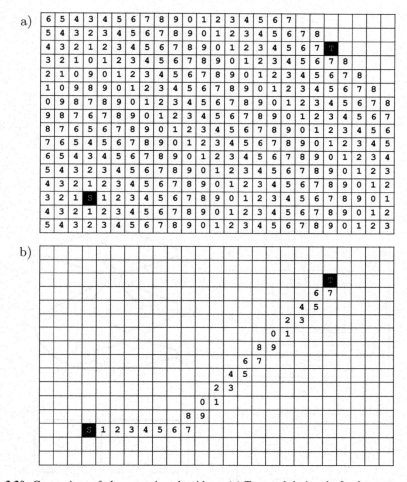

Fig. 3.30 Comparison of planar routing algorithms. (**a**) Traversal during the Lee's maze routing algorithm. The initial and target positions are denoted with S and T. The numbers indicate the first digit of the distance from the source. Due to exhaustive traversal, the entire layout is almost traversed until the target is found. (**b**) In the A^* algorithm, the Euclidean distance is considered during the routing process. Fewer nodes are traversed as compared to maze routing.

3.6 Summary

Early computing systems were sufficiently uncomplicated to be manually designed by several people. No systematic design process was necessary. The complexity of modern integrated systems, however, requires cooperation and close interactions among many people and a variety of advanced design tools and algorithms. Abstraction is a technique to decompose the design process into multiple, fairly independent layers. This layered design process greatly reduces the amount of

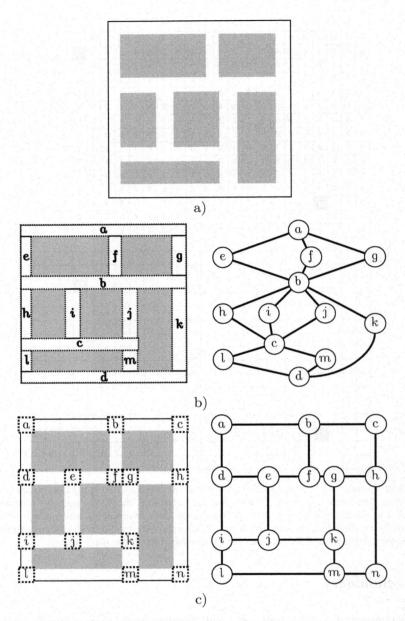

Fig. 3.31 Graph representation of a layout floorplan. a) Original floorplan, b) channel connectivity graph, and c) switchbox connectivity graph.

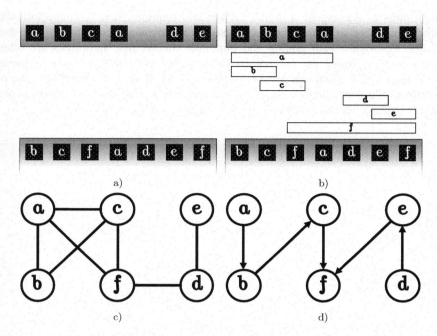

Fig. 3.32 Constraint graphs for channel routing of six nets. a) Arrangement of nets. The terminals are located on opposite sides of the channel. b) Net ranges for generating a horizontal constraint graph. A net range stretches between the leftmost terminal and rightmost terminal of a particular net. c) Horizontal constraint graph, and d) vertical constraint graph.

information that is simultaneously considered. Solutions at a particular abstraction layer depend less on the other abstraction layers and thus can be generalized, greatly enhancing the design process.

The design of VLSI systems can be divided into four abstraction layers, namely, register transfer, gate, circuit, and physical. At each abstraction layer, a VLSI system can be represented as a network. At the RTL, the system is viewed as interconnected registers and combinatorial logic blocks. By representing the RTL system as a timing graph, clock skew scheduling can be performed to synchronize the data flow within the network. At the gate layer, each register and logic block are decomposed into a network of logic gates. Logic circuits are represented and processed using graph-based OBDD and AIG to efficiently verify the functionality of a logic network. At the circuit layer, the logic circuits are viewed as a network of transistors. Electrical circuits are often represented as weighted graphs where the weight of the edges represents the conductance of a corresponding wire. A matrix representation of a graph enables the analysis of the voltages and currents at any node within an electrical circuit. At the physical layer, the network of gates and circuit blocks is embedded into a physical layout. Graph-based algorithms are widely utilized to convert a circuit into a layout. Circuit partitioning uses k-cut algorithms to decompose a digital system into multiple parts for more efficient processing. Tree structures are widely used in floorplanning, including O-tree and B-tree topologies. Path finding and minimum Steiner tree algorithms are often applied to determine the optimal interconnect routes. The flexibility of graphs combined with the inherent network structure of VLSI systems enables the use of graphs as a primary method to model and optimize a wide range of design issues across multiple abstraction layers in VLSI circuits and systems.

Chapter 4
Synchronization in VLSI

During the past decades, VLSI systems have undergone significant increases in computational performance, driven primarily by three factors, technology scaling, advances in circuit design, and evolution of computer architectures. Due to technology scaling, the switching delay of the transistors is significantly reduced, accelerating the combinatorial logic speed of the IC's. Furthermore, greater numbers of devices can be placed within the IC, providing much larger computational resources and fewer chip-to-chip constraints. Circuit design techniques have also significantly advanced, greatly elevating the performance of VLSI systems. For example, carry lookahead adders, such as Kogge-Stone Adder [293] and Brent-Kung Adder [294], enabled the addition of two n-bit numbers in $O\left(\log_2(n)\right)$ time, replacing the ripple carry adder which exhibits linear time complexity (see Fig. 4.1). Examples of computational speedup due to advances in computer architecture include instruction pipelining, branch prediction, and multicore processors. While accelerating the computational speed of integrated circuits, these factors have significantly increased the complexity of VLSI systems. A wide range of supporting infrastructural circuitry is necessary to ensure correct functionality while improving the performance of VLSI systems. This support circuitry may include power and ground distribution networks, signal buffers, thermal sensors, task schedulers, and self-test circuitry.

The clock distribution network is one of the primary infrastructural circuits in VLSI systems, distributing a periodic waveform to synchronize the data flow within a synchronous system [230]. The computational performance of a synchronous IC largely depends upon the clock distribution network. If the clock signal is not delivered at the required time, the data propagates through the logic network in an incorrect order, producing erroneous results. Precise delivery of the clock signal is therefore a vital part of any synchronous VLSI system.

The movement of data within a circuit is synchronized by a clock signal delivered to each register. The clock distribution network is the backbone of any synchronous VLSI system, distributing this periodic electronic signal to every

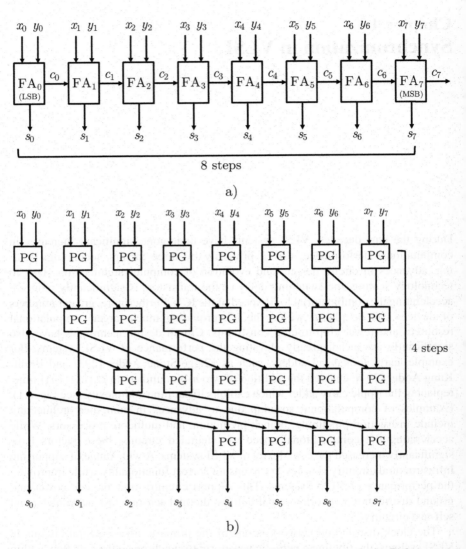

Fig. 4.1 Adder circuits. a) Eight bit ripple carry adder exhibiting linear complexity. Eight steps are required to perform the computation. b) Eight bit Kogge-Stone adder [295] exhibiting logarithmic complexity. Four steps are required to complete the operation.

synchronous element. Several features distinguish the clock distribution network from other VLSI subsystems [232]. The clock signal typically has the highest on-chip frequency and determines the maximum data processing speed. The network is also the longest on-chip signal driving a highly capacitive system, traveling across a circuit block or often across the entire IC. The clock network also exhibits the largest on-chip fanout, since all of the synchronous elements require a clock signal.

Synchronization presents a significant challenge in the sequential circuit design process ever since the inception of digital computing. For example, the Z1, the earliest programmable binary computer, built in 1938, suffered from poor synchronization due to mechanical stress in the device components [296]. The advent of fully electronic machines, such as the ENIAC in 1945 [19], eliminated the issue of mechanical stress, gradually enhancing the reliability of computers.

The architecture of early computer systems was relatively simple for clock networks despite being designed using *ad hoc* approaches. The rapid increase in circuit complexity exposed the need for a more systematic approach to the design of synchronous systems. Seminal studies of synchronous circuit architectures were presented in the 1950's, primarily by David A. Huffman, Edward F. Moore, and George H. Mealy [62, 297]. Unlike combinatorial (static) circuits, the output of the switching circuits, described in [297], depends not only upon the inputs, but also on the present state of the system. A graph theoretic basis for sequential circuits, described in [298, 299], further advanced the design of sequential circuits. These models enabled powerful graph-based techniques, such as a signal transition matrix and graph reduction methodologies [299, 300]. Subsequent developments in the synchronous circuit design process drastically increased the speed of computers, alongside advancements in other areas of computer development such as logic design [301, 302] and programming [303]. In 1945, the ENIAC operated at 100 kHz [304]. The ILLIAC IV, completed in 1966, operated at 25 MHz, performing a billion floating point operations per second [305].

Mealy or Moore finite state machines were sufficiently accurate for those systems where the delay of the signal lines is significantly smaller than the clock period. Soon, however, higher clock frequencies exposed certain practical issues in synchronous systems. Path delays became comparable to the clock period and therefore required special consideration to ensure correct operation. Electrical circuit analysis, such as Modified Nodal Analysis (MNA), can be used to analyze the timing of these sequential logic systems. Circuit-level techniques however exhibit poor scalability, requiring prohibitive runtime for the analysis of large systems [306]. Graph-based static timing analysis (STA) has therefore been developed to evaluate system timing without requiring electrical simulation. The program evaluation and review technique (PERT), developed by the United States Navy in 1958 [307], is regarded as the first tool to perform STA of integrated circuits. Interestingly, PERT was originally a project management tool for managing the flow of the Polaris nuclear missile program [308]. The technique was adopted for STA of logic systems in 1965 [309].

Unequal path delays within a clock distribution network can produce clock skew – different arrival times of the clock signal between sequentially-adjacent registers [230]. Clock skew is a significant performance and reliability issue in synchronous systems. In 1965, the issue of clock skew was first mentioned in the literature [310–312], where the authors noted the existence of inhomogeneous clock arrival times in synchronous circuits. In [311], one of the earliest analyses of the bounds on a clock period was presented, where system performance was noted as limited by clock skew, logic and interconnect delay, and propagation delay uncertainty. In 1969, Cotten first presented an analysis of the clock period in a multistage pipeline while considering clock skew [313]. Increasing the clock period, however, remained the primary method for managing clock skew, ignoring clock skew induced race conditions which are independent of clock frequency [230]. The operating frequency of the ILLIAC IV, for example, was limited to 5 MHz due to, in no small part, system clock skew. Similar discussions are reported in [314, 315].

The severity of the effect of clock skew affects the clocking architectures. Constraining the design was the primary strategy to minimize clock skew in early computers. For example, in [316], the propagation delay of all of the data paths are (nearly) equalized, and the clock pulse generator is placed at the physical center of the computer. With the clock signal distributed radially, the effects of clock skew are mitigated by removing all possibilities of race conditions [316]. The concept of delay equalization within a clock network was developed into a symmetric H-tree topology, introduced in [317]. Ensuring zero clock skew, however, requires significant on-chip resources. A less stringent, globally asynchronous, locally synchronous (GALS) clocking paradigm was introduced in 1984 [318]. By splitting an integrated circuit into separate clock regions, the delay from a clock source to each register is reduced, typically producing less clock skew. The transfer of data among the separate clock domains is established by an asynchronous communication protocol [319].

In the 1980's, several graph-based EDA tools emerged that consider nonzero path delays. For example, in [320], a data path delay analysis tool is presented. If the clock skew of the target system exceeded a specified value, the user is warned. In 1984, a critical path weighting methodology for on-chip layout optimization is described [321]. The data path and clock delay are analyzed and the layout is accordingly adjusted. Other notable timing analysis tools include Crystal [322], CELTIC [323], and SCALD [234]. All of these STA tools were developed during the early 1980's, allowed VLSI circuits of growing complexity to be efficiently verified for static timing violations.

Since the discovery of clock skew, clock skew has been viewed as a deleterious effect that required containment. Elimination of clock skew is, however, not necessary for the correct functionality of a synchronous system. Since 1989 [238], design techniques incorporating nonzero clock skew have been explored. A timing verification algorithm is presented in [324], where the delay of the clock signal is considered. Clock skew optimization, *i.e.*, intentional adjustment of clock skew to improve the delay characteristics of a synchronous system, is discussed in [231–233, 325–327]. Algorithms for synthesizing the clock tree layout are presented in [235, 328]. An important role in these techniques is played by graph theory, introduced into clock distribution network design and analysis during the 1990's [237, 240, 329–332]. In this chapter, graph-based methods for the design of clock distribution networks are described. An overview of graph-based timing analysis is presented in Section 4.1. Clock skew scheduling is discussed in Section 4.2. Clock tree layout synthesis is discussed in Section 4.3.

4.1   Graph-based timing analysis

A synchronous circuit is a sequential logic system where the data flow is coordinated by a clock signal. Unlike combinatorial circuits, where data processing starts immediately after arrival of an input signal, sequential circuits store data in memory units. Examples of typical memory elements in CMOS are flip flops and latches. In flip flops, the release of a datum is triggered by a rising or falling edge of the clock signal (edge sensitive), whereas in latches, the level of the clock signal triggers the release of the datum. The arrival of the clock signal at a memory element initiates a new clock period for a data path. The released datum propagates through the combinatorial logic subcircuit toward the next memory unit, where the datum is stored until the arrival of the next clock signal.

A sequential circuit consists of four primary components:

1. Registers that store intermediate data and release the data upon the arrival of a clock signal. The two primary types of registers are flip flops and latches where each element releases the stored datum upon sensing, respectively, an edge and level of the clock signal.
2. Clock generator circuit, such as a phase locked loop (PLL) [333] utilizing a voltage controlled oscillator (VCO) to produce a clock signal, as illustrated in Fig. 4.2.
3. Clock distribution network – an interconnect network connecting the clock generator to the registers. Due to the nonideal characteristics of the network, the clock signal is delivered with a delay that varies depending upon the location of the registers.
4. Combinatorial logic that performs the computation.

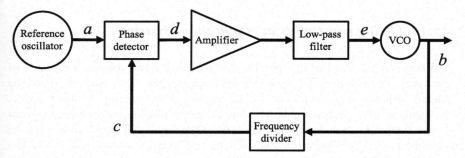

Fig. 4.2 Structure of a phase locked loop (PLL). a) A low frequency oscillator generates a reference periodic signal [334]. This signal exhibits low variations in response to environmental conditions, such as the temperature. b) A high frequency voltage controlled oscillator (VCO), such as a relaxation oscillator [335] or Pierce oscillator [336], generates a high frequency signal. The output of the VCO exhibits high sensitivity to parameter variations. c) The frequency of the VCO output is downscaled by a frequency divider. d) The phase detector compares the phase of signal c to the phase of the reference oscillator. e) The change in average phase difference is converted into the input voltage of the VCO, thereby maintaining a constant high frequency at the PLL output.

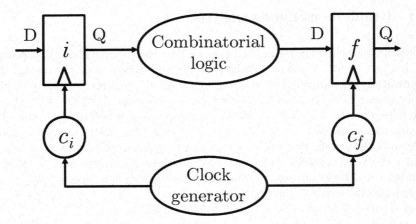

Fig. 4.3 Local data path. The datum enters at input D of register i and is stored until the clock signal c_i arrives at register i at time t_i. The datum is released from terminal Q and propagates through the combinatorial logic toward input D of register f. The datum is stored until the clock signal c_f arrives at register f at time t_f. The datum captured during the previous clock period is released at terminal Q of register f.

A circuit model of a sequential circuit is shown in Fig. 4.3. An input datum enters the sequential circuit and is stored in a register. Upon arrival of the clock signal, the datum is released and the signal propagates through combinatorial logic until the signal arrives at the next register. The process continues with each new clock signal received by the registers. Note that the delay of the clock signal to the different registers is not necessarily uniform.

4.1.1   Timing constraints in synchronous systems

A datum is initially stored in register i until the clock signal triggers the release of the datum. For example, the clock signal arrives at register i at time $t = 0$ and the released datum propagates through the combinatorial logic circuit. At time $t = d_{i,f}$, the datum arrives at register f and is stored until the clock signal arrives at register f. For the purpose of clock skew scheduling, the sequential circuit model can be simplified into a directed multigraph G with self-loops, as shown in Fig. 4.4. The nodes of G represent the registers, and the edges represent the combinatorial logic within the data paths. Maximum and minimum delays are assigned to edges within the graph, representing lower and upper bounds on the propagation delay.

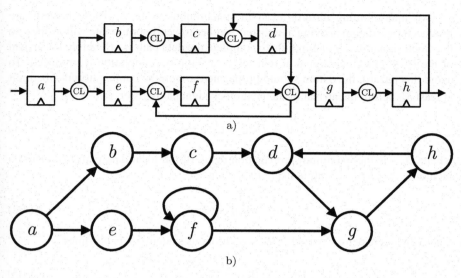

Fig. 4.4 Conversion of a sequential logic circuit into a timing graph. The clock distribution network is not shown for clarity. a) Sequential logic circuit, and b) equivalent multigraph. CL stands for combinatorial logic.

4.1.1.1 Local timing constraints

The constraints described in this section consider only sequentially-adjacent registers and are therefore referred to as *local timing constraints* [239]. Consider a properly functioning local data path consisting of initial register i, combinatorial logic, and final register f, as shown in Fig. 4.3. At time t_n^i, the nth clock signal arrives at register i. The input of register i remains constant for at least δ_h^i, the hold time of register i. The hold time is the minimum time *after* arrival of the clock signal during which the register input is maintained constant to ensure correct transfer of the datum between the input and output of the register. After the clock-to-output delay t_{C-Q}^i of register i, the datum stored in register i is released and propagates through the combinatorial logic. After propagating through the combinatorial logic path, the datum arrives at register f. Upon arrival, the datum awaits the clock signal of period $(n + 1)$ for at least δ_s^f, the setup time of register f. The setup time is the minimum time *before* the arrival of the clock signal during which the register input is maintained constant to ensure correct transfer of data. At time t_{n+1}^f, the clock signal of the next period $(n + 1)$ arrives at register f. After delay t_{C-Q}^f, the datum is released to the next combinatorial logic block.

Two major timing hazards exist in synchronous systems. These hazards are produced by a datum arriving at register f either too early or too late with respect to the clock signal. If the datum fails to complete the data path (i, f) before the arrival of the $(n + 1)^{\text{th}}$ clock signal at register f, a setup time violation is produced. To avoid this hazard, the datum should propagate through the combinatorial subcircuit at least δ_s^f before the arrival of the next clock signal at register f,

$$t_n^i + D_{i,f} \leq t_{n+1}^f - \delta_s^f, \tag{4.1}$$

where $D_{i,f}$ is the maximum propagation delay of the local data path from i to f, including the clock-to-output delay t_{C-Q}^i. Since

$$t_{n+1}^f = t_n^f + T_{CP}, \tag{4.2}$$

where T_{CP} is the clock period of the synchronous system, (4.1) becomes

$$s_{i,f} \leq T_{CP} - D_{i,f} - \delta_s^f, \tag{4.3}$$

where $s_{i,f}$ is the *clock skew* of the local data path (i, f),

$$s_{i,f} = t_n^i - t_n^f. \tag{4.4}$$

A positive clock skew indicates that the clock signal arrives at register i *after* arriving at register f. Conversely, a negative clock skew indicates that the clock signal arrives at register i *before* arriving at register f. Constraint (4.3) determines the upper bound $u_{i,f}$ on the clock skew of data path (i, f),

$$u_{i,f} = T_{CP} - D_{i,f} - \delta_s^f. \tag{4.5}$$

A clock skew greater than $u_{i,f}$ produces *zero clocking* (or a setup time violation) [233] – a data hazard preventing the correct transfer of the datum to the next combinatorial data path.

Alternatively, if the datum arrives at register f too early, the transfer of the preceding datum from the input to the output of f is disrupted. The combinatorial circuit should therefore be completed at least δ_h^f after the nth clock signal arrives at f,

$$t_n^i + d_{i,f} \geq t_n^f + \delta_h^f. \tag{4.6}$$

The clock skew is therefore bound by

$$s_{i,f} \geq -d_{i,f} + \delta_h^f, \tag{4.7}$$

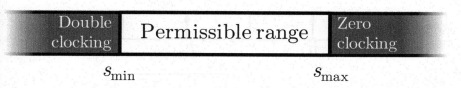

Fig. 4.5 Permissible range of clock skew $[s_{min}, s_{max}]$ for a local data path. A clock skew smaller than s_{min} produces a race condition (double clocking), whereas a clock skew greater than s_{max} produces a clock period violation (zero clocking). To ensure correct functionality, the clock skew should be within the permissible range.

where $d_{i,f}$ is the minimum propagation delay of the local data path from i to f, including the clock-to-output delay t_{C-Q}^i. The lower bound $l_{i,f}$ on clock skew is therefore

$$l_{i,f} = -d_{i,f} + \delta_h^f. \tag{4.8}$$

A clock skew less than $l_{i,f}$ produces *double clocking* (or a hold time violation) [233] – a data hazard, where incorrect data is transferred to the next combinatorial data path. Note that the lower bound on clock skew does not depend upon the clock period.

The permissible range (PR) of clock skew $PR_{i,f} = [l_{i,f}, u_{i,f}]$ is the range of clock skew that ensures correct functionality of a data path (i, f) [239]. This concept is illustrated in Fig. 4.5. Exceeding the lower bound produces a race condition [233]; the datum stored in f is not released in time and is overwritten by the output of i, as illustrated in Figs. 4.6a to 4.6c. Exceeding the upper bound produces zero clocking, where the clock signal arrives at register f before the datum passes through the combinatorial logic, as shown in Fig. 4.6e. Correct timing of a local data path is ensured when the clock skew is within the permissible range [237], as shown in Fig. 4.6d.

Data path delay uncertainty greatly affects the timing characteristics of a synchronous system. Observe from (4.3) and (4.7), that increasing the *data skew* — the difference between the maximum and minimum data propagation delay — narrows the PR of a data path,

$$u_{i,f} - l_{i,f} = T_{CP} - \delta_s^f - \delta_h^f - DS_{i,f}, \tag{4.9}$$

where $DS_{i,f} = D_{i,f} - d_{i,f}$ is the data skew of data path (i, f). Since the width of a PR cannot be negative, the minimum feasible clock period for a single data path is

$$T_{CP}^{min} = \delta_s^f + \delta_h^f + DS_{i,f}. \tag{4.10}$$

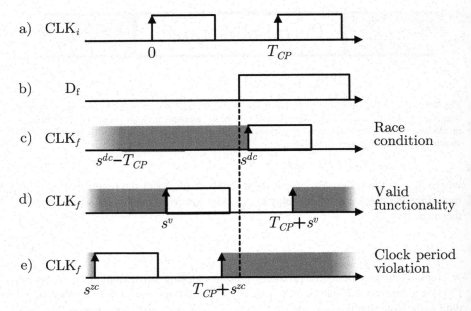

Fig. 4.6 Functionality of a local data path shown in 4.3 for different clock skews. a) Clock signal at initial register. b) Arrival of datum at final register. c) Clock skew $s^{dc} < s_{min}$, producing a race condition (double clocking). The datum arrives at register f before the clock signal of the same period arrives at register f. d) A valid clock skew s^v, where $s_{min} \leq s^v \leq s_{max}$. e) Clock skew $s^{zc} > s_{max}$, producing a clock period violation (zero clocking). The datum arrives at register f after the clock signal of the next period arrives at register f.

The clock period cannot be further reduced with clock skew scheduling. Expression (4.10) is however not the only constraint on the clock period. Global clock period constraints also exist, as discussed in the upcoming section.

4.1.1.2   Global timing constraints

Equations (4.5) and (4.8) describe the PR for sequentially-adjacent registers. Satisfying the local timing constraints is necessary but does not guarantee correct functionality of a synchronous system. Global timing constraints also exist that require non-adjacent registers to be considered. Two topologies within a graph influence the synchronization process – reconvergent global data paths and logic cycles [231]. To better understand the effects of these topologies on synchronization, it is necessary to review the effects of a serial connection of registers on the PR.

Serial data path.

The clock skew within a synchronous system shares many properties with voltage within an electrical circuit. For example, the clock skew describes the difference in arrival time of a clock signal, while voltage describes the difference in electric potential. Another example of the similarity of these two concepts is the addition of skew (voltage) along a path within an underlying graph. Consider three sequentially-adjacent registers, R_1, R_2, and R_3. The clock skew $s_{1,3}$ between R_1 and R_3 is

$$s_{1,3} = s_{1,2} + s_{2,3}. \tag{4.11}$$

More generally, the clock skew between registers R_1 and R_n connected by path $p = \{r_1, \ldots, r_n\}$ is the sum of the clock skews along the path p [230],

$$s_{1,n} = \sum_{i=1}^{n-1} s_{i,i+1}. \tag{4.12}$$

The upper (u_p) and lower (l_p) bounds on clock skew are therefore a sum of the bounds of each local data path along path p,

$$u_p = nT_{CP} - \sum_{(i \to f) \in p} \left[\delta_s^f + D_{i,f} \right], \tag{4.13}$$

$$l_p = - \sum_{(i \to f) \in p} \left[-\delta_h^i + d_{i,f} \right]. \tag{4.14}$$

The resulting PR is wider than the PR of each local data path in p. The PR of the non-adjacent registers is therefore rarely discussed in the literature, since the PR of a sequentially-adjacent local data path is more restrictive. In reconvergent data paths and logic cycles, however, the PR of the non-adjacent registers affects the synchronization process.

Reconvergent (parallel) paths.

Systems with parallel paths are common in logic circuits. A generalized example of this topology is shown in Fig. 4.7a. Registers d and c represent, respectively, divergent and convergent registers. Those paths with the smallest and greatest propagation delay are referred to, respectively, as the short and long path. Reconvergent paths shrink the PR of the data path and impose constraints on the minimum clock period [231],

$$T_{CP}^{d,c} = \frac{D_L - d_S + \delta_c^s + \delta_c^h}{|m - n + 1|}, \tag{4.15}$$

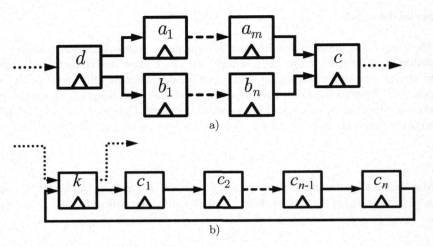

Fig. 4.7 Examples of global data path constraints in sequential circuits. a) Reconvergent path with long path $[d, a_1, \ldots, a_n, c]$ and short path $[d, b_1, \ldots, b_m, c]$, and b) cyclic path $[i, c_1, \ldots, c_n, i]$.

where $D_{d,c}$ is the maximum delay of the long path with m registers, and $d_{d,c}$ is the minimum delay of the short path with n registers.

Furthermore, the topology of a reconvergent path limits the clock skew between divergent and convergent registers. Satisfying the PR for one of the parallel paths is not sufficient to ensure correct operation, since the PR for a different parallel path may be violated. The equivalent PR for a reconvergent path $(d \rightsquigarrow c)$ is therefore the intersection of the PR for each parallel path p,

$$l_{d,c} = \max_{p \in (d \rightsquigarrow c)} (l_{d,c}^p) \tag{4.16}$$

$$u_{d,c} = \min_{p \in (d \rightsquigarrow c)} (u_{d,c}^p). \tag{4.17}$$

Special consideration is required to analyze parallel paths with feedback. The clock skew between any two registers exhibits antisymmetry, *i.e.*, clock skew in a feedback path is the negative of the clock skew of the forward data path [230],

$$s_{i,j} = -s_{j,i}. \tag{4.18}$$

The PR of a feedback path is therefore

$$l_{j,i} = -u_{i,j} \tag{4.19}$$

$$u_{j,i} = -l_{i,j}. \tag{4.20}$$

Therefore, if feedback $(c \rightsquigarrow d)$ exists in a reconvergent path, the clock skew bounds are converted into the equivalent PR of path $(d \rightsquigarrow c)$ before determining the intersection of the PR,

$$l_{d,c} = \max\left(\max_{p\in(d\rightsquigarrow c)}\left(l_{d,c}^p\right), -\min_{p\in(c\rightsquigarrow d)}\left(u_{c,d}^p\right)\right) \tag{4.21}$$

$$u_{d,c} = \min\left(\min_{p\in(d\rightsquigarrow c)}\left(u_{d,c}^p\right), -\max_{p\in(c\rightsquigarrow d)}\left(l_{c,d}^p\right)\right). \tag{4.22}$$

Cyclic data paths.

An example of a sequential circuit containing cycle $(k \rightsquigarrow k)$ with n nodes is shown in Fig. 4.7b. The datum enters register k, returning to the same register in n cycles. The cycle traversal is completed in time nT_{CP}, limiting the clock period to the average propagation delay of the datapath [231],

$$T_{CP}^{kk} = \frac{1}{n} \sum_{(i,j)\in(k\rightsquigarrow k)} (D_{i,j} + \delta_j^s). \tag{4.23}$$

To determine the lower bound on the clock period, all reconvergent paths and cycle paths need to be determined. To find all reconvergent paths, all simple paths between each divergent and convergent register also need to be determined. A simple path between two vertices can be found in linear time [337]. A common algorithm for finding all cycles within a graph is proposed by Johnson [338], where the search is completed in $O((|V| + |E|)(c + 1))$ time, where c is the number of cycles. The number of cycles and reconvergent paths within a graph can however be prohibitively large, up to $n!$ in a complete graph. Although this number is significantly smaller in practical graphs, this requirement limits the maximum size of a circuit for which the permissible range can be accurately determined. An effective method for controlling the size of a circuit is the GALS design paradigm [339]. By decomposing the circuits into separate clock domains, the permissible range can be efficiently determined within each partition.

4.1.1.3 Constraint graph

For timing graph $G = (V_G, E_G)$, clock skew constraints (4.7) and (4.3) produce $2|E_G|$ inequalities that describe the PR of each data path. These constraints can be transformed into a system of inequalities,

$$\begin{bmatrix} Y_d^T \\ -Y_d^T \end{bmatrix} \mathbf{t} - \begin{bmatrix} U \\ -L \end{bmatrix} \leq \mathbf{0}_{2|E_G|}, \tag{4.24}$$

where $\mathbf{0}_n$ is $n \times 1$ zero vector, $\mathbf{t}$ is the vector of clock arrival times, Y_d is the directed incidence matrix of a timing graph G, and $U \in \mathbb{R}^{|E_G|}$ and $L \in \mathbb{R}^{|E_G|}$ are vectors describing, respectively, the upper bound $u_{i,f}$ and lower bound $l_{i,f}$ on the clock skew for each data path (i, f). The system expressed by (4.24) is the *system of difference constraints* [107],

$$Ax \le b, \tag{4.25}$$

where A is the coefficient matrix whose rows contain one 1 and one -1, x is the vector of variables, and b is the vector of constraints. These systems consist of inequalities of the form,

$$x_i - x_j \le b_{i,j}. \tag{4.26}$$

A system of difference constraints can be efficiently described using a *constraint graph* [107]. Fundamentally, a constraint graph is a directed graph $G_c = (V \cup \{v_0\}, E \cup E_0)$. Set V is a set of variables in $\mathbf{x}$. An edge $(i, j) \in E$ connects two nodes if there exists the inequality,

$$x_j - x_i \le b_{j,i}. \tag{4.27}$$

The weight of edge (i, j) is $b_{j,i}$. An additional node v_0 is added to the node set. Edges in set E_0 have zero weight and connect v_0 to each node in V.

In the context of synchronization, this general definition can be transformed into a more specific definition of a *timing constraint graph*. The timing constraint graph $G_C = (V_G \cup \{v_0\}, E_0 \cup E_l \cup E_u, e_l, e_u)$ is derived from the timing graph G and depicts the minimum and maximum clock skew constraints for each data path [331]. The node set of a constraint graph is identical to the node set of a timing graph G with an added node v_0. The set E_l is a copy of the edge set of the timing graph. The weight function $e_L : E_l \to \mathbb{R}$ associates each edge within E_l with the double clocking constraint (4.7),

$$e_l(i, f) = -l_{i,f} = d_{i,f} - \delta_h^f. \tag{4.28}$$

Set E_u is comprised of the reversed edges of E_G,

$$E_u \equiv \{(f, i)|(i, f) \in E_G\}. \tag{4.29}$$

The function $e_u : E_u \to \mathbb{R}$ associates each edge within set E_u with the zero clocking constraint (4.3),

$$e_u(i, f) = u_{f,i} = T_{CP} - D_{f,i} - \delta_s^i. \tag{4.30}$$

To illustrate the construction of the timing constraint graph, consider the sequential circuit depicted in Fig. 4.8a. Four inequalities describe the zero clocking

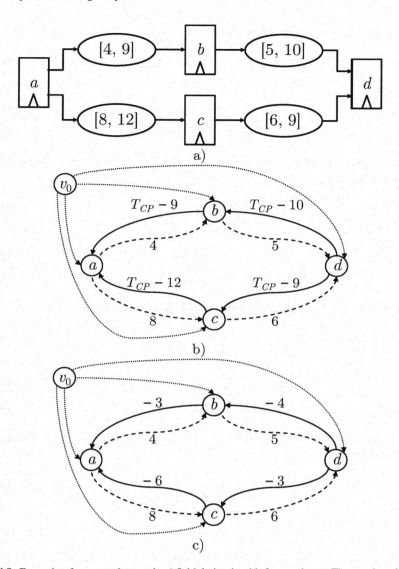

Fig. 4.8 Example of a constraint graph, a) Initial circuit with four registers. The numbers in the square brackets denote the maximum and minimum propagation delay in arbitrary time units (tu). b) Constraint graph. The solid edges belong to set E_u and denote the setup time constraint. Observe that the weight of these edges is a function of the clock period T_{CP}. The dashed edges belong to set E_l and denote the hold time constraint. The dotted edges denote the zero weight edges connecting v_0 with all other nodes. c) Constraint graph for $T_{CP} = 6$ tu. Observe that zero cycle $[a, b, d, c, a]$ is produced, indicating that no further reduction in clock period is possible without modifying the data paths.

constraints (ignoring the setup and hold time),

$$t_a - t_b \geq -4 \Rightarrow t_b - t_a \leq 4, \tag{4.31}$$

$$t_a - t_c \geq -8 \Rightarrow t_c - t_a \leq 8, \tag{4.32}$$

$$t_b - t_d \geq -5 \Rightarrow t_d - t_b \leq 5, \tag{4.33}$$

$$t_c - t_d \geq -6 \Rightarrow t_d - t_c \leq 6. \tag{4.34}$$

These equations form the edge set $E_l = \{(a, b), (a, c), (b, d), (c, d)\}$. The weight of these edges are

$$e_l(a, b) = 4, \tag{4.35}$$

$$e_l(a, c) = 8, \tag{4.36}$$

$$e_l(b, d) = 5, \tag{4.37}$$

$$e_l(c, d) = 6. \tag{4.38}$$

Another four inequalities describe the zero clocking constraint,

$$t_a - t_b \leq T_{CP} - 9, \tag{4.39}$$

$$t_a - t_c \leq T_{CP} - 12, \tag{4.40}$$

$$t_b - t_d \leq T_{CP} - 10, \tag{4.41}$$

$$t_c - t_d \leq T_{CP} - 9. \tag{4.42}$$

These equations form the edge set $E_u = \{(b, a), (c, a), (d, b), (d, c)\}$. The weight of these edges is

$$e_u(b, a) = T_{CP} - 9, \tag{4.43}$$

$$e_u(c, a) = T_{CP} - 12, \tag{4.44}$$

$$e_u(d, b) = T_{CP} - 10, \tag{4.45}$$

$$e_u(d, c) = T_{CP} - 9. \tag{4.46}$$

A set of zero weight edges E_0 connecting v_0 with each node is added to the graph. The resulting constraint graph is depicted in Fig. 4.8b.

A feasible solution to a system of difference equations can be found by finding shortest path from node v_0 to each of the nodes within the constraint graph [107]. Consider the constraint graph for $T_{CP} = 6$ tu, depicted in Fig. 4.8c. The shortest paths to each node within the graph are

$$t_a = -9, \tag{4.47}$$

$$t_b = -5, \tag{4.48}$$

$$t_c = -3, \tag{4.49}$$

$$t_d = 0. \tag{4.50}$$

The permissible range of each data path is

$$PR_{a,b} = [-4, -3], \tag{4.51}$$

$$PR_{a,c} = [-8, -6], \tag{4.52}$$

$$PR_{b,d} = [-5, -4], \tag{4.53}$$

$$PR_{c,d} = [-6, -3]. \tag{4.54}$$

The clock skew in each local data path is within the corresponding PR. Recall from Subsection 2.7.1, that if a negative cycle exists within a graph, a shortest path does not exist since the cost of a traversal can be made arbitrarily small by repeatedly traversing the negative cycle. As demonstrated in [107], a feasible solution to (4.24) only exists if no negative cycle exists within the constraint graph. A negative cycle can be detected using the Bellman-Ford algorithm [331], as discussed in Subsection 2.7.1.4.

Note that the upper bound on clock skew is a function of the clock period. A minimum clock period T_{CP}^{min} therefore exists that satisfies (4.24) while producing a zero weight cycle [331]. Observe that cycle $[a, b, d, c, a]$ has zero weight in the constraint graph depicted in Fig. 4.8c. A clock period of 6 tu is therefore minimum for the specific circuit and cannot be further reduced without modifying the data paths. Using the delay insertion method, discussed in Subsection 4.2.2, a smaller clock period can be produced by adding delay to selected data paths.

4.2   Clock skew scheduling

Clock skew scheduling is the process of determining the individual clock skew for each local data path to enhance the characteristics of a system of sequential logic. Three primary quality metrics of a sequential logic circuit exist, robustness, performance, and power dissipation. All of these metrics can be enhanced with clock skew scheduling.

4.2.1   Robustness

Improved reliability of a synchronous system against timing violations is a major advantage of clock skew scheduling. To illustrate the importance of clock skew

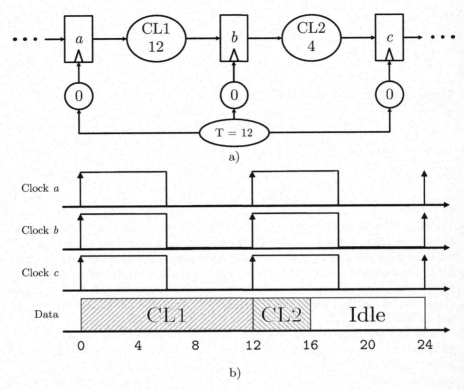

Fig. 4.9 An illustrative example of a system with three flip flops exhibiting zero clock skew. For simplicity, the internal register delays, namely, clock-to-output delay, setup time, and hold time, are neglected. a) Topology of the system. CL and T refer, respectively, to the combinatorial logic and clock period. The numbers indicate delays in time units (tu). b) Data flow within the system. At time $t = 0$, a datum is released from flip flop a to CL1. After completing CL1, the datum reaches flip flop b. The clock signal triggers the release of the data into CL2. After 4 tu, the datum reaches flip flop c and is stored in flip flop c for 8 tu.

scheduling, consider a data path consisting of three registers, a, b, and c, and two combinatorial logic blocks, CL1 and CL2, as illustrated in Fig. 4.9a. The clock period T is 12 time units (tu), barely sufficient for completing the CL1. Observe, however, that the CL2 is completed in only 4 tu. An *idle* time $t_{idle} = 8$ tu therefore exists during which a datum is stored in register c until it is released after the clock signal arrives at register c, as depicted in Fig. 4.9b. Assuming the setup time, clock to output time, and hold time of the registers are negligible, the PR of (a, b) and (b, c) are, respectively, $[-12, 0]$ tu and $[-4, 8]$ tu. Observe that a zero clock skew requires data path (a, b) to operate at the edge of the PR. In a practical system, the actual arrival time of the clock signal may be different from the predicted arrival time due to parameter variations, such as manufacturing defects, temperature fluctuations, and electromagnetic interference. The clock skew between registers a and b may therefore become positive, shifting the clock skew beyond the PR. Alternatively, the delay of CL1 can exceed 12 tu, thus failing to deliver the datum to register b before the arrival of the clock signal. To mitigate this issue, the clock period can be increased to accommodate variations in propagation delay. The speed of a synchronous system is however reduced.

With clock skew scheduling, the 8 tu of idle time in (b, c) can be exploited to increase the time allocated for CL1. Consider the topology shown in Fig. 4.10a. The clock arrives at register b delayed by 4 tu, producing clock skews, $s_{a,b} = -4$ tu and $s_{b,c} = 4$ tu. The skew $s_{a,b}$ is now farther from the edge of the PR. The time allocated for CL1 is increased to $T - s_{a,b} = 16$ tu, while the time allocated for CL2 is reduced to $T - s_{b,c} = 8$ tu. The idle time of CL2 in the zero skew example is effectively split between CL1 and CL2. Clock skew scheduling is therefore often called "cycle stealing," since part of the cycle is stolen from the fast path (b, c) and is given to the slow path (a, b) [324, 340]. Process variations, affecting the propagation delay of CL1, are therefore less likely to produce a timing violation. The improved tolerance to process variations may greatly increase the manufacturing yield of an integrated circuit [231]. Note that since the clock period is unchanged, the overall system performance is unaffected.

The primary objective of robustness driven clock skew scheduling is shifting the clock skew towards the center of the PR [239]. A deviation of the clock skew from the center of the PR is

$$f(\mathbf{s}) = ||\mathbf{s} - \mathbf{s}^*||, \tag{4.55}$$

where $\mathbf{s} \in \mathbb{R}^{|E|}$ is the vector of clock skew for each local data path, and each entry in $\mathbf{s}^* \in \mathbb{R}^{|E|}$ is at the center of the PR for the corresponding data path. By minimizing $(f(\mathbf{s}))^2$, the clock skew can be chosen to maximize the robustness of the synchronous system.

Several constraints exist that prevent an arbitrary adjustment of $\mathbf{s}$. The clock skew of each data path is limited by the PR, as described by (4.5) and (4.8). The clock skew between the terminal registers of a circuit module is often set to zero [231],

$$s_{i,j} = 0 \forall i, j \in B_{I/O}, \tag{4.56}$$

where $B_{I/O}$ is the set of those registers within a module connected to the I/O terminals. The primary motivation for this constraint is easier synchronization of the module with the rest of the IC [230].

Cyclic paths within the timing graph prevent shifting the clock skew of each datapath towards the center of the PR. The sum of the clock skews along cyclic path $(k \rightsquigarrow k)$ is zero due to the linear dependence among the clock skews [231],

$$\sum_{(i \rightsquigarrow j) \in (k \rightsquigarrow k)} s_{i,j} = 0. \tag{4.57}$$

The number of cycles within the graph can be prohibitively large. For example, the number of cycles in a complete directed graph of degree n is

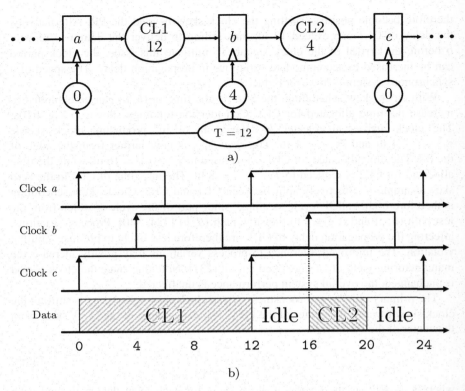

a)

b)

Fig. 4.10 A system exploiting clock skew to enhance robustness. a) Topology of the system. The clock signal arriving to register b is delayed by 4 tu. b) Data flow within the system. After completing CL1, the datum reaches flip flop b at time 12 tu and is stored for 4 tu in register b. Similarly, after completing CL2, the datum is stored for 4 tu in register c.

$$\sum_{i=1}^{n-1} \binom{n}{n-i+1} (n-i)!.$$ (4.58)

Only a subset of cycles is however necessary to consider, as depicted in Fig. 4.4. Three cycles (excluding the self-loop at node f) are present in the system, namely $[dgfeabcd]$, $[dghd]$, and $[dhgfeabcd]$, forming three linear equations,

$$s_{d,g} + s_{g,f} + s_{f,e} + s_{e,a} + s_{a,b} + s_{b,c} + s_{c,d} = 0$$ (4.59)

$$s_{d,g} + s_{g,h} + s_{h,d} = 0$$ (4.60)

$$s_{d,h} + s_{h,g} + s_{g,f} + s_{f,e} + s_{e,a} + s_{a,b} + s_{b,c} + s_{c,d} = 0.$$ (4.61)

Equation (4.61) can be expressed as the sum of (4.59) and (4.60), and therefore does not impose additional constraints on the clock skew within a system. Observe the

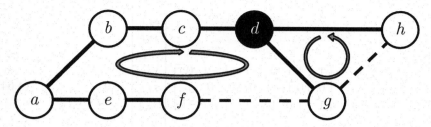

Fig. 4.11 Determining the cycle basis of a sequential circuit based on a spanning tree. Three cycles are present in the circuit, $abcdgfea$, $dhgd$, and $abcdhgfea$. The spanning tree separates the edges into two groups, basis edges (solid line) belonging to the spanning tree, and chords (dash line), the remaining nodes. Each of the chords (hg and fg) corresponds to an independent cycle (respectively, $abcdgfea$ and $dhgd$), forming a cycle basis of the graph. The cycle $abcdhgfea$ can be expressed as a combination of independent cycles and is therefore not included in the cycle basis.

similarity of (4.57) with Kirchhoff's voltage law, where the sum of the voltage drops along the cycle is zero.

A *cycle basis* – minimum subset of linearly independent cycles – is sufficient to analyze the clock skew constraints induced by cycles within a graph. The cycle basis for graph G is shown in Fig. 4.11. To determine the cycle basis, a spanning tree algorithm is used. An arbitrary node is initially selected as the root. The graph is traversed while ignoring the direction of the edges (*i.e.*, traversal in the opposite direction of an edge is allowed), skipping the visited nodes until all of the vertices are traversed. Many algorithms for traversal exist [341], including fundamental breadth-first search (BFS) [111] and depth-first search (DFS) [342]. Edges within and outside the resulting tree are referred to, respectively, as basis and chord edges. Each chord corresponds to a distinct independent cycle. The number of basis edges is $n_b = |V| - 1$. The number of chords (and independent cycles) is therefore

$$n_c = |E| - |V| + 1. \tag{4.62}$$

The cycle basis is efficiently expressed as the circuit connectivity matrix $B \in \mathbb{R}^{(n_c \times |E|)}$. By arbitrarily choosing the direction of each cycle, an entry for edge e within cycle c is

$$b_{c,e} = \begin{cases} 1, \text{ if } e \text{ follows } c, & \text{(4.63a)} \\ -1, \text{ if } e \text{ opposes } c, & \text{(4.63b)} \\ 0, \text{ if } e \text{ does not belong to } c. & \text{(4.63c)} \end{cases}$$

For example, the clock skew constraint due to data path cycles for the system shown in Fig. 4.4 is

$$\begin{bmatrix} 1 & 1 & 1 & 1 & 0 & -1 & -1 & -1 & 0 \\ 0 & 0 & 0 & 1 & -1 & 0 & 0 & 0 & 1 \\ 1 & 1 & 1 & 0 & 1 & -1 & -1 & -1 & -1 \end{bmatrix} \begin{bmatrix} s_{a,b} \\ s_{b,c} \\ s_{c,d} \\ s_{d,g} \\ s_{d,h} \\ s_{a,e} \\ s_{e,f} \\ s_{f,g} \\ s_{g,h} \end{bmatrix} = \mathbf{0}. \qquad (4.64)$$

Clock skew scheduling to maximize the robustness of the sequential system can therefore be expressed as a quadratic programming problem,

Minimize

$$||\mathbf{s} - \mathbf{s} * ||^2, \qquad (4.65)$$

subject to

$$B\mathbf{s} = \mathbf{0} \qquad (4.66)$$

$$l_{i,f} \leq s_{i,f} \leq u_{i,f} \forall (i, f) \in E_G. \qquad (4.67)$$

Constraint (4.67) maintains the clock skew of each data path within the PR. The optimized clock skew schedule is converted into a schedule of clock arrival times $\mathbf{T_{CD}}$ by choosing the reference node, similar to the process of choosing a ground potential in the analysis of electrical circuits.

The main feature of the quadratic programming problem (4.67) is minimization of the cumulative squared distance between the clock skew and the center of the PR [241]. Sensitivity to timing variations however typically varies among data paths. Furthermore, process parameter variations limit the ability to precisely estimate the maximum $D_{i,j}$ and minimum $d_{i,j}$ delays of a data path (i, j). Bounds on delays, $d_{i,j}$ and $D_{i,j}$, can therefore be statistically modeled, respectively, as $\tilde{d}_{i,j}$ and $\tilde{D}_{i,j}$, where the tilde denotes a random variable. The bounds on clock skew, (4.3) and (4.7), are therefore random variables and can be expressed as

$$\tilde{u}_{i,f} = T_{CP} - \tilde{D}_{i,f} - \delta_s^f, \qquad (4.68)$$

$$\tilde{l}_{i,f} = -\tilde{d}_{i,f} + \delta_h^f. \qquad (4.69)$$

The probability of producing a clock skew outside the PR is $P(\tilde{l}_{i,f} \leq s_{i,f} \leq \tilde{u}_{i,f})$. With this formulation, the probability of a timing violation can be explicitly considered during the optimization process. Several algorithms exist that utilize this formulation to minimize the probability of a timing violation. In [343], the propagation delay of each data path is modeled as a Gaussian random variable. The accuracy of the delay estimates is improved by eliminating *false paths* [344], *i.e.*, logic paths that are never traversed (sensitized) by a data signal. The clock skew scheduling algorithm in [343] minimizes the maximum probability of a timing failure among all data paths, achieving up to a 53% improvement in yield. The accuracy of Gaussian statistical modeling in timing analysis may however suffer due to a variety of issues such as the topological and spatial correlation between the device parameters and nonlinear delay models [345]. The algorithm described in [343] is generalized in [346] to support arbitrary probability distributions; achieving, on average, an improvement of 17.7% over [343].

4.2.2   Performance

Clock skew within the system shown in Fig. 4.10 is adjusted to reduce the likelihood of a timing violation. Both of the data paths exhibit an idle time of 4 tu. If the likelihood of process variations is relatively low, the system performance can be enhanced by increasing the clock frequency by reducing the idle time. Consider the example illustrated in Fig. 4.12. The clock period of the system can be reduced by 4 tu, eliminating the idle time. The throughput of the system is increased by 50% by reducing the clock period by 4 tu, from $T = 12$ tu to $T = 8$ tu.

Although clock skew scheduling reduces the clock period below the critical path delay, the PR of all of the data paths is narrowed. A lower bound on the clock period therefore exists that prevents an arbitrary increase in the clock frequency. Recall that a feasible clock skew schedule exists if the corresponding constraint graph does not contain a positive cycle. The minimum clock period is found by finding a clock period that produces a zero weight directed cycle in the constraint graph [107, 331]. An example of a constraint graph for the data path depicted in Fig. 4.10 is shown in Fig. 4.10. A zero weight cycle (i, f, i) is produced when $T_{CP} = 5$ tu. After finding the minimum clock period, a feasible clock schedule can be produced by solving a set of linear inequalities (4.24).

Although the zero weight cycle within a constraint graph yields a minimum clock period for a specific circuit, further reduction in the clock period is possible by modifying the data paths with an intentional delay [231, 347]. To illustrate this effect, consider the data path shown in Fig. 4.13a. Two data paths, CL_1 and CL_2, connect two registers, i and f. The minimum and maximum delay of the paths is, respectively, $[d_1, D_1] = [4, 10]$ tu and $[d_2, D_2] = [8, 14]$ tu. Assuming the internal register delays are negligible, the minimum clock period satisfying (4.5) and (4.8) is 10 tu, achieved by inducing negative clock skew $s_{i,f} = -4$ tu, as illustrated in Fig. 4.13b. Suppose a delay element of 4 tu is intentionally inserted into data path CL_1, yielding delay $[d_1', D_1'] = [8, 14]$. The minimum achievable clock period with the modified topology is 6 tu, produced by creating a negative clock skew $s_{i,f} = -8$ tu, as depicted in Figs. 4.13c and 4.13d. Observe that the delay of data path CL_1 is

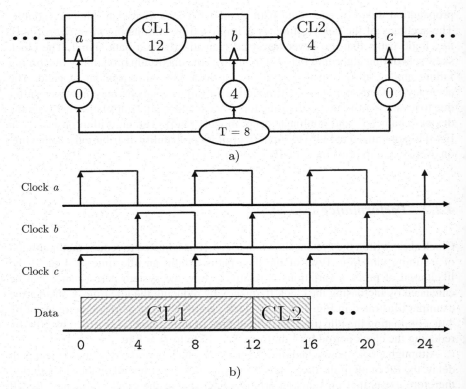

Fig. 4.12 System exploiting clock skew to enhance performance. a) Topology of the system. The clock signal arriving at register *b* is delayed by 4 tu, similar to the topology shown in Fig. 4.10a. The clock period *T* is reduced to 8 tu. b) Data flow within the system. The idle time within the system is eliminated to enhance the system speed.

aligned with the delay of data path CL_2. The total delay uncertainty is therefore reduced, reducing the minimum clock period [231].

Different algorithms are proposed in the literature that support delay insertion to reduce a clock period. A constraint graph can be used to determine whether a circuit may benefit from delay insertion. Recall that the edge set of a constraint graph consists of edges denoting the setup time constraint E_u and hold time constraint E_l. According to [348], if a zero weight cycle at T_{CP}^1 contains an edge $(i, j) \in E_l$, there exists a delay insertion strategy achieving a smaller clock period where $T_{CP}^2 < T_{CP}^1$. A counterexample is however shown in Fig. 4.14. A 3 tu delay is inserted in data path (a, b), achieving a clock period of 5 tu. Further reductions in the clock period are not possible since the maximum delay uncertainty within the circuit is 5 tu. No zero weight cycle can however be found without the hold time constraint edges. This algorithm in [348] does not consider clock period limitations due to reconvergent paths, as discussed in Subsection 4.1.1. This limitation is overcome in [349] where the clock period is minimized along with the inserted delay.

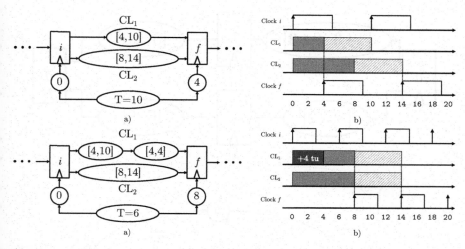

Fig. 4.13 Illustrative example of performance enhancement due to delay insertion. a) Two parallel data paths between registers i and f. The numbers in the brackets denote the minimum and maximum delay of the data paths. b) Timing diagram of data and clock signals in circuit (a). The minimum clock period is 10 tu. c) An additional 4 tu delay is inserted into data path CL_1. d) Timing diagram of data and clock signals in circuit (c). The minimum clock period is reduced to 6 tu by reducing the delay uncertainty.

4.2.2.1   Wave pipelining

Several issues pertaining to clock period minimization exist. Observe that the propagation delay of the data paths depicted in Fig. 4.13c is greater than the clock period. A second datum is therefore released before the first datum is completely processed. This phenomenon is called *wave pipelining* and is depicted in Fig. 4.15 [350]. Observe that multiple data travel at different stages within the same data path. Unlike traditional synchronous systems where the data signals are temporally separated, the signals are spatially separated within a wave pipelined data path. To reduce the clock period below the propagation delay, a combinatorial circuit should propagate multiple data [351]. *Data skew* — the difference between the maximum and minimum propagation delay — is an important factor affecting the performance of wave pipelined systems. Similar to clock skew that limits the maximum clock frequency, the data skew determines the maximum rate of wave pipelining (and, hence, the clock frequency), as illustrated in Fig. 4.16. Another important consideration during the delay insertion process is the cost of the additional delay elements. The clock skew scheduling algorithm proposed in [347] minimizes not only the clock period but also the number of inserted delay elements.

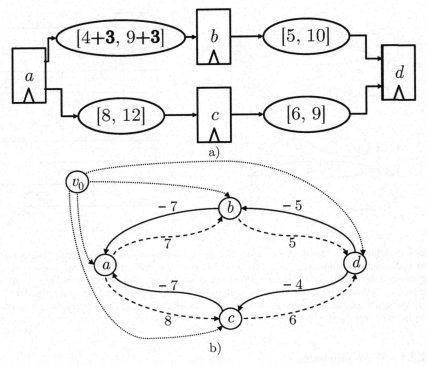

Fig. 4.14 Counterexample illustrating minimization of clock period with delay insertion where the zero weight cycles contain edges belonging to E_l. a) Circuit with inserted delay (from Fig. 4.8). A delay of 3 tu is inserted into data path (a, b). b) The constraint graph for $T_{CP} = 5$ tu. Due to delay uncertainty $D_{b,d} - d_{b,d} = 5$ tu, reducing the clock period below 5 tu produces a timing violation. Zero weight cycles $[a, b, a]$ and $[b, d, b]$ however both contain edges from set E_l.

4.2.3 Power

Clock skew scheduling is a technique to adjust the arrival time of the clock signals at the registers [233]. This technique is extended with the delay insertion method discussed in the previous section. The propagation delay of the data paths is intentionally increased to raise the clock frequency. Adjustment of the propagation delay along the data and clock paths can also reduce the power consumption of a synchronous system [352, 353].

Two dominant sources of power dissipation in modern IC's are switching activity and leakage current [232]. Certain design parameters affect the power dissipation of an IC. The dynamic power dissipated by a switching circuit is [354]

$$P_{dyn} = \alpha C_L V_{DD}^2 f, \qquad (4.70)$$

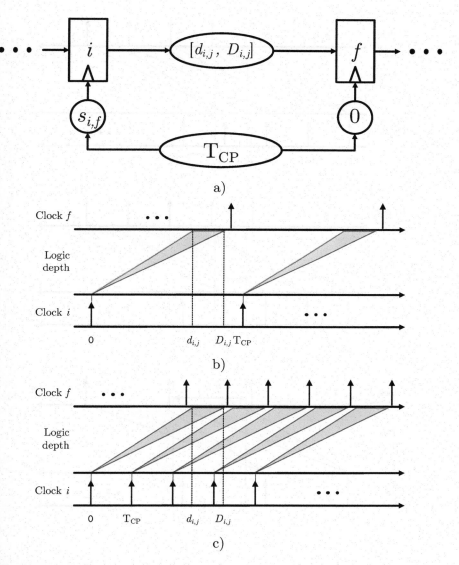

Fig. 4.15 Timing diagrams of conventional pipelining and wave pipelining. a) Example data path. b) Propagation of data during conventional pipelining. The triangular curves depict the minimum and maximum propagation delay of the datum. Only a single datum is processed during a clock period. c) Propagation of data during wave pipelining. Multiple data are processed during a clock period. Observe that the clock period is smaller than the propagation delay of the data path.

where C_L is the load capacitance, V_{DD} is the supply voltage, f is the clock frequency, and α is the activity factor representing the probability of switching during a cycle. Note that $\alpha = 1$ for a clock signal since switching occurs during

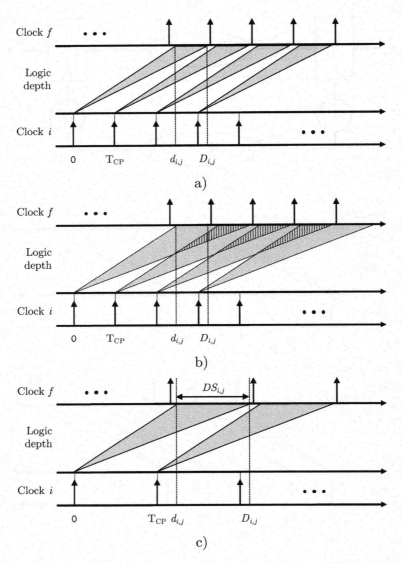

Fig. 4.16 Effect of data skew on maximum rate of pipelining in an example data path, shown in Fig. 4.15a. a) Correct data flow at high clock frequency and small data skew. b) Data flow at high clock frequency and large data skew. The diagonal cross hatched pattern illustrates data collisions. c) Correct data flow achieved by lowering the clock frequency.

every clock period. The load capacitance C_L is typically comprised of the gate capacitance and interconnect capacitance. Assuming the interconnect capacitance of the local interconnects is negligible, the dynamic power is proportional to the area WL of the transistors,

$$P_{dyn} \propto \alpha W L V_{DD}^2 f. \tag{4.71}$$

Two major types of leakage current in modern IC's are subthreshold leakage current and gate leakage current [232]. The subthreshold leakage power P_{subth} is dissipated due to a small current from the source and drain of a transistor when the device is in the cutoff state. P_{subth} is a function of device dimensions and voltage [232],

$$P_{subth} \propto \frac{W}{L} e^{-V_{th}/V_T}, \tag{4.72}$$

where

$$V_T = \frac{k_B T}{q} \tag{4.73}$$

is the thermal voltage, k_B is the Boltzmann constant, T is the temperature, and q is the electric charge of an electron (elementary charge). Gate leakage power P_{gl} is dissipated by the electrons tunneling through the gate oxide, producing a leakage current. P_{gl} is proportional to the total gate area [355],

$$P_{gl} \propto W L. \tag{4.74}$$

Other factors affecting the leakage current are technology or environmental parameters not directly controlled during the IC design process, such as the doping concentration, carrier mobility, ambient temperature, and dielectric permittivity and thickness.

Expressions (4.71), (4.72), and (4.74) indicate that to reduce power dissipation, the supply voltage and transistor width should be reduced while the threshold voltage should be increased. These modifications however directly degrade the speed of the logic circuitry [356]. A tradeoff therefore exists between power dissipation and propagation delay. Observe, however, that reducing the speed of the combinatorial logic affects the overall speed of a synchronous system only if the combinatorial logic is a part of a critical path. Consider the example system shown in Fig. 4.13a. Since data path b is a critical path, slowing data path b requires increasing the clock period, degrading the overall system speed. In contrast, the speed of data path a can be reduced without affecting the clock period.

Different methods exist that modify one or several components of (4.71), (4.72), and (4.74) thereby reducing the power dissipation. In a multiple supply voltage technique [357], the non-critical paths within a combinatorial circuit are driven by a smaller supply voltage. The transistors in CL2, shown in Fig. 4.10, can therefore be connected to a smaller voltage, $V_{DD}^{low} < V_{DD}^{high}$. Lowering the supply voltage is highly effective in alleviating dynamic power due to the quadratic relationship between the supply voltage and dynamic power (see (4.71)). Producing different supply voltages within a circuit however requires voltage converters [358, 359] or additional power distribution networks [75]. Multiple threshold voltage technique often accompanies a multiple supply voltage technique [360, 361]. By increasing the threshold voltage along a non-critical path, the subthreshold leakage power of a circuit can be drastically reduced, at the cost of greater delay. Gate sizing is another approach [362], trading speed for power consumption. By reducing the width of a transistor, the propagation delay is increased while the leakage and switching currents are reduced.

Consider the data path shown in Fig. 4.9. Local data path (b, c) is a non-critical path exhibiting idle time, since the propagation delay is smaller than the clock period. The speed and power consumption of (b, c) can therefore be reduced without affecting the overall performance of the circuit. The majority of the data paths in a practical synchronous system are non-critical, with more than 65% of the data paths at least twice faster than the slowest paths [352]. This feature of practical systems indicates the significant potential for reducing power dissipation in integrated systems. Furthermore, the propagation delay of a clock signal (affecting the clock skew) can be adjusted by gate sizing, further reducing the power consumption.

A power-aware clock skew scheduling algorithm is presented in [363]. The idle time of each data path is initially calculated. The size and threshold voltage of the transistors along each data path exhibiting an idle time is adjusted to reduce the power consumption of the circuit, while maintaining a constant system-level clock frequency. The algorithm specifically targets leakage power, achieving, on average, an 18.8% reduction. The dynamic power is likely also lowered due to the smaller capacitive load of the circuit. An algorithm for minimizing the power dissipated by a sequential system based on changing the supply voltage is presented in [364]. A discrete set of supply voltages is considered available. The associated delays are precomputed for each gate to obtain a function $\bar{P}_e : \mathbb{R} \to \mathbb{R}$, mapping the delay of a gate to a dissipated power. Power optimization can therefore be performed by optimizing the delay of the data paths. The solution is discretized to adapt the solution to a set of available supply voltages. An average of 9% reduction in power is achieved in the benchmark circuits.

Several limitations exist that limit the reduction in power consumption by voltage and frequency scaling. First, the additional delay $\Delta D_{i,j}$ introduced into the data path (i, j) by downsizing or voltage scaling should not exceed the idle time available within the data path,

$$T_{CP} - s_{i,j} \geq D_{i,j} + \Delta D_{i,j} + \delta_s. \tag{4.75}$$

Another limitation is the accuracy of the delay models. Due to supply and threshold voltage variations combined with parameter variations, precise control of the additional delay of the data paths is extremely difficult to manage [365]. Delay uncertainty should therefore be considered during the optimization process. Furthermore, the switching process in CMOS dissipates short-circuit power [352] due to current flowing from the power supply to ground while both the pull-up and pull-down networks operate in the conducting state. The short-circuit power increases with a smaller load capacitance, lower threshold voltage, greater switching time, and higher supply voltage [232]. Decreasing the size of the logic gates reduces the load capacitance and switching speed, significantly increasing the short-circuit power. Any power savings from downsizing the gates is therefore reduced if the short-circuit power is not considered during the sizing process.

4.3 Clock tree synthesis

Many topologies to enhance clock distribution networks have been discussed in the literature. Symmetric tree structures, such as an H-tree [236], equalize the distance traversed by the clock signal to each register, effectively producing zero clock skew (see Fig. 4.17). The regular structure of the tree however limits the application of a symmetric clock tree to highly regular layouts[366]. A modification of the H-tree is introduced in [235], where the shape of the clock tree is deformed while the distance traveled by the clock signal to each of the registers is maintained equal (see Fig. 4.17). Additional on-chip resources are however required to equalize the delay of the clock signal to different registers. The most prominent non-tree type of clock distribution network is a mesh topology [232], depicted in Fig. 4.17. The mesh structure provides a low impedance path to the clock sinks, thereby equalizing the delay from the clock source to each clock sink [367]. This structure, however, consumes large on-chip area and metal resources as compared to a tree topology.

The objective of these topologies is to minimize or eliminate any clock skew within a circuit. However, as discussed in the previous section, zero skew systems typically exhibit suboptimal performance and require significant on-chip resources or additional circuitry to minimize the clock skew. With *useful* clock skew, the overhead of the clock distribution network can be significantly reduced [239, 328]. The clock skew scheduling process specifies the time of arrival of a clock signal to each register within a network. To fully utilize a set of optimal clock arrival times, a clock distribution network satisfying the prescribed clock arrival time schedule T_{CD} is necessary.

A buffered asymmetric tree topology is highly suitable for clock distribution networks due to the flexibility of the layout, control of the arrival times, and smaller area overhead [368]. The clock tree synthesis process typically consists of two steps - topological and embedding [328]. The objective of the topological step is an abstract representation of the clock tree that achieves the clock arrival time at the registers while lowering the area overhead, such as wire length and buffer insertion. This abstract representation is converted into a physical layout during the embedding step. The layout parameters, such as the length and width of the interconnects, are tuned to distribute the clock signal to the specific on-chip location of the registers.

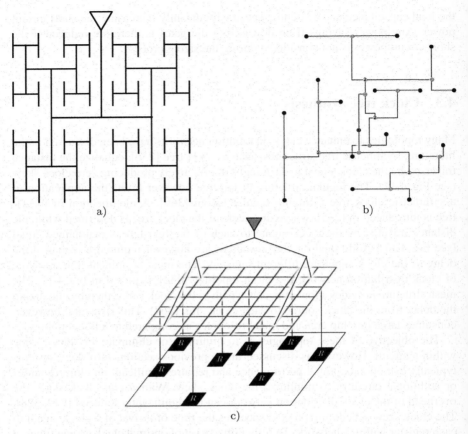

Fig. 4.17 Example of clock tree topologies that minimize clock skew. a) Symmetric H-tree, b) Delay-matched tree. The propagation delay of the clock signal from the clock source to each sink is equalized, producing zero clock skew. c) Mesh clock distribution network. The interconnects are placed between the leaves of the tree to provide a low impedance path between the leaves of the clock tree, reducing the clock skew.

4.3.1 Clock tree topology

A circuit model of a buffered tree-based clock distribution network, commonly referred to as clock tree, is shown in Fig. 4.18a. The connection among the buffers and registers constitute a directed tree graph. The unique root buffer B_0 is connected to a clock generator. The internal nodes of the clock tree correspond to the buffers amplifying the clock signal to mitigate attenuation and noise. Each register corresponds to a leaf within a clock tree. The number of leaves within the clock tree is therefore equal to the number of registers within the circuit.

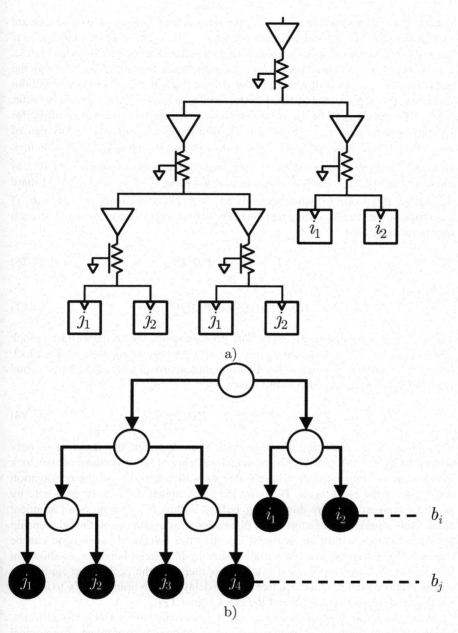

Fig. 4.18 Buffered clock tree topology for a sequential circuit with six registers. a) Circuit representation. The interconnect impedance is modeled as a distributed impedance, and b) graph representation. The registers and buffers are represented by, respectively, filled and empty circles.

One of the earliest works discussing the fundamental features of a clock tree are [368] and [240]. An equivalent graph model $T = (V_B \cup V, E_B)$ of a clock tree is utilized for the analysis of clock trees, where V_B is the set of buffers (internal nodes), V is the set of registers (leaf nodes), and E_B is the set of branches (edges) within the clock tree. This model is illustrated in Fig. 4.18b. The leaf nodes (*i.e.*, nodes without successors) are represented by filled circles, while the hollow circles represent buffer nodes. The arrival time of the clock signal to a register i is the sum of all buffer and interconnect delays along the path P_i from the clock source to i. Any pair of registers (i, j) is connected to the clock source via a common path $P_{i,j}^*$. The unique paths, $P_{i,j}^i$ and $P_{i,j}^j$, connect the corresponding register to the common path $P_{i,j}^*$, as illustrated in Fig. 4.19. Paths $P_i = [a, b, c, e, g, i]$ and $P_j = [a, b, d, f, h, j]$ share the section $P_{i,j}^* = [a, b]$, while sections $P_{i,j}^i = [c, e, g, i]$ and $P_{i,j}^j = [d, f, h, j]$ are unique to the corresponding registers. The arrival time of the clock signal to each register is therefore

$$t_i = PD(P_{i,j}^*) + PD(P_{i,j}^i), \qquad (4.76)$$

$$t_j = PD(P_{i,j}^*) + PD(P_{i,j}^j), \qquad (4.77)$$

where PD is the propagation delay. The primary challenge of the buffered clock tree is realizing the clock skew determined during the scheduling process. The clock skew $s_{i,j}$ between registers (i, j) is the difference in arrival time of the clock signal from the clock source to each of the leaves,

$$s_{i,j} = PD(P_{i,j}^i) - PD(P_{i,j}^j). \qquad (4.78)$$

Expression (4.78) indicates that the clock skew between the registers is only controlled by the difference in the propagation delay of those portions of the clock tree unique to both registers. Observe that only the *difference* in the propagation delay affects the clock skew. Delaying the clock signal delivery to each gate by an equal amount (e.g., by downsizing buffers within $P_{i,j}^*$) produces an identical clock skew among the registers. This phenomenon is analogous to the relationship between voltages within an electrical circuit – the voltage at every node can be increased by a constant amount without affecting the circuit behavior, as shown in Fig. 3.24. Clock tree topological synthesis is therefore the problem of finding an abstract representation of a clock tree and the delay of each branch in E_B to achieve the required relative arrival time of the clock signal $\mathbf{T_{CD}}$.

Many clock tree topologies can satisfy a given schedule of clock arrival times. Clock topology synthesis is often viewed as an optimization problem. Topological optimization of the clock tree is presented in [240]. The branching depth b_i of node i is the number of buffers connecting the clock source to the node. Assuming the delay of a buffer is Δ_b and the interconnect delay is negligible, the delay between the clock generator and an arbitrary node i is $b_i \Delta_b$. In this case, the clock skew

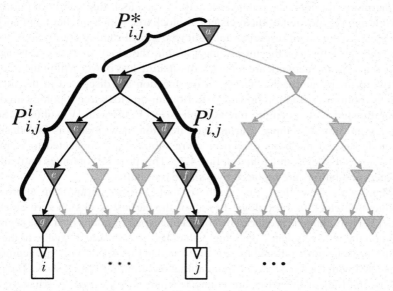

Fig. 4.19 Shared and unique paths within a clock tree. The clock signal travels from the root of the clock tree to registers i and j, traversing path $P^*_{i,j} = [a, b]$. The signal bifurcates at buffer b, producing paths $P^i_{i,j} = [c, e, g, i]$ and $P^j_{i,j} = [d, f, h, j]$ that are unique for, respectively, i and j. The difference in the clock arrival time of i and j is the difference in the propagation delay of $P^i_{i,j}$ and $P^j_{i,j}$, and does not depend on $P^*_{i,j}$.

between registers i and f is only the difference in the depth of the register within the clock tree,

$$s_{i,f} = (b_i - b_f)\Delta_b. \tag{4.79}$$

Substituting (4.79) into (4.3) and (4.7) yields

$$s_{i,f} = (b_i - b_f)\Delta_b \geq -d_{i,f} \tag{4.80}$$

$$s_{i,f} = (b_i - b_f)\Delta_b \leq T_{CP} - D_{i,f}. \tag{4.81}$$

By varying the depth parameters b_i and b_f, the clock skew in an abstract tree can approximate the skew in the clock skew schedule. The clock tree topology can therefore be viewed as a mixed-integer programming problem that ensures that the difference in the buffer delay satisfies the clock skew schedule.

Topology generation [240] assumes no information is available regarding the location of the registers. The quality of the clock tree topology is however greatly influenced by the location of the sinks. Incorporating the location information into the synthesis process can therefore significantly enhance the quality of the topology.

Furthermore, the total interconnect length of the clock tree can be minimized by considering the location of the registers. A clock tree with a smaller length occupies less area, saving metal resources and relieving global routing congestion [369]. By reducing the length of the interconnect within the clock tree, the resistance and capacitance of the network are reduced, dissipating less power within the tree [232].

Two primary approaches based on the location are discussed in the literature; namely, bottom-up and top-down [370]. Bottom-up clock tree topology algorithms generally start with an empty forest graph $F = (V, \varnothing)$ with $|V|$ subtrees. Sets V_B and E_B are initially empty. During each iteration, new node v is introduced into the node set and the roots of two or more subtrees are connected to v, merging into a new subtree, as illustrated in Fig. 4.20. This process continues until all of the registers and buffers are replaced by a single root buffer.

Many zero skew clock tree synthesis tools utilize a bottom-up approach for producing a clock tree topology. In one of the earliest works on zero clock skew trees [371], a subtree is produced by connecting a pair of points by a wire segment. A point on the wire segment equalizing path delay from the clock source to the leaves is chosen as the root of the subtree, as illustrated in Fig. 4.17b. The wire segments are connected with each other while maintaining the minimum clock skew. The process repeats until all of the wires are connected into a single tree. A balanced binary tree topology is generated from $n = 2^k$ register locations, where $k \in \mathbb{N}$, since each node has exactly two children and all registers within a tree are on the same level. Different heuristics can be applied to choose the pairs of points to be connected, such as finding the closest pairs of points or creating a Hamiltonian cycle through the points. Application of the algorithm to an arbitrary number of registers (*i.e.*, $n \neq 2^k$, $k \in \mathbb{N}$) is not discussed here.

A different bottom-up technique is presented in [372]. The register located farthest from the clock source is initially connected. The clock tree is next extended to connect an additional register. This process repeats until all of the registers are connected to the clock tree. Unlike [371], this procedure can handle an arbitrary number of registers. Further improvements in the clock tree topology are presented in [373, 374]. A nearest neighbor graph is produced from a set of register locations, as shown in Fig. 4.21. Edge (p, q) indicates that the point q is the closest neighbor for node p. The weight of each edge is the distance between the endpoints. During each iteration, edge $e = (v_1, v_2)$ with the smallest weight is chosen. Edge e and endpoints are contracted into a single node v. The process repeats until all of the edges are contracted.

An important extension to zero skew topology generation techniques is presented in [375], where the clock tree is generated for the prescribed skew, e.g., the skew determined during the clock skew scheduling process. Unlike purely spatial techniques which choose the closest nodes during topology synthesis, the unequal clock arrival times prescribed by a clock skew schedule require additional consideration. Consider four registers, a, b, c, and d, arbitrarily placed within a layout. Suppose four registers, a, b, c, and d, arbitrarily placed within a layout have the following relationship among the arrival times,

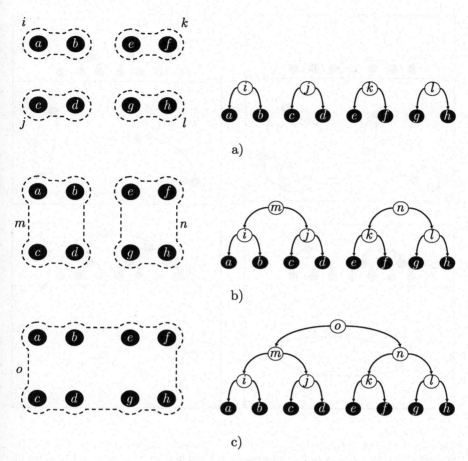

Fig. 4.20 Bottom-up construction of clock tree topology. a) The individual registers are initially grouped into several groups. For each group, a new internal node is added to the tree. The nodes within each group become children of the corresponding parent node. b) Similar to registers, groups are merged into larger groups, producing a new parent node for each group. c) The tree is completed after all of the groups are merged into a single group and the root node is added.

$$t_a \gg t_b \gg t_c \gg t_d. \tag{4.82}$$

The clock tree generated while only considering the location of the registers would deliver the clock signal to node d too early. Additional delay along the clock path to nodes a, b, and c is therefore required, as illustrated in Fig. 4.22a. The delay can be provided using delay elements or wire snaking. Both of these options however require additional on-chip resources. An alternative clock tree topology is shown in Fig. 4.22b. The delay from the clock source to node d is smaller than the delay to nodes a, b, and c. This topology better matches the prescribed clock arrival time

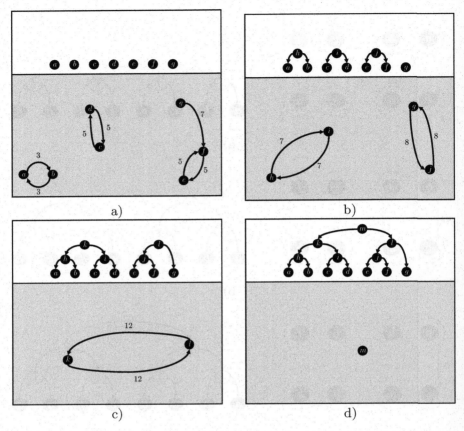

Fig. 4.21 Clock tree topological synthesis based on a nearest neighbor (NN) graph [373]. The abstract clock tree topology is shown above the NN graph. If an edge (p, q) exists within the NN graph, the distance between p and q is smaller than the distance between p and any other node within the NN graph. a) Initial clock tree topology and NN graph. The tree is initially a forest of $|V|$ singular trees. b) Edges (a, b), (c, d), and (e, f) have been contracted, producing, respectively, nodes h, i, and j. The position of the new nodes ensures minimum clock skew between the registers. Those registers located closest to each other become siblings within the clock tree. c) Edges (h, i) and (g, j) are contracted, producing, respectively, parent nodes k and l. d) Edge (k, l) is contracted into node m. The only remaining node m becomes the root of the clock tree, completing the topological synthesis process.

than a balanced topology, requiring fewer delay elements and less wire snaking. In [375], the cost of merging is based on both the position and clock arrival times. 69% fewer buffers and 60% less interconnect are required, on average, after minimizing the merging cost during the clock tree topological synthesis process.

An alternative to a bottom-up approach is a top-down method, where the clock tree is generated by repeatedly splitting the set of registers into clusters (see Chapter 11). Each cluster is recursively divided until the clusters contain a single

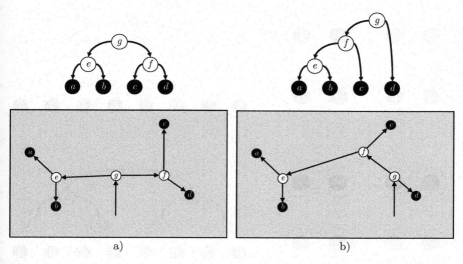

Fig. 4.22 Reduction in clock distribution overhead by modifying the topology. a) Balanced clock tree topology matches the length from the clock source to the registers. b) Imbalanced clock tree topology intentionally delaying the arrival of the clock signal to specific registers.

register. This process effectively partitions circuit graph G into multiple parts, as described in Subsection 3.5.1 and illustrated in Fig. 4.23.

The method of means and medians is one of the oldest top-down techniques for topological synthesis [376]. The set of register locations is recursively split into clusters based on the location. This method however requires a greater total wirelength as compared to bottom-up techniques. A similar technique is presented in [377], where the load capacitance of the registers within a cluster is considered when partitioning the registers. The difference in the total load capacitance of the registers within each subset is minimized, thereby producing a smaller difference in delay.

A top-down clock tree synthesis algorithm utilizing useful skew is described in [370]. The precise clock arrival times are assumed unknown, but the PR of each data path is determined from the circuit topology. The clock tree topology is generated by recursively bipartitioning a cluster S into two subclusters, S_1 and S_2. The clock skew between sequentially-adjacent registers is minimized during the clustering process by employing the following heuristic,

$$W_{1,2} = a \frac{P R_{1,2}}{N_{1\to2} + N_{2\to1}} + b|N_{1\to2} - N_{2\to1}|, \qquad (4.83)$$

where $N_{1\to2}$ ($N_{2\to1}$) refer to the number of data paths starting in S_1 (S_2) and ending in S_2 (S_1), a and b are the weight parameters, and $P R_{1,2}$ is the intersection of the PR of all edges connecting S_1 and S_2,

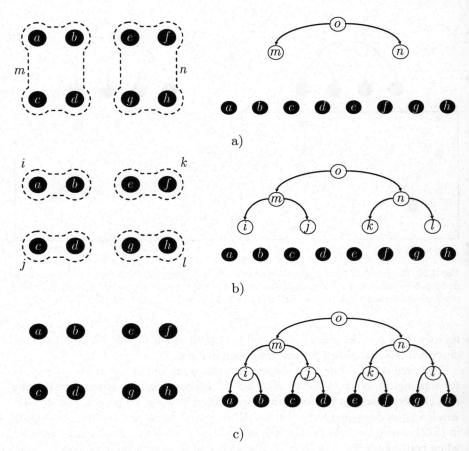

Fig. 4.23 Top-down clock tree topological synthesis process. a) The process commences with assigning root node (*o*). The set of nodes is initially split into several subsets. Each subset is associated with a child of the root node. b) Subsets are further divided into smaller subsets, and a child node corresponding to a subset is added to the tree. c) The tree is completed after all of the sets are reduced to the individual registers. These registers become the leaves of the tree.

$$PR_{1,2} \equiv \bigcup_{\forall (i,j) \in E_G, i \in S_1, j \in S_2} PR_{i,j}. \qquad (4.84)$$

By maximizing heuristic (4.83), the set of registers is recursively split into smaller clusters, producing a clock tree.

Top-down clustering for a prescribed clock skew schedule is described in [66] (see Chapter 11). Each register within the layout is described using a triple (x, y, wt), where x and y describe the position of the register, t is the clock arrival time, and w is the weight parameter characterizing the importance of the clock signal. During the tree topological synthesis process, the registers are recursively clustered based on the location and arrival times. Clustering with a smaller w produces trees which prioritize the location over the arrival time, while a larger w minimizes the difference in arrival times within the clusters.

4.3.2 Clock tree embedding

The abstract clock tree topology $T = (V_B \cup V, E_B)$ described in the previous section specifies the interconnections among the clock generators, buffers, and registers. The exact position and wiring are determined during the embedding stage. Early efforts on clock tree synthesis were based solely on minimizing the amount of metal resources utilizing such techniques as minimum spanning trees and Steiner minimum trees [376]. With the increase in clock frequency, however, clock skew has became the primary performance limitation. After development of the H-tree [236], considerable research effort has been devoted to zero skew clock tree synthesis. This problem is significantly more restrictive than producing a clock network that satisfies the upper bound (4.3), and lower bound (4.7), on clock skew. Later works which describe clock tree synthesis permit a bounded clock skew [378] or utilize useful skew [328]. Clock trees for a prescribed clock skew schedule have also been explored [66, 375], such as described in Chapter 11.

4.3.3 Method of means and medians

One of the earliest works on asymmetric zero skew clock trees is described in [376], where the difference in the distance from the clock source is reduced by the method of means and medians. In each set of points, $\{(x_1, y_1), \ldots, (x_n, y_n)\}, \in S$, the clock signal is routed from the clock source towards the center of mass (x_c, y_c) of the registers within the cluster,

$$x_c = \frac{\sum_{i=1}^{n}}{x_i}, \tag{4.85}$$

$$y_c = \frac{\sum_{i=1}^{n}}{y_i}. \tag{4.86}$$

The set of points is split into two subsets based on the location, and the clock distribution network is extended to the center of mass of each subset. This process is repeated until the individual nets are connected to the clock tree, as illustrated in Fig. 4.24. With this technique, clock skew is reduced to below 200 ps, at the time constituting approximately 20% of the typical clock period (at frequency of about one gigahertz).

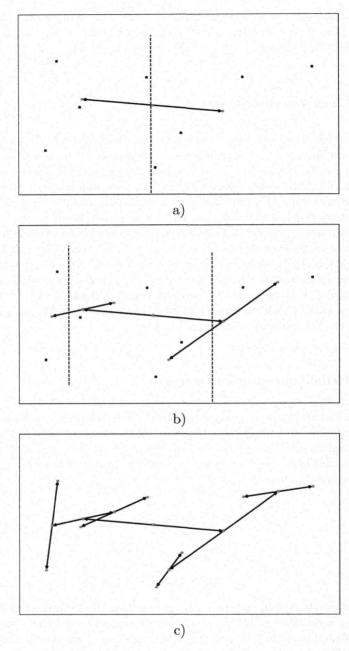

Fig. 4.24 Clock tree synthesis using method of means and medians [376]. a) Eight registers are split into two equal sets based on the x-coordinate. b) Each set is further divided into a pair of smaller sets. c) Final clock tree.

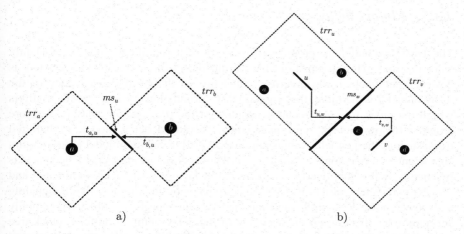

Fig. 4.25 Deferred merge embedding for clock tree synthesis. a) Tilted rectangular regions (TRR) for two registers. The signal routed from a merging segment has an equal delay to both of the cells. b) The segments are merged in subsequent stages based on TRR until the root buffer is reached.

4.3.4   Deferred merge embedding

Further reductions in skew have been achieved by applying the deferred merge embedding (DME) algorithm that has become the standard approach for producing zero skew clock trees [329]. In the original formulation of DME, the propagation delay of an interconnect is assumed directly proportional to the length [329]. With this assumption, the goal of DME is to produce a clock tree that equalizes the length of the wire from the clock source to each of the clock sinks. The registers are processed in a bottom-up order, starting with the registers at the lowest level of a tree. Assuming the interconnects follow a Manhattan geometry and the delay is proportional to the wire length, a tilted square region around each register exists. All points on the boundary of this region are equally delayed from the register. Points located at the intersection of two regions produce a merging segment – a diagonal line whose points have the same delay from each register. Consider merging segment ms_u between registers a and b, as illustrated in Fig. 4.25a. Delays $t_{a,u}$ and $t_{b,u}$ from the merging segment to the respective registers are equal, *i.e.*, $t_{a,u} = t_{b,u}$.

During early iterations of DME, a set $M_1 = \{m_1^1, \ldots, m_1^{2n}\}$ of $2n$ merging segments is produced from $4n$ registers. Similar to registers, for each distance d, a tilted rectangular region exists around each merging segment. During a subsequent iteration, a set of n new merging segments $M_2 = \{m_2^1, \ldots, m_2^n\}$ is produced from $2n$ previous merging segments, as illustrated in Fig. 4.25b. This process repeats until a binary tree of merging segments is produced. During the second part of the algorithm, the exact locations are determined in a top-down order. The merging points of each merging segment are selected to ensure the minimum total wirelength, thereby producing a clock tree.

4.3.5   Elmore delay

The DME methodology produces a tree whose leaves are located equidistantly from the root. The propagation delay from the clock source to the registers is however not likely to be precisely equal due to the relative inaccuracy of the delay model. Furthermore, the original DME formulation does not consider buffers within the clock tree. An accurate estimate of the propagation delay is therefore required to lower clock skew uncertainty. Advanced delay models of buffered trees, such as the Penfield-Rubinstein model [379] or Sakurai model [380], require significant computational resources. Furthermore, achieving perfect accuracy is practically impossible due to a wide range of factors, such as environmental and process parameters variations [381], electromagnetic interference [382], signaling and power noise [383–385], and interaction with repeaters [386]. A commonly accepted tradeoff between the accuracy of the delay within a buffered clock tree is exemplified by the Elmore delay model [153].The Elmore delay from an internal node u to a descendant node v is [373, 377, 387]

$$t_{u,v} = \sum_{e \in u \rightsquigarrow v} r_e(\frac{c_e}{2} + C_v), \tag{4.87}$$

where C_v is the total capacitance of the subtree rooted at node v, recursively defined as

$$C_v = \begin{cases} C_{L_v}, & \text{if } v \text{ is the leaf node} \quad (4.88a) \\ c_v + \sum_{(v,w) \in E_B} c_{v,w} + C_w, & \text{if } v \text{ is the internal node,} \quad (4.88b) \end{cases}$$

and r_e and c_e denote, respectively, the interconnect resistance and interconnect capacitance of edge e. A generalization of DME to buffered RC trees based on the Elmore delay model [153] is proposed in [377, 387]. A more accurate delay model based on π interconnect model is used in [374, 388], achieving further reductions in clock skew and wirelength.

4.3.6   Bounded skew tree

Producing zero skew requires additional interconnect for balancing the path lengths. The zero skew requirement is however excessively strict since only two fundamental constraints, namely, (4.7) and (4.3) need to be satisfied. A bounded skew tree (BST) is proposed in [389, 390], where a zero skew constraint is replaced with a global nonzero skew constraint s_{max}. Due to the range of available clock skews, the merging segments produced during the zero skew routing process are transformed into octilinear merging regions. Consider merging region mr_u between registers

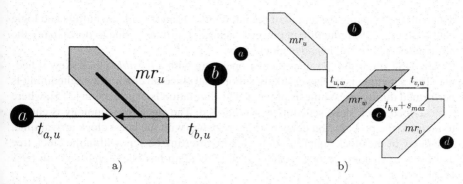

Fig. 4.26 Merging regions during bounded skew tree synthesis. a) Merging region mr_u due to two registers. $t_{u,a}$ and $t_{u,b}$, denote the propagation delay of a clock signal from mr_u to the respective register. $t_{u,a}$ and $t_{u,b}$ differ by no more than s_{max}. b) Regions are merged in subsequent stages, minimizing the interconnect length while maintaining the skew below s_{max}.

a and b, as illustrated in Figs. 4.26a and 4.26b. The propagation delay from the merging region mr_u to registers a and b differs by no more than s_{max}, and the length of interconnect connecting a, b, and mr_u is minimized in a Manhattan geometry,

$$t_{a,u} - t_{b,u} \leq s_{max}. \tag{4.89}$$

The octilinear regions offer additional flexibility in clock tree construction, enabling a significant reduction in wirelength. In [391], the bounded skew is expressed as upper and lower bounds on the length of the interconnect connecting a register with the clock source. Notably, the total wirelength produced by these bounded skew tree algorithms approaches the total wirelength of a Steiner minimal tree as the skew bound increases [390]. Practical considerations for bounded skew clock tree synthesis are discussed in [378], including unequal propagation speed of a signal on a different layer and obstacles within the layout. Further improvements in bounded skew clock tree synthesis are reported in [392], where dynamic programming is applied to find the optimal merging point within the merging regions.

4.3.7   Useful skew tree

The primary limitation of the BST algorithm is the use of a global bound on clock skew. Within a sequential circuit, however, the PR of each data path is typically different. The global skew bound used in BST algorithms is effectively an intersection of all PR within a system [328], excessively restricting the layout of the clock tree. The DME methodology has been extended to produce a clock tree which exploits useful clock skew. The earliest work on producing layouts utilizing useful skew is presented in [370]. Negative clock skew [393] is imposed on certain data path by adjusting the length of the interconnects and sizing the buffers and logic gates. Up to a 22% reduction in power is achieved, primarily due to downsizing the logic gates enabled by negative clock skew.

The UST algorithm produces a tree operating with a *feasible* clock skew schedule. This schedule is, however, likely suboptimal since no clock skew scheduling is performed during the synthesis process. This limitation is overcome in [375], where a prescribed clock skew schedule is synthesized. The clock tree topology is chosen to increase the propagation delay to those registers with the latest clock arrival time. The delay of the clock signal is controlled by inserting buffers within the clock tree and by wire snaking. A 60% reduction in wirelength has been reported with 69% fewer buffers.

A crucial assumption in DME is arbitrary placement of the merging segments. In certain practical systems, placing the wire intersection within the merging segments is not feasible due to manufacturing constraints. This limitation is overcome in the QuCTS algorithm [66], where the intersection of the interconnects is constrained to a discrete set of points. Similar to [375], a clock skew schedule is first produced to determine the optimal clock frequency while maximizing robustness. A more complete description of QuCTS is provided in Chapter 11.

4.4 Summary

Most modern high performance integrated systems are synchronous, employing a clock signal to synchronize the flow of data within an IC. The clock signal has the largest fanout and operates at the highest frequency in an IC. Along with power and ground, the clock signal is distributed using the largest on-chip network. Sophisticated graph-based methods for the design and analysis of clock distribution networks have been developed.

Due to the finite propagation speed of a signal within an interconnect, simultaneously delivering the clock signal to each gate is a complicated task. Clock skew therefore exists within each data path. The clock skew imposes timing constraints on clock signal delivery, such as zero clocking, double clocking, and minimum clock period, as discussed in Subsection 4.1.1. Timing graphs and constraint graphs efficiently represent these constraints, enabling powerful graph algorithms, such as cycle bases or spanning trees, to be used during the clock distribution network synthesis process.

During the clock skew scheduling process, introduced in Section 4.2, the clock arrival time of each register is adjusted to ensure the local timing constraints are satisfied. Different system characteristics can be enhanced with clock skew scheduling. The robustness of a system with respect to process and environmental parameter variations (and, therefore, the manufacturing yield) can be improved by shifting the clock skew towards the center of the permissible range of each local data path. By exploiting the idle time within each local data path, the system clock frequency can be increased, thereby enhancing the performance. The non-critical paths identified during the scheduling process can be intentionally slowed thereby reducing power dissipation with no effect on overall system performance.

The schedule of clock arrival times determined with clock skew scheduling is realized by constricting a buffered clock tree. During the topological clock tree synthesis process, an abstract structure of the tree is determined, as described in Section 4.3. Different approaches to constructing an abstract clock tree exist, such as recursive top-down bipartitioning or bottom-up merging based on location. The resulting abstract clock tree undergoes the clock tree embedding process, where the layout of a clock tree is determined. Common clock tree embedding methods include method of means and medians (MMM) and deferred merge embedding (DME).

Chapter 5
Circuit analysis

High level design of VLSI systems assumes correct functionality of the underlying electrical circuits. Digital systems, for example, utilize clearly distinguishable binary signals. At lower abstraction layers, however, the electrical signals behave similar to analog signals. The electrical waveforms therefore satisfy stringent requirements, such as propagation delay, slew rate, and power dissipation, to satisfy signal integrity requirements. To evaluate these waveforms, circuit level analysis of VLSI systems is necessary. Due to the significant increase in the speed and complexity of integrated systems, accurate and computationally efficient circuit analysis has gained critical importance over the past decades.

Since the establishment of a mathematical structure for circuit theory in 1827 by G. S. Ohm [394], different methods for circuit analysis have been reported in the literature. A graph theoretic basis for circuit analysis was described in 1847 by G. R. Kirchhoff [256] by postulating two laws governing the current and voltage relationship within an arbitrary electrical circuit. By the late 19th century, the theory of transient and alternating current was developed, incorporating the concepts of capacitance and inductance into the circuit analysis process [395]. Nonlinear circuit theory emerged in the early 20th century, driven by the advent of nonlinear devices, particularly vacuum tubes [396].

Entering the era of integration, the need for accurate analysis of complex systems motivated the development of circuit simulation tools. Tensor analysis of electrical circuits, pioneered in 1934 by G. Kron [397], was a crucial precursor of early circuit simulators. The Transistor Analysis Program (TAP), developed in 1959 [398], is considered the earliest circuit simulation program [399]. Based on TAP, more advanced simulation tools were developed, including NET1 in 1963 [400] and SCEPTRE in 1967 [401], capable of handling a wide range of circuits, including both passive and nonlinear components. Important advancements in numerical integration, driven primarily by H. Shichman [402, 403], were vital in creating CIRcuit analysis PACkage (CIRPAC) [403] which exhibited an order of magnitude speedup as compared to other simulators of the time.

R. Bairamkulov, E. G. Friedman, *Graphs in VLSI*,
https://doi.org/10.1007/978-3-031-11047-4_5

Fig. 5.1 An example of AC analysis of a low pass filter using SLIC [261]. a) Target circuit, and b) SLIC code. Resistor R1, inductor L1, and capacitor C1 are described on lines 4 to 6. Note the similarity with SPICE syntax. The TEMP statement specifies the operating temperature of the circuit. The GAIN statement specifies the output (03 and 00) and input (01 and 00) ports and the type of analysis (e.g., AC voltage transfer function). The range of frequencies is 0.1 to 100 MHz with ten points per decade, as specified in the FREQ statement on line 3. The input is terminated with the END statement.

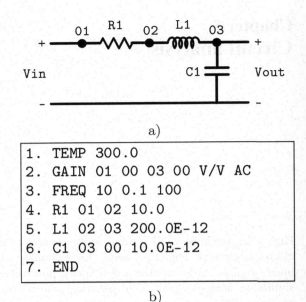

a)

```
1. TEMP 300.0
2. GAIN 01 00 03 00 V/V AC
3. FREQ 10 0.1 100
4. R1 01 02 10.0
5. L1 02 03 200.0E-12
6. C1 03 00 10.0E-12
7. END
```

b)

The application of sparse matrix analysis to circuit simulation was a crucial advancement in the Advanced STatistical Analysis Program (ASTAP), developed in IBM in 1971 [404], significantly reducing memory requirements. Variable time step integration further improved accuracy and runtime by increasing or reducing the time resolution if the rate of change in the parameters is, respectively, high or low [404]. Other notable circuit simulators of the early 1970's include Computer Analysis of Nonlinear Circuits, Excluding Radiation (CANCER) [260], and Simulator for Linear Integrated Circuits (SLIC) [261], which utilized advanced linear algebraic methods for linearization, numerical stability, and accuracy control. An important feature of CANCER and SLIC was the user friendly input description language that contributed to the widespread adoption of these tools in both the industrial and academic communities. An example of a circuit described in the SLIC language is shown in Fig. 5.1. Note the similarity with current circuit simulation tools. Since an electrical circuit is fundamentally a graph, only connectivity information is required to describe a circuit, enabling efficient textual representation of the system.

The popularity of CANCER in the academic community motivated the development in 1973 of the open source Simulation Program with Integrated Circuit Emphasis, commonly known today as SPICE [399]. The second version of SPICE, released in 1975 [50], became the worldwide standard for circuit simulation. The success of SPICE2 can be largely attributed to the applicability of the tool to a wide range of linear and nonlinear circuits. This crucial feature of SPICE2 is achieved by utilizing modified nodal analysis (MNA), a robust method for numerical circuit analysis [405].

5.1 Modified nodal analysis

First presented in 1975 [64], MNA is a versatile method for analyzing linear circuits. The impedances, current sources, voltage sources, and nonlinear devices within a circuit are described in matrix form. If the conductance of each wire, voltage of each voltage source, and current of each current source is known, the potential difference across each edge (i.e., a circuit element, such as a resistor or current source) can be determined.

Suppose a circuit is represented by a directed multigraph $G = (V, E)$, the direction of the edges is arbitrary chosen, and the edge set is composed of five subsets,

$$E = E_v \cup E_i \cup E_r \cup E_c \cup E_l, \tag{5.1}$$

each representing, respectively, independent voltage sources, independent current sources, resistors, capacitors, and inductors. Recall from Subsection 3.4.1 that Y_d is the incidence matrix of a directed graph where an entry is

$$y_{n,e} = \begin{cases} 1, \text{ if the positive terminal of element } e \text{ connected to node } n & (5.2a) \\ -1, \text{ if the negative terminal of element } e \text{ connected to node } n & (5.2b) \\ 0, \text{ otherwise.} & (5.2c) \end{cases}$$

The elements within the network can be ordered such that

$$Y_d = \begin{bmatrix} Y_v & Y_i & Y_r & Y_c & Y_l \end{bmatrix}, \tag{5.3}$$

$$\mathbf{v} = \begin{bmatrix} \mathbf{v}_v \\ \mathbf{v}_i \\ \mathbf{v}_r \\ \mathbf{v}_c \\ \mathbf{v}_l \end{bmatrix}, \tag{5.4}$$

$$\mathbf{i} = \begin{bmatrix} \mathbf{i}_v \\ \mathbf{i}_i \\ \mathbf{i}_r \\ \mathbf{i}_c \\ \mathbf{i}_l \end{bmatrix}, \tag{5.5}$$

where $\mathbf{v} \in \mathbb{R}^{|E|}$ and $\mathbf{i} \in \mathbb{R}^{|E|}$ are vectors of, respectively, the voltage across and current through the corresponding element, and subscripts v, i, r, c, and l indicate the type of circuit element, respectively, the independent voltage and current sources, resistors, capacitors, and inductors. The elements of $\mathbf{i}_i \in \mathbb{R}^{|E_i|}$

represent the current through the independent current sources and are known *a priori*. The remaining current and voltage vectors are related via the following relationships [406],

$$\mathbf{i}_r = \mathcal{G}\mathbf{v}_r, \tag{5.6}$$

$$\mathbf{i}_c = \mathcal{C}\frac{d}{dt}\mathbf{v}_c, \tag{5.7}$$

$$\mathbf{v}_l = \mathcal{L}\frac{d}{dt}\mathbf{i}_l, \tag{5.8}$$

where $\mathcal{G} \in \mathbb{R}^{|E_r|\times|E_r|}$ and $\mathcal{C} \in \mathbb{R}^{|E_c|\times|E_c|}$ are diagonal matrices representing, respectively, the conductance and capacitance of the respective elements, and $\mathcal{L} \in \mathbb{R}^{|E_l|\times|E_l|}$ is the inductance matrix representing the self- and mutual inductance within a circuit. Note that $\mathcal{L}$ is a diagonal matrix if the mutual inductances are ignored.

The primary equation governing the static analysis of circuits without dependent sources can be formulated as

$$\begin{bmatrix} G & Y_v \\ Y_v^T & 0 \end{bmatrix} \mathbf{e} = Y_i \mathbf{i}_i, \tag{5.9}$$

where $\mathbf{e} \in \mathbb{R}^{|V|}$ is the vector of voltage at each node, and $G = Y_g \mathcal{G} Y_g^T$ is the conductance matrix of a resistive network.

By constructing and solving (5.9), the steady state voltage at each node can be determined. Practical VLSI circuits however contain circuit elements that display transient behavior. These elements include linear primitives, such as capacitors and inductors, and nonlinear elements, such as transistors and memristors. To model the behavior of these elements, numerical differentiation is applied. Each transient element is replaced by an equivalent circuit element called a *companion model* that includes resistors and independent sources. For example, the transient current $i_C(t)$ through a capacitor as a function of time t is

$$i_C(t) = C\frac{dv_C(t)}{dt}, \tag{5.10}$$

where C is the capacitance and v_C is the voltage across the capacitor. Discretization by the Backward Euler method yields

$$i_C(t^k) = \frac{C}{h}v_C(t^k) - \frac{C}{h}v_C(t^{k-1}), \tag{5.11}$$

where t^{k-1} and t^k are consecutive discrete time instants, and h is the time step. This expression is equivalent to

$$i_C(t^k) = g_{eq}v_C(t^k) + i_{eq}, \tag{5.12}$$

where g_{eq} is the equivalent instantaneous conductance of the capacitor,

$$g_{eq} = \frac{C}{h}, \tag{5.13}$$

and i_{eq} is the equivalent current source across the capacitor,

$$i_{eq} = -\frac{C}{h}v_C(t^{k-1}). \tag{5.14}$$

During transient analysis, a capacitor is replaced by an equivalent companion model, as shown in Fig. 5.2a. During each time step, the transient parameters within the model are adjusted, modeling the instantaneous behavior of the element.

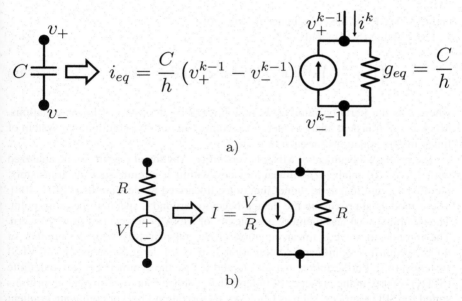

Fig. 5.2 Companion models for transient analysis of circuits. a) Capacitor model, and b) independent voltage source model

In matrix form, the companion models transform (5.9) into

$$\begin{bmatrix} A & Y_v \\ Y_v^T & 0 \end{bmatrix} \mathbf{x} + \begin{bmatrix} C & 0 & 0 \\ 0 & L & 0 \\ 0 & 0 & 0 \end{bmatrix} \frac{d}{dt} \mathbf{x} = \begin{bmatrix} Y_i \mathbf{i}_i \\ 0 \end{bmatrix}, \tag{5.15}$$

or, in a more compact form,

$$\tilde{G}\mathbf{x} + \tilde{C}\dot{\mathbf{x}} = \mathbf{b} \tag{5.16}$$

where

$$A = \begin{bmatrix} G & Y_l \\ -Y_l^T & 0 \end{bmatrix}, \tag{5.17}$$

$$\mathbf{x} = \begin{bmatrix} \mathbf{e} \\ \mathbf{i}_l \\ \mathbf{i}_v \end{bmatrix}, \tag{5.18}$$

and $C = Y_c C Y_c^T$ is the capacitance matrix [406].

Discretizing (5.15) yields

$$\left(\tilde{G} + \frac{2}{h}\tilde{C} \right) \mathbf{x}^k = \mathbf{b}^k + \mathbf{b}^{k-1} - \mathbf{x}^{k-1} \left(\tilde{G} - \frac{2}{h}\tilde{C} \right), \tag{5.19}$$

where k is the iteration number. Equation (5.19) is a system of linear equations. MNA-based transient analysis therefore requires an iterative solution of a system of linear matrix equations for each time step.

The primary advantage of MNA is versatility. Any linear circuit can be analyzed with MNA. To analyze nonlinear devices within a circuit, such as transistors, memristors, and magnetic tunnel junctions, linearized models are used [407, 408]. These models approximate the device behavior around a specific operating point. The computational and memory complexity of MNA is however of great concern. The runtime to solve a linear equation grows superlinearly with the number of nodes, requiring significant computational time for large systems, as in VLSI systems [145]. Furthermore, matrices $\tilde{G}$ and $\tilde{C}$ lose the symmetric positive definite (SPD) property in the presence of independent voltage sources. Efficient algorithms suited for SPD matrices, such as Cholesky factorization [409] or conjugate gradient method [410], can therefore no longer be used to solve (5.15), requiring more expensive algorithms such as LU factorization [411].

To preserve the SPD property, those circuit elements producing the voltage source and additional nodes within the network can be transformed using a Norton equivalent circuit to eliminate the voltage source and any associated rows and columns. An example of a Norton equivalent of an independent voltage source connected in series with a resistor is shown in Fig. 5.2b [412]. By eliminating the independent voltage sources, (5.15) becomes

$$\begin{bmatrix} G & Y_l \\ -Y_l^T & 0 \end{bmatrix} \mathbf{e} + \begin{bmatrix} C & 0 \\ 0 & L \end{bmatrix} \frac{d}{dt} \mathbf{e} = Y_i \mathbf{i}_i, \tag{5.20}$$

yielding the SPD matrix $\left(\tilde{G} + \frac{2}{h}\tilde{C} \right)$ in (5.19) [406].

Despite restoring the SPD property, the analysis of large circuits requires significant computational resources. Direct linear matrix solvers place excessive demand on the memory during computations of extremely large networks. Alternative methods have been introduced to circumvent the superlinear complexity of the circuit analysis process. For example, a significant structural similarity exists between linear electrical circuits and finite element discretization of partial differential equations (PDE). Methods for accelerated solution of PDEs are therefore often applicable to the analysis of linear circuits. Many approaches utilize graph theory to accelerate the analysis process. A variety of techniques for fast circuit analysis is described in the upcoming sections.

5.2   Iterative numerical methods

Both DC and transient forms of MNA can be represented as a standard system of linear equations [65],

$$A\mathbf{x} = \mathbf{b}, \tag{5.21}$$

where $\mathbf{x}$ is the vector representing the voltage at each node and the current through each voltage source within a network, and $\mathbf{b}$ is the vector of the current being injected and the voltage sources. Network models of modern ICs are prohibitively large, disallowing the use of direct solution methods such as LU factorization or Cholesky factorization [411]. Iterative solvers, such as the conjugate gradient (CG) method [410] or generalized minimal residual method (GMRES) [413], should therefore be used to circumvent this limitation.

Reformulating (5.21) yields

$$\mathbf{b} - A\mathbf{x} = 0. \tag{5.22}$$

If vector $\mathbf{x}$ is replaced by vector $\mathbf{x}'$, (5.21) becomes

$$\mathbf{b} - A\mathbf{x}' = \mathbf{r}, \tag{5.23}$$

where $\mathbf{r}$ is called a residual. Observe that the norm of the residual $||\mathbf{r}||$ becomes smaller if $\mathbf{x}'$ is close to $\mathbf{x}$. Iterative linear equation solvers attempt to minimize $||\mathbf{r}||$ by iteratively adjusting vector $\mathbf{x}'$, thereby closely approximating the exact solution $\mathbf{x}$.

Classic iterative algorithms are *stationary* methods [414] that represent a system of equations as

$$\mathbf{x}^k = B\mathbf{x}^{k-1} - \mathbf{c}, \tag{5.24}$$

where $\mathbf{x}^k$ is the approximation of the solution after the kth iteration, and matrix B and vector $\mathbf{c}$ depend upon the chosen iterative method. Note that matrix B and vector $\mathbf{c}$ remain invariant during the iterative process, hence the methods are called stationary [415]. Classical stationary methods operate by *splitting* matrix A into two matrices [415],

$$A = M - N. \tag{5.25}$$

The kth iteration of these methods is

$$\mathbf{x}^k = M^{-1}\left(N\mathbf{x}^{k-1} - b\right). \tag{5.26}$$

Common iterative methods suggest different methods for splitting matrix A [416],

$$M = \begin{cases} D, & \text{Jacobi method} & (5.27a) \\ D - E, & \text{Forward Gauss-Seidel method} & (5.27b) \\ D - F, & \text{Backward Gauss-Seidel method} & (5.27c) \\ \dfrac{1}{\omega}D - E, & \text{Successive OverRelaxation (SOR) method,} & (5.27d) \end{cases}$$

where ω is the relaxation parameter, E and F are, respectively, the strictly lower and strictly upper triangular parts of A, and D is the diagonal part of A.

Advanced iterative methods are typically not stationary, i.e., the terms of (5.24) are not maintained constant. The CG method [410] is one of the most common non-stationary iterative method for solving systems of the form described by (5.21) when matrix A is symmetric positive definite (SPD). The algorithm is based upon the observation that the exact solution $\mathbf{x}$ minimizes a convex function,

$$f(\mathbf{x}) = \frac{1}{2}\mathbf{x}^T A\mathbf{x} - \mathbf{b}^T \mathbf{x}. \tag{5.28}$$

The solution of (5.21) is determined by minimizing $f(\mathbf{x})$ via gradient descent [417]. The gradient of function $f(\mathbf{x})$ is

$$\nabla f(\mathbf{x}) = A\mathbf{x} - \mathbf{b}. \tag{5.29}$$

The solution is found by iteratively shifting $f(\mathbf{x})$ in the direction of steepest descent,

$$\mathbf{x}^{k+1} = \mathbf{x}^{k-1} - \alpha^k \left(A\mathbf{x}^{k-1} - \mathbf{b} \right), \tag{5.30}$$

or, alternatively,

$$\mathbf{x}^k = \mathbf{x}^{k-1} + \alpha^k \mathbf{r}^{k-1}, \tag{5.31}$$

where α^k is the step size during iteration k.

A limitation of the CG method is the limited applicability of the algorithm, since only SPD matrices are considered. Furthermore, the upper bound on the number of CG iterations before converging is equal to the size of the matrix, impractical for large systems. A large variety of solvers is proposed that partially or fully overcome these limitations, including the BIConjugate Gradient (BICG), BIConjugate Gradient STABilized (BICGSTAB), MINimal RESidual (MINRES), and Generalized Minimal RESidual methods (GMRES) [418].

Preconditioning is often used to accelerate the convergence of the iterative methods. During preconditioning, a linear matrix equation is transformed into

$$\tilde{A}\tilde{\mathbf{x}} = \tilde{\mathbf{b}} \tag{5.32}$$

where

$$\tilde{A} = M_1^{-1} A M_2^{-1}, \tag{5.33}$$

$$\tilde{\mathbf{x}} = M_2 \mathbf{x}, \tag{5.34}$$

$$\tilde{\mathbf{b}} = M_1^{-1} \mathbf{b}, \tag{5.35}$$

and $M = M_1 M_2$ is a nonsingular matrix called a *preconditioner* [418]. After solving (5.32), (5.34) is solved to determine $\mathbf{x}$. Proper choice of the preconditioner accelerates the convergence of the iterative solvers, ensuring that determining M and solving (5.32) and (5.34) are more efficient than solving the original system. The computational cost of producing the preconditioner should however be small since the difficulty in producing matrix M negates the computational benefits.

Matrix M, described in (5.25), and (5.27a) to (5.27d) can be used as a preconditioner during the circuit analysis process [416]. Several features make methods (5.27a) to (5.27d) attractive. Systems involving matrix M are relatively easy to solve, since M is either diagonal or triangular. Systems produced in practical VLSI systems are sparse diagonally dominant matrices. The diagonal elements of a Laplacian matrix of an underlying graph are typically larger than the non-diagonal elements. The number of nodes of an underlying circuit graph is proportional to the number of edges, since most nodes are only connected to the immediate neighbors. Systems composed of sparse diagonally dominant matrices are well suited for preconditioning using split matrices, significantly accelerating the convergence process [418].

Other popular preconditioning approaches exist that include incomplete factorization and approximate inverse. Incomplete LU (ILU) factorization, for example, is based on approximating $A \approx \tilde{L}\tilde{U}$ [416], where $\tilde{L}$ and $\tilde{U}$ are sparse upper and lower triangular matrices, yielding a preconditioned system,

$$\tilde{L}^{-1}A\tilde{U}^{-1}(\tilde{U}\mathbf{x}) = \tilde{L}^{-1}\mathbf{b}. \tag{5.36}$$

Incomplete Cholesky decomposition is a similar procedure restricted to positive definite matrices where the sparse approximation $A \approx R^T R$ is determined. The SParse Approximate Inverse (SPAI) [419] preconditioner exhibits performance superior to incomplete factorization methods [420] when applied to diagonally dominant problems.

The iterative methods and preconditioners described in this section are considered general purpose, effectively handling a wide range of problems while significantly reducing the memory requirements. Superior performance can however be achieved by applying advanced analysis methods, exploiting special features of practical circuit graphs, such as sparsity, smoothness, and graph partitioning. The upcoming subsections describe enhancements to MNA-based circuit analysis, including domain decomposition, multigrids, and hierarchical matrices.

5.2.1 Domain decomposition

Due to the superlinear complexity of linear system solvers, the divide-and-conquer approach [421] can be effective in tackling these problems. Two advantages make divide-and-conquer algorithms particularly attractive for circuit analysis. If solving a problem requires $O(n^p)$ time, where n is the problem size and $p > 1$, decomposing the problem into m sequentially solved parts yields a runtime of $O\left(\frac{n^p}{m^{p-1}}\right)$, assuming negligible computational overhead. Due to the superlinear complexity of linear system solvers, i.e., $p > 1$, this approach can be effective in reducing the computational burden. Furthermore, these m parts can be processed in parallel, further reducing the runtime.

Domain decomposition (DD) is one of the most successful divide-and-conquer strategies for circuit analysis. The main principle of the DD technique is partitioning a circuit graph $G = (V, E)$ into multiple subgraphs $G_i = (V_i, E_i), i \in \{1, \ldots, m\}$ and a subgraph of interface nodes $G_0 = (V_0, E_0)$. An illustrative example is shown in Fig. 5.3. Observe that the interface nodes do not belong to any subgraph, and no node belongs to more than one partition. Note that this limitation not only prohibits conduction between the subgraphs, but also forbids capacitive and

Fig. 5.3 Domain decomposition process within a grid. a) A large mesh is divided into $m = pq$ subdomains $\{\Omega_1, \ldots, \Omega_{pq}\}$ (gray nodes) and interfaces (black nodes). b) Connectivity within domain i is described by matrix A_i, while connections with the interface are described in E_i and F_i. A_Γ encodes the connectivity within the interface. The dimensions of A_Γ are typically smaller than the dimensions of A_i.

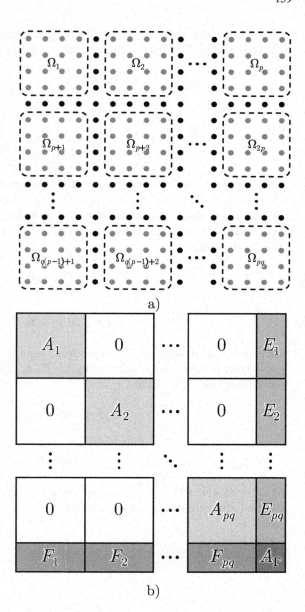

inductive coupling between subdomains. The standard linear equation of (5.21) can therefore be represented as an "arrowhead matrix" [422],

$$
\begin{bmatrix}
A_1 & 0 & \dots & 0 & E_1 \\
0 & A_2 & \dots & 0 & E_2 \\
\vdots & \vdots & \ddots & \vdots & \vdots \\
0 & 0 & \dots & A_m & E_m \\
F_1 & F_2 & \dots & F_m & A_0
\end{bmatrix}
\begin{bmatrix}
\mathbf{x}_1 \\
\mathbf{x}_2 \\
\vdots \\
\mathbf{x}_m \\
\mathbf{x}_0
\end{bmatrix}
=
\begin{bmatrix}
\mathbf{b}_1 \\
\mathbf{b}_2 \\
\vdots \\
\mathbf{b}_m \\
\mathbf{b}_0
\end{bmatrix},
\tag{5.37}
$$

where A_i represents the connectivity within subgraph G_i, E_i and F_i represent the connectivity between G_i and interface G_0, A_0 represents the connectivity among the interface nodes, $\mathbf{x}_i$ denotes the unknown voltages and currents within G_i, and $\mathbf{b}_i$ is the vector of current and voltage sources connected to G_i. Equation (5.37) can be split into two parts,

$$
A\mathbf{x} + E\mathbf{x}_0 = \mathbf{b},
\tag{5.38}
$$

$$
F\mathbf{x} + A_0\mathbf{x}_0 = \mathbf{b}_0,
\tag{5.39}
$$

where A is a block diagonal matrix produced from subgraph matrices A_i, and F, E, $\mathbf{x}$, and $\mathbf{b}$ are produced by concatenating F_i, E_i, $\mathbf{x}_i$, and $\mathbf{b}_i$. Solving (5.38) for $\mathbf{x}$ and substituting into (5.39) yields

$$
\mathbf{x} = A^{-1}(\mathbf{b} - E\mathbf{x}_0),
\tag{5.40}
$$

$$
(A_0 - FA^{-1}E)\mathbf{x}_0 = \mathbf{b}_0 - FA^{-1}\mathbf{b}.
\tag{5.41}
$$

To solve (5.41), $P = A^{-1}E$ and $\mathbf{q} = A^{-1}\mathbf{b}$ are determined. Due to the block diagonal structure of A, these equations can be decomposed into m independent equations,

$$
P_i = A_i^{-1}E_i,
\tag{5.42}
$$

and

$$
\mathbf{q}_i = A_i^{-1}\mathbf{b}_i.
\tag{5.43}
$$

Matrix P and vector q are substituted into (5.41), yielding

$$
(A_0 - FP)\mathbf{x}_0 = \mathbf{b}_0 - F\mathbf{q}.
\tag{5.44}
$$

Since the size of the interface set $|V_0|$ is typically much smaller than the size of any subgraph $|V_i|$, (5.44) requires relatively small computational resources. Equation (5.40) is transformed into

$$\mathbf{x} = \mathbf{q} - P\mathbf{x}_0. \qquad (5.45)$$

Similar to (5.42) and (5.43), (5.45) can be decomposed into m independent systems,

$$\mathbf{x}_i = \mathbf{q}_i - P_i\mathbf{x}_0. \qquad (5.46)$$

Due to the mutual independence of these expressions, (5.42), (5.43), and (5.46) can be solved in parallel. Furthermore, due to the small dimensions (i.e., relatively few interface nodes), the total runtime to solve m systems is smaller than the runtime to solve the original system, yielding additional performance improvement.

One of the earliest applications of domain decomposition in circuit analysis is discussed in [423]. A DRAM system composed of 130,000 transistors was successfully analyzed using 27 connected workstations operating in parallel. On-chip power delivery system analysis using domain decomposition is proposed in [422]. Domain decomposition combined with direct LU factorization of the subdomains achieved the maximum performance in case studies, completing a DC analysis of a ten million node system in 450 seconds.

The major advantage of domain decomposition is parallelization. Increasing the number of subdomains however increases the size of the interface graph G_0, potentially negating any performance gains. Overlapping domain decomposition modifies the original non-overlapping technique by allowing partitions to overlap [424], as illustrated in Fig. 5.4. The combined analysis of overlapping domains is based on the Schwarz method [425] and is used in [426] to complete the analysis of a power grid with 192 million nodes in five minutes while utilizing 1,200 processors operating in parallel.

5.2.2 $\mathcal{H}$-matrix

Another divide-and-conquer approach to circuit analysis is the application of the hierarchical matrix ($\mathcal{H}$-matrix) technique. Assume the target graph $G = (V, E)$ is described by a nodal analysis matrix A. The method commences with hierarchical clustering of the entries within matrix A, yielding cluster tree T, as illustrated in Fig. 5.5d. The first step of top-down clustering [427] splits matrix A into multiple submatrices,

$$A = \begin{bmatrix} A_{1,1} & \cdots & A_{1,m} \\ \vdots & \ddots & \vdots \\ A_{m,1} & \cdots & A_{m,m} \end{bmatrix}. \qquad (5.47)$$

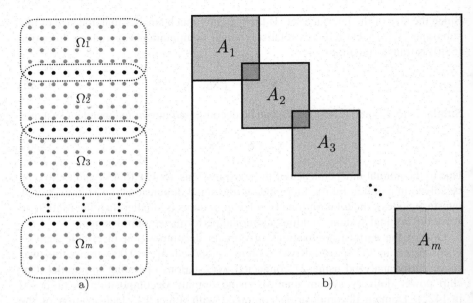

Fig. 5.4 Overlapping domain decomposition technique. a) Overlapping domains within the mesh. The black dots represent shared nodes. b) Resulting MNA matrix with overlapping sections corresponding to shared nodes.

Blocks $A_{i,j} \in \mathbb{R}^{p \times q}$ become the children of the root node of cluster tree T. The diagonal blocks $A_{i,i}$, $i \in \{1, \ldots, m\}$ are typically full rank matrices, while the off diagonal blocks are rank deficient. If rank $k_{i,j}$ of block $A_{i,j}$ is smaller than the specified threshold $k_{\min}$, the matrix can be efficiently factorized as the product of two small matrices,

$$A_{i,j} = M N^T, \qquad (5.48)$$

where $M \in \mathbb{R}^{p \times k}$, $N \in \mathbb{R}^{q \times k}$, and $k \ll p, q$. Any block within T is split if the size of the block is greater than the specified minimum size $m_{\min}$ and if the rank of the matrix is greater than $k_{\min}$. Otherwise, the block is not split and is stored in factored form.

The main purpose of an $\mathcal{H}$-matrix is a cluster tree representation of matrix A. The resulting block matrix is illustrated in Fig. 5.5c. Those leaves stored in factored form require relatively low processing runtime. These features enable an efficient approximation of the LU factorization and inverse of matrix A, yielding significant improvement in runtime and memory. For example, the complexity of LU factorization is reduced from $O(n^3)$ to $O(n (\log n)^2)$. Partial element equivalent circuit analysis of a power supply layout is presented in [428], achieving up to four orders of magnitude speedup and up to a 50 fold reduction in memory requirements. The compatibility of the $\mathcal{H}$-matrix with finite element analysis

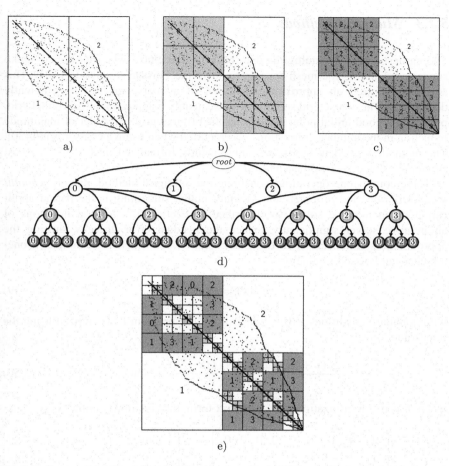

Fig. 5.5 Construction of an $\mathcal{H}$-matrix of a sparse matrix describing a circuit with 256 nodes and 471 edges. The black dots represent the sparsity pattern of the original matrix. a) The original matrix is divided into four submatrices. b) Submatrices 1 and 2 are significantly rank deficient and are therefore not divided. Submatrices 0 and 3 are further split into four submatrices. c) $\mathcal{H}$-matrix after the third iteration. d) Cluster tree T after three iterations. e) Final $\mathcal{H}$-matrix after five iterations. The densest regions along the diagonal are split until the minimum submatrix size is achieved.

enables efficient fine grained thermal analysis of a three-dimensional IC with more than a million discrete points [429].

5.2.3   *Multigrid methods*

The earliest works on multigrids date back to the 1960's when R. P. Fedorenko suggested doubling the mesh spacing to solve the Poisson's equation [430]. This approach allowed an approximate solution of a coarse grid to be efficiently determined. The coarse grid solution is subsequently mapped onto the original grid, providing a good initial point for solving the original grid. An order of magnitude reduction in the number of iterations was reported when using a coarser grid [430]. The method, applied in [430], was subsequently formalized in the 1970's by A. Brandt [431] and W. Hackbusch [432].

Three cornerstone operations constitute the multigrid method, namely *smoothing*, *restriction*, and *prolongation*. Fundamentally, smoothing is the partial application of an iterative solver, as described in Section 5.2. Several iterations of an iterative solver significantly reduce the residual error, thereby shifting the approximate solution closer to the exact solution. Consider, for example, a one-dimensional Poisson's equation,

$$f''(x) = \sin(x), \tag{5.49}$$

with boundary conditions $f(0) = f(1) = 0$. The problem is discretized using the trapezoidal rule,

$$\frac{f(x+h) - 2f(x) + f(x-h)}{h^2} = \sin(x), \tag{5.50}$$

where h is the discretization step. In matrix form, the equation becomes

$$\begin{bmatrix} 1 & 0 & 0 & 0 & \dots & 0 & 0 & 0 & 0 \\ -1 & 2 & -1 & 0 & \dots & 0 & 0 & 0 & 0 \\ 0 & -1 & 2 & -1 & \dots & 0 & 0 & 0 & 0 \\ \vdots & \vdots & \vdots & \vdots & \ddots & \vdots & \vdots & \vdots & \vdots \\ 0 & 0 & 0 & 0 & \dots & 0 & -1 & 2 & -1 \\ 0 & 0 & 0 & 0 & \dots & 0 & 0 & 0 & 1 \end{bmatrix} \mathbf{x} = \begin{bmatrix} \sin(0) \\ \sin(h) \\ \sin(2h) \\ \vdots \\ \sin(1-h) \\ \sin(1) \end{bmatrix}. \tag{5.51}$$

The system is solved using the Loose GMRES (LMGRES) method [433]. Convergence of the algorithm is depicted in Fig. 5.6. Observe that any difference between the adjacent points is quickly reduced, i.e., the high frequency errors are significantly dampened during each iteration.

Eliminating low frequency errors however requires significantly longer runtime. To overcome this issue, a restriction step is performed to coarsen the domain of the system. The frequency of the error is therefore effectively increased, permitting efficient elimination by smoothing. Formally, restriction of function $f : E \rightarrow F$ is function $f|_A : A \rightarrow F$, where $A \subset E$ and $f(x) = f|_A(x) \forall x \in A$. In the context of multigrids, this operation is effectively coarsening, reducing the number of points

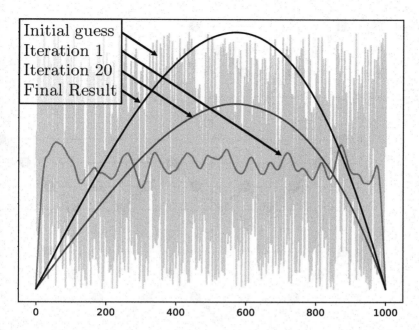

Fig. 5.6 Convergence of Poisson's equation using LMGRES method. The one-dimensional space is discretized using 1,001 points. The initial guess is a vector of random numbers. After the first iteration, the high frequency components of the initial vector are significantly reduced, producing a smoother curve. To eliminate the remaining low frequency components, additional iterations are necessary. In this example, the solution is achieved after 62 iterations.

within the grid. Suppose the grid is described as a graph $G_0 = (V_0, E_0)$, and function $v : V \to \mathbb{R}$ maps each node to the node voltage. The goal of the coarsening operation is a reduced version of an initial grid $G_1 = (V_1 \subset V_0, E_1)$, where the voltage within the original system $v_0(n)$ is equal to the voltage within coarse system $v_1(n)$ at any node $n \in V_1$. To recover the original solution from the approximate coarse solution, a prolongation operation is performed. Using interpolation, the solution of a coarse grid is mapped onto a fine grid. The resulting vector is typically a close approximation of a solution requiring few iterations to converge.

A single restriction procedure is often insufficient for significant acceleration. The restriction process is therefore repeatedly applied to further reduce the size of the mesh. Multiple prolongation operations are therefore necessary to recover the original grid. These procedures are formalized in *cycles* where a system undergoes a series of restrictions, prolongations, and smoothing. The most common cycles are V-cycle, W-cycle, and F-cycle [434], as illustrated in Fig. 5.7. These techniques tradeoff robustness with computational speed. The V-cycle is typically faster than the other cycle types but may fail to converge to a correct result. In contrast, the F-cycle requires more operations but is highly robust and accurate.

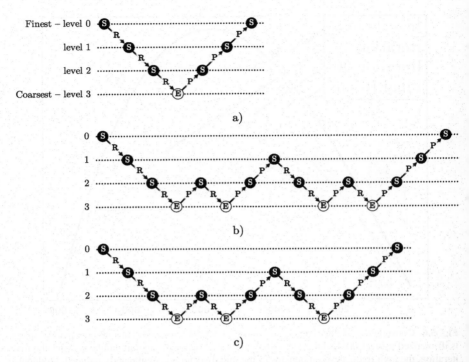

Fig. 5.7 Multigrid cycles. a) V-cycle, b) W-cycle, and c) F-cycle. 'S', 'R', and 'P' denote, respectively, the smoothing, restriction, and prolongation operations. At the coarsest level, a system is typically solved using an exact solver. This operation is denoted by 'E.'

One of the earliest applications of geometric multigrids to VLSI systems is described in [435]. A relatively straightforward application of geometric multigrids in [435] yields a 300 fold improvement in runtime with a peak error of over 20%. Geometric multigrids are highly suitable for analyzing regular physical layouts, supporting the efficient analysis of circuits with tens of millions of nodes [436].

The theoretical computational complexity of the geometric multigrid is $O(|V|)$ [437], an attractive feature in the analysis of large systems. A major limitation of geometric multigrids is reliance on the structural regularity of the problem domain. The algebraic multigrid (AMG) is an important generalization of geometric multigrids where no structural information is required for restriction and prolongation. One of the earliest applications of AMG to circuit analysis is proposed in [438], where a 16 to 20 fold improvement in runtime is achieved. Another notable result is PowerRush, an AMG-based DC and transient simulator exhibiting linear complexity [439, 440]. Up to nine levels of grid reduction are reported in [439], completing the analysis in 169 seconds after reducing a circuit with 38 million nodes to 264 nodes. The efficiency of AMG simulation enables large scale circuit optimization. For example, in [441], decoupling capacitor allocation is performed using AMG, optimizing circuits with up to a million nodes.

Restriction, smoothing, and prolongation are highly parallelizable due to the small number of steps with few dependencies [442]. These features enable GPU-based geometric multigrid acceleration of the circuit analysis process, achieving up to two orders of magnitude speedup [442, 443].

5.3 Non-MNA techniques

MNA and associated enhancements enable the efficient analysis of a wide range of complex circuits. Alternative techniques however exist that avoid MNA-based equations, often yielding superior performance as compared to MNA-based methods. Three techniques are presented in this section, namely, scattering parameters, random walks, and lattice graph analysis.

5.3.1 Scattering parameters

The detailed structure of an IC component is often unknown. This situation frequently occurs in two cases. If the components of the integrated system are supplied by a third party vendor, the internal structure of the components is treated as intellectual property (IP) and is typically not described or is purposely obfuscated [444, 445]. The structure of a component can also be highly complex, complicating the construction of a distributed model [75]. A scattering parameter (S parameter) model is often utilized in these cases, characterizing the frequency response of a circuit to input stimuli without revealing the internal structure (a black box). Examples of an S parameter model with two and n ports are depicted in Fig. 5.8. Parameters a_k and b_k correspond to normalized power waves [446],

$$a_k = \frac{1}{2g_k} \left(V_k + I_k Z_k \right), \tag{5.52}$$

$$b_k = \frac{1}{2g_k} \left(V_k - I_k Z_k^* \right), \tag{5.53}$$

where Z_k is the reference impedance at port k. Z_k^* denotes the complex conjugate of Z_k and

$$g_k = \sqrt{|\Re(Z_k)|}. \tag{5.54}$$

By measuring the response b_m of a circuit at port m in response to a unit power wave at port k, scattering parameter $s_{k,m}$ is

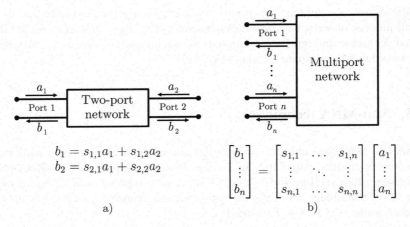

Fig. 5.8 Scattering parameter (S parameter) model of a component. a) Two port network, and b) multiport network. The component is treated as a black box with no knowledge of the internal structure, as opposed to a grey or white box utilizing, respectively, partial or complete structural information [447]. By applying stimuli at different ports of the components, the response of the system at each port is determined. The relationship between an excitation at port i and the response at port j is described by S parameters.

$$s_{k,m} = \frac{b_m}{a_k}.$$ (5.55)

In a multiport network, scattering parameter matrix S is produced that describes the relationship among the signals at different ports,

$$\begin{bmatrix} b_1 \\ b_2 \\ \vdots \\ b_n \end{bmatrix} = \begin{bmatrix} s_{1,1} & s_{1,2} & \cdots & s_{1,n} \\ s_{2,1} & s_{2,2} & \cdots & s_{2,n} \\ \vdots & \vdots & \ddots & \vdots \\ s_{n,1} & s_{n,2} & \cdots & s_{n,n} \end{bmatrix} \begin{bmatrix} a_1 \\ a_2 \\ \vdots \\ a_n \end{bmatrix}.$$ (5.56)

Note that any scattering parameter is a function of frequency. Measurements should therefore be performed at different frequencies to evaluate the response over the entire bandwidth of interest.

A major advantage of the S parameter model is the applicability of the model to an arbitrary system. The S parameter model requires no information describing the internal structure of the system. Furthermore, based on an S parameter matrix, other electromagnetic characteristics of a system can be determined [446]. Based on open circuit impedance (Z) parameters, for example, crucial parameters can be determined such as the self- and mutual inductances within a network [332].

$$Z = G_0^{-1} (I - S)^{-1} \left(S Z_0 - Z_0^* \right) G_0,$$ (5.57)

where I is the identity matrix,

$$G_0 = \begin{bmatrix} g_1 & 0 & \cdots & 0 \\ 0 & g_2 & \cdots & 0 \\ \vdots & \vdots & \ddots & \vdots \\ 0 & 0 & \cdots & g_n \end{bmatrix}, \tag{5.58}$$

and

$$Z_0 = \begin{bmatrix} Z_1 & 0 & \cdots & 0 \\ 0 & Z_2 & \cdots & 0 \\ \vdots & \vdots & \ddots & \vdots \\ 0 & 0 & \cdots & Z_n \end{bmatrix}. \tag{5.59}$$

Other parameters widely used during the design of analog circuits, such as Y, $ABCD$, or h parameters [446], can also be derived from the S parameters.

5.3.2   Random walks

A random walk is a stochastic process that describes a succession of steps of an object within a mathematical space [448]. A classic example of a random walk is a random walk along a one-dimensional integer axis, as illustrated in Fig. 5.9a. The particle is initially at position 0 and, every time step, the particle moves in a random direction. Different types of space and probability distributions of the transitions in a random walk exist, such as a discrete two-dimensional space, continuous two-dimensional space with a variable step length (such as Lévy Flight [449]), or a biased, continuous walk in three-dimensional space (see Figs. 5.9b to 5.9d). Common issues relating to a random walk include the expected distance of an object from the source after n steps, probability of a return to the origin after n steps, and probability of reaching a before reaching b, where a and b are arbitrary points within the space.

Manifestations of a random walk in physical systems have been studied before this term was first coined. In 1880, Lord Rayleigh studied the amplitude of oscillations due to multiple strings vibrating at the same frequency with a random phase [450]. This problem is analogous to a random walk on a one-dimensional axis. The erratic movement of dust particles, what will later be called Brownian motion, was discovered as early as 1784 by the Dutch scientist, J. Ingen-Housz [451]. A formal study of random walks has been applied to different physical phenomena, including diffusion in molecular physics [452], genetic drift in genetics [453], and measuring certain features of the World Wide Web [454, 455].

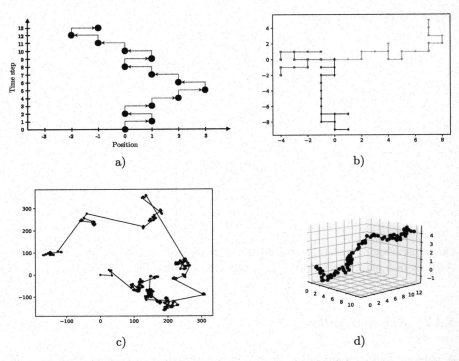

Fig. 5.9 Examples of a random walk. a) Discrete time one-dimensional random walk on an integer axis. b) Three unbiased random walks within a two-dimensional integer grid. c) Unbiased random walk within a continuous two-dimensional space. The direction of the step is uniformly random. The step size follows a Cauchy distribution. This type of random walk is commonly referred to as Lévy Flight [449]. d) Biased random walk within a three-dimensional integer space. The probability of a transition toward $+\infty$ is greater than the probability of a transition toward $-\infty$ along the x, y, and z axes.

One of the most extensively studied spaces of a random walk is a graph, where a particle moves towards the neighboring vertex at each time step. The probability of moving from vertex a toward vertex b is proportional to the weight of the edge (a, b). The analogy between a random walk and an electrical network was studied by C. St. J. A. Nash-Williams in [456]. The random walk equivalent of the effective resistance between a and b is the commute time between nodes a and b, i.e., the expected number of steps in a random walk starting at a, visiting b, and returning to a (see Fig. 5.10). The conductance of a resistor is equivalent to the weight of an edge within a network. The probability of a transition along a specific edge is proportional to the weight of that edge. A particle moving in a random walk is therefore less likely to transition along a high resistance edge (or path).

The analogy between electrical circuits and random walks can be exploited in the analysis of electrical circuits. Different simulation tools based on random walks have been explored in the literature. By performing a random walk experiment

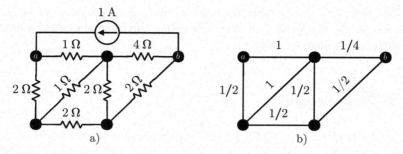

Fig. 5.10 Effective resistance between arbitrary nodes a and b is equivalent to the expected number of steps for a random walk to visit node b while starting and returning to node a, a) electrical circuit, and b) equivalent graph. The probability of transitioning towards a neighboring node is proportional to the conductance of the corresponding edge.

multiple times, the average number of steps converges towards the commute time which corresponds to the effective resistance. The earliest application of a random walk to linear circuit analysis is described in [70]. A major advantage highlighted in [70] is the linear relationship between circuit size and computational complexity. Different circuit simulation tools have been described in the literature, achieving a significant speedup as compared to conventional circuit analyses [70, 457–459]. Another aspect is the locality of the random walk. If the target nodes are located close to each other within a network, the random walk is more likely to terminate while exploring only a small portion of the network. This result is highly desirable when studying system perturbations, since only the affected portion of the system is analyzed. This advantage is exploited in [460] in the analysis of incremental changes in power grids.

A major issue pertaining to random walk-based simulation tools is the number of random walk experiments required to achieve a sufficiently small error. The error ε of a random walk is inversely proportional to the square root of the number of experiments M,

$$\varepsilon \propto \frac{1}{\sqrt{M}}. \tag{5.60}$$

To reduce the error by 50%, the number of experiments should be increased four fold. To overcome this issues, the 'importance sampling' technique is introduced in [461], significantly improving the speed of convergence. Another challenge of random walk-based tools is the possibility of excessively long walks, negating any computational speedup [70]. This issue can be eliminated by limiting the length of the random walk. The accuracy of the solution is however degraded by limiting the length.

Random walks are found in a wide variety of VLSI applications. A sensitivity analysis of VLSI power networks [462], for example, is a notable application, where the critical parameters affecting a power grid are evaluated. Matrix preconditioning based on random walks is described in [463], significantly accelerating the circuit analysis process. Other notable applications of random walks in VLSI include modeling of thermal behavior [464], decoupling capacitor placement [465], and electromigration analysis [466].

5.3.3 Lattice graph

Due to the large scale of VLSI systems, the physical structures are often highly regular, composed of millions to billions of identical elements distributed within a system. The on-chip power grid is a prominent example of a regular structure composed of two or more layers of identical interconnects. An example of a power grid is shown in Fig. 5.11a. Grids are highly reliable due to the many redundant paths. The number of paths connecting two corners of a grid is [467]

$$\binom{x+y}{x} = \frac{(x+y)!}{x!y!}, \tag{5.61}$$

where x and y are, respectively, the horizontal and vertical dimensions of a grid. The number of paths in (5.61) grows superlinearly with x and y, yielding a high degree of redundancy even in relatively small grids. The failure of a single or multiple wire segments can be tolerated since the remaining wires provide the necessary connections. Additional benefits of a grid include shielding and decoupling that reduce parasitic capacitive and inductive coupling in global clock and data lines [468]. A grid structured power network can be modeled as a two-dimensional resistive lattice, as shown in Fig. 5.11b [469]. Depending upon the metal pitch and die size, the dimensions of a grid can vary from hundreds to tens of thousands of segments [72]. The large dimensions enable the use of infinite mesh methods for analyzing grid structured power networks.

Multidimensional mesh structures assuming infinite mesh dimensions have been extensively studied in the literature. In 1936, W. H. McCrea studied the following problem [470],

> "In a rectangular lattice, at every time instant a point P moves from one lattice point to one of the neighboring points. Each adjacent point has equal probability of being selected. Determine the probability that the particular boundary point is ultimately reached."

Variations of this problem on different two- and three-dimensional lattices were solved by W. H. McCrea and F. J. W. Whipple in 1940 [471], where different finite and infinite rectangular lattices are analyzed. A notable result is an expression for the average flow of particles between the source and target points within an infinite two-dimensional lattice,

$$G(x, y) = \frac{2}{\pi} \int_0^\pi \frac{1 - \cos(\lambda y) \exp(-\mu|x|)}{\sinh(\mu)} d\lambda, \tag{5.62}$$

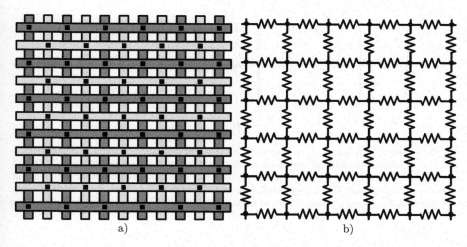

Fig. 5.11 On-chip power grid. a) Layout of power (dark grey) and ground (light grey) distribution networks, and b) power distribution network modeled as a resistive lattice.

where $\cos(\lambda) + \cosh(\mu) = 2$, and x and y denote the number of resistors separating the source and target points in, respectively, the horizontal and vertical direction.

The link between random walk and circuit theory was not widely recognized in the 1940's. An electrical formulation of the problem solved by W. H. McCrea and F. J. W. Whipple in [471] is

Determine the effective resistance between two arbitrary points (x_0, y_0) and (x, y) within a two dimensional grid of resistors with resistance r (see Fig. 5.11b)

An easier problem of determining the effective resistance between *adjacent* nodes in an infinite resistive lattice was solved in 1949 [472] based on the principles of symmetry and superposition. Suppose current $4i$ is injected at an arbitrary node a, as illustrated in Fig. 5.12a. Due to symmetry, the current through each of the four adjacent branches is i. Now withdraw current $4i$ from neighboring node b. The current though each adjacent branch is also i, as shown in Fig. 5.12b. By superimposing these solutions, the current through a resistor connecting a and b is $2i$, as illustrated in Fig. 5.12c. The effective resistance is found by equating the voltage drop across resistor ab with the voltage drop across the effective resistance of the grid,

$$4i \, R_{eff} = 2ir, \tag{5.63}$$

yielding

$$R_{eff} = \frac{r}{2}. \tag{5.64}$$

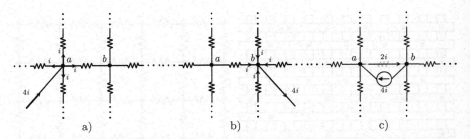

Fig. 5.12 Superposition applied to determine the effective resistance between adjacent nodes a and b in an infinite two-dimensional resistive lattice. a) Current $4i$ is injected into node a. Due to symmetry, the current through each adjacent resistor is i, flowing away from node a. b) Current $4i$ is drawn from node b. Due to symmetry, the current through each adjacent resistor is i, flowing towards node b. c) Superposition of current injection and withdrawal. The current through resistor ab is $2i$.

Despite the relative simplicity of the problem for adjacent nodes, the general problem requires advanced mathematical methods. Different alternative solutions to the problem of determining the effective resistance within a grid have been presented in the literature [472] in the context of operational calculus [473], discrete analytic functions [474], partial differential equations [475–477], random walks [471, 478], and lattice Green's function [479]. Notable examples include the expressions by A. Stöhr [475],

$$R(x, y) = -\frac{1}{2\pi} \int_0^\infty \left[\left(1 - \frac{t}{\zeta}\right)^{x+y} \left(1 - \frac{t}{\zeta^3}\right)^{x-y} (1 - \zeta t)^{-x+y} \left(1 - \zeta^3 t\right)^{-x-y} \right] \frac{dt}{t}, \quad (5.65)$$

$$\zeta = e^{\frac{2\pi i}{8}}, \quad (5.66)$$

and

$$R(x, y) = -\frac{1}{\pi} \int_0^\pi \frac{\left[\lambda - \sqrt{\lambda^2 - 1}\right]^y \cos x\theta}{\sqrt{\lambda^2 - 1}} d\theta, \quad (5.67)$$

$$\lambda = 2 - \cos\theta, \quad (5.68)$$

F. Spitzer [478],

$$R(x, y) = \frac{1}{8\pi^2} \int_{-\pi}^{\pi} \int_{-\pi}^{\pi} \frac{1 - \cos(x\alpha + y\beta)}{1 - \frac{1}{2}(\cos\alpha + \cos\beta)} d\alpha \, d\beta, \quad (5.69)$$

B. van der Pol [473],

$$R(x, y) = \frac{1}{2\pi} \int\limits_{0}^{\infty} \left[1 - \left(\frac{t+i}{t-i}\right)^{x+y} \left(\frac{t-1}{t+1}\right)^{|x-y|} \right] \frac{dt}{t}, \tag{5.70}$$

and W. H. McCrea and F. J. W. Whipple [471], later rediscovered by G. Venezian [476] and J. Cserti [479],

$$R(x, y) = \frac{1}{\pi} \int\limits_{0}^{\pi} \frac{1 - e^{-x\mu} \cos y\lambda}{\sinh \mu} d\beta, \tag{5.71}$$

$$\cosh \mu + \cos \lambda = 2. \tag{5.72}$$

Expressions (5.65) to (5.71) describe uniform resistive lattices. Many practical VLSI grids are anisotropic, *i.e.*, the resistance along the horizontal dimension is not the same as the resistance along the vertical dimension. An expression for the resistance within an infinite anisotropic resistive grid is presented in [469],

$$R(x, y, k) = \frac{kr}{\pi} \int_{0}^{\pi} \frac{2 - e^{-|x|\alpha} \cos y\beta}{\sinh \alpha} d\beta, \tag{5.73}$$

where k is the ratio of the horizontal resistance to the vertical resistance, and

$$k + 1 = k \cos \beta + \cosh \alpha. \tag{5.74}$$

This result has significant value for the analysis of power grids. To determine the equivalent resistance within an $M \times N$ grid using MNA, a solution of the linear equation of size $MN \times MN$ is necessary, requiring prohibitive computational time. In contrast, the effective resistance between two nodes within a grid can be found in constant time, assuming these nodes are sufficiently far from the grid boundaries. A linear complexity, IR voltage drop analysis algorithm is introduced in [480]. The contribution of the voltage sources and current loads to IR voltage drops is evaluated separately based on the effective resistance computed in constant time. The solutions are superimposed to determine the total IR voltage drop within a circuit. The solution is further accelerated by observing that the IR voltage drop contribution of distant voltage sources and loads is negligible. By restricting the analysis to the vicinity of a node, the runtime can be drastically reduced while maintaining the error below 0.5%.

5.4 Summary

Due to the stringent performance requirements of modern VLSI systems, the demand for accurate circuit analysis has drastically increased over the past decades. The immense complexity of modern VLSI systems however makes standard circuit analysis based on MNA impractical. A wide range of algorithms have been proposed to reduce the runtime of the circuit analysis process while maintaining sufficient accuracy. The most prominent techniques are described in this chapter.

Domain decomposition methods split a circuit into multiple independent domains, thereby reducing the computational complexity and enabling parallelization. In the $\mathcal{H}$-matrix representation, the sparsity of practical matrices is exploited to produce a cluster tree, enabling efficient algorithms with less memory requirements and lower computational complexity. Using multigrid techniques, a solution is initially approximated using a coarse version of the system. The solution is subsequently determined after interpolation and smoothing operations.

Alternative circuit analysis techniques attempt to accelerate the circuit analysis process by avoiding costly MNA-based analysis. A complex or obfuscated circuit can be represented by a multiport network model, efficiently described by S parameters. In random walk-based methods, the voltage within a grid is determined statistically, yielding linear computational complexity at a fixed accuracy. The infinite lattice model can often be used to analyze large grids, often encountered in on-chip power distribution systems.

Common circuit analysis methods are discussed in this chapter. These methods enhance traditional MNA processes or follow alternative approaches. Despite the immense potential, few of these techniques are used in mainstream circuit analysis methodologies. Further research is required to improve the versatility and performance of the advanced circuit analysis techniques discussed in this chapter. For example, a significant limitation of infinite lattice analysis is poor accuracy near the boundaries of the grid. This limitation is overcome by applying the image method [71] and infinity mirror technique [72], described, respectively, in Chapters 6 and 7.

Chapter 6
Effective resistance of truncated infinite mesh structures

A mesh structure is an important topology for modeling a variety of physical and mathematical phenomena. The structure consists of regularly placed nodes within a multidimensional space and connected with resistors to adjacent nodes. Despite the theoretical nature of an infinite mesh structure, a variety of practical examples exist, where the size and regularity support the assumption of an infinite grid. For example, the resistance of a large uniform conducting sheet can be modeled as a resistive grid [65, 145, 476], enabling the use of an infinite resistive grid to model, for example, substrate noise [72, 481]. A mesh structure is prevalent in modern integrated circuits, particularly in power and ground distribution networks [381] and clock distribution networks [66]. The power and ground delivery networks typically consist of layered perpendicular metal interconnects [72, 232]. A typical on-chip delivery network structure is shown in Fig.6.1a. During the analysis process, power supply and ground networks are typically analyzed separately [75], as shown in Fig.6.1b. The resulting grid can be modeled as a resistive mesh, as shown in Fig.6.1c.

Analysis of power delivery noise in power grids is an important problem in VLSI systems. Conventional nodal analysis tools typically exhibit superlinear computational complexity, resulting in significant simulation time. An alternative approach for the analysis of power delivery grid circuits is proposed in [480]. To simplify the analysis, the resistive mesh is reduced to an equivalent effective resistance where the grid is assumed to be infinitely large. The primary benefit of this approach is significantly lower complexity, independent of grid size. The main drawback, however, is higher error in proximity of the grid boundaries due to the assumption of an infinite grid.

This chapter aims to bridge this gap. The effective resistance of a large resistive grid near the edges and corners is modeled as a truncated infinite mesh. In this chapter, the infinite mesh truncated along a single dimension is called a half-plane mesh (Fig. 6.2a), while an infinite mesh truncated along two orthogonal dimensions is called a quarter-plane mesh (Fig. 6.2b).

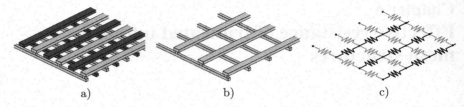

Fig. 6.1 Two layer power and ground network mesh modeling process. a) Original view of two-layer mesh. The light and dark gray segments are connected to, respectively, power and ground. b) A simplified model with the ground mesh removed. c) Equivalent resistive mesh of the power network.

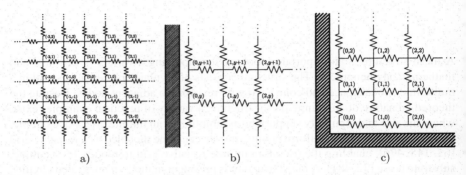

Fig. 6.2 Portions of two-dimensional infinite resistive structures. a) Portion of fully infinite mesh near the origin. b) Portion of half-plane mesh near the edge. c) Portion of quarter-plane mesh near the corner.

By utilizing the image and superposition methods, exact integral and approximate closed-form expressions for half- and quarter-plane meshes are presented. A brief historical review of the infinite resistive mesh analysis is presented in Section 6.1. The electric potential in an infinite mesh is studied in Section 6.2. The image method for electric circuits and a derivation of the exact integral equations are described in Section 6.3, followed by a derivation of the closed-form expressions in Section 6.4. The accuracy of the results is discussed in Section 6.5. The findings are summarized in Section 6.6.

6.1 Historical perspective

Determining the effective resistance between two nodes of an infinite resistive mesh, also known as a Liebman mesh [482], is a classical problem. The objective is to determine an effective two-port resistance given a two-dimensional network with identical resistors between adjacent nodes, as shown in Figs. 6.2a and 6.2b.

The problem has been studied from a variety of perspectives. An intuitive solution for determining the effective resistance between adjacent nodes within a mesh is described in [482, 483], where superposition of the current sources and symmetry are used to determine the voltage between adjacent nodes. The first general solution for this problem was published in 1940 [471], where the probability of reaching a specific node within a lattice during a random walk is determined, a process closely related to finding the effective resistance within a grid [484]. A solution, specific to electrical circuits, was published in 1950 [473], where a two-dimensional elliptic wave partial differential equation is applied to an infinite lattice. Several later works have been published describing alternative methods to solve this problem, including Fourier Transform [476, 477], Green's function [479] and graph theory [485].

Several extensions and variations of solutions to this problem have been published. In [477, 479], the problem is solved for a multi-dimensional grid, and triangular and hexagonal infinite lattices. Regular and semi-regular polyhedric structures as well as multi-dimensional cubes are described in [486], and an infinite cylindrical grid is considered in [487]. More practical considerations are included in [469], where a solution for an infinite grid with unequal horizontal and vertical resistances is provided.

Despite the problem being well studied, little attention has been devoted to the effects of truncations on the effective resistance. One version for determining the effective resistance in an infinite mesh is provided in [485],

$$R_{eff}(x, y) = \frac{1}{\pi i} \int_0^\pi \frac{1 - e^{x \cos^{-1}(2 - \cos(\alpha))} \cos(y\alpha)}{\sqrt{1 - (2 - \cos(\alpha))^2}} d\alpha. \tag{6.1}$$

The accuracy of (6.1) is compared with numerical analysis of a large resistive mesh. The relative error of the effective resistance near the edges and corners is shown, respectively, in Figs. 6.3 and 6.4. Due to the assumptions of symmetry and regularity, the effective resistance is more accurately evaluated near the center of the grid, where the effect of the boundaries is less significant. Near the edges and corners, however, the error of (6.1) can reach 40%, limiting the applicability of the integral expression.

6.2   Electric potential in an infinite mesh

The solution proposed in this chapter is based on modifying the methods described in [469, 485]. An alternative Green's function-based approach is presented in Appendix A. Consider a fully infinite anisotropic resistive mesh. Let the horizontal and vertical resistances be, respectively, $r_x = r$ and $r_y = kr$. Assign coordinates to each node, inject current I into node (x_0, y_0), and let the current exit at a node infinitely far from the injection node. Denote the potential at node (x, y) due to

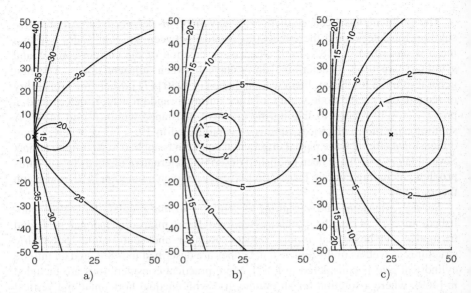

Fig. 6.3 Relative error (in per cent) of the effective resistance expression of an infinite grid (6.1) [485] within the proximity of the grid edge. The actual resistance is determined using a nodal analysis between node $(x_0, 0)$ and node (x, y) for a) $x_0 = 0$, b) $x_0 = 10$, and c) $x_0 = 25$. The grid dimensions are 101×201. The point $(x_0, 0)$ is indicated by the $\times$-mark.

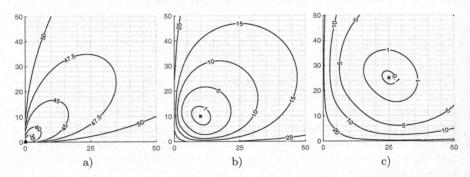

Fig. 6.4 Relative error (in per cent) of the effective resistance expression of an infinite grid (6.1) [485] within the proximity of the grid corner. The actual resistance is determined using a nodal analysis between node (x_0, y_0) and node (x, y) for a) $x_0 = y_0 = 0$, b) $x_0 = y_0 = 10$, and c) $x_0 = y_0 = 25$. The grid dimensions are 101×101. The point (x_0, y_0) is indicated by the $\times$-mark.

current I injected at (x_0, y_0) as $\phi_{x_0,y_0}(x, y)$. Three important properties of this potential exist. First, if the current source is moved by distance (a, b), the potential distribution across the grid is also moved by the same distance,

$$\phi_{x_0,y_0}(x, y) = \phi_{x_0+a,y_0+b}(x + a, y + b). \tag{6.2}$$

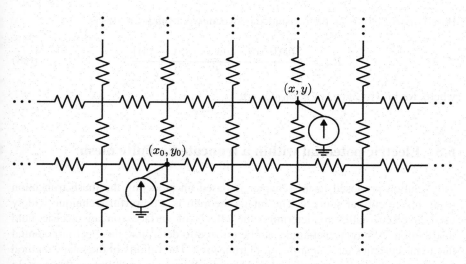

Fig. 6.5 Current injection into an infinite resistive mesh

Another important property is symmetry, *i.e.*, the current source and probe coordinates can be swapped,

$$\phi_{x_0,y_0}(x, y) = \phi_{x_0,y}(x, y_0) = \phi_{x,y}(x_0, y_0) = \phi_{x,y}(x_0, y_0). \tag{6.3}$$

From these properties, note that

$$\phi_{x_0,y_0}(x, y) = \phi_{-x_0,y_0}(-x, y) = \phi_{-x_0,-y_0}(-x, -y) = \phi_{x_0,-y_0}(x, -y). \tag{6.4}$$

To evaluate the effective resistance R_{eff} between nodes (x_0, y_0) and (x, y), the two current sources can be superimposed, as shown in Fig. 6.5. Knowing the voltage drop between these nodes allows the effective resistance to be determined,

$$R_\infty = \frac{V_{x_0,y_0} - V_{x,y}}{I}, \tag{6.5}$$

where V_{x_0,y_0} and $V_{x,y}$ are the voltage, respectively, at (x_0, y_0) and (x, y). V_{x_0,y_0} and $V_{x,y}$, in turn, can be expressed as the superposition of the potentials due to multiple current sources,

$$V_{x_0,y_0} = \phi_{x_0,y_0}(x_0, y_0) - \phi_{x,y}(x_0, y_0), \tag{6.6}$$

$$V_{x,y} = \phi_{x_0,y_0}(x, y) - \phi_{x,y}(x, y). \tag{6.7}$$

Based on (6.2) to (6.7), the resistance between two arbitrary nodes is

$$R_\infty = \frac{2\big(\phi(0, 0) - \phi(x - x_0, y - y_0)\big)}{I}, \tag{6.8}$$

where, for brevity, $\phi(x, y) = \phi_{0,0}(x, y)$.

6.3   Electric potential within a truncated infinite mesh

The solution proposed in this chapter is based on modeling the mesh truncation using image current sources. The image theorem is a powerful technique widely used in electrostatics to determine the effects of surfaces on an electric field distribution. A similar technique can be utilized to determine the electric potential due to the current source near the mesh truncation. The validity of the image method for a truncated mesh is established in Appendix B using the uniqueness theorem. In Subsection 6.3.1, the potentials of a fully infinite grid determined in Section 6.2 are superimposed to model the behavior of a truncated grid. In Subsection 6.3.2, the integral expression for the effective resistance in a half-plane and quarter-plane grid is presented.

6.3.1   Modeling truncation with image

Consider the case where an infinite grid of resistors is truncated at $x = 0$, removing all of the nodes with negative coordinates, as shown in Fig. 6.2. The assumption of symmetry along the x-axis becomes invalid, making the solutions reported in [471] to [479] inapplicable for a truncated mesh.

To circumvent this limitation, truncation can be replaced with another topology modification which satisfies the boundary conditions, (B.1) and (B.2). The truncated mesh structures are modeled as a fully infinite mesh with boundary conditions. The condition for the half-place mesh is

$$\phi(0, y) - \phi(-1, y) = 0, \tag{6.9}$$

i.e., the current flowing through the grid edge is zero. Similarly, a quarter-plane mesh is modeled as a fully infinite mesh with the following boundary conditions,

$$\phi(0, y) - \phi(-1, y) = 0, \tag{6.10}$$

$$\phi(x, 0) - \phi(x, -1) = 0. \tag{6.11}$$

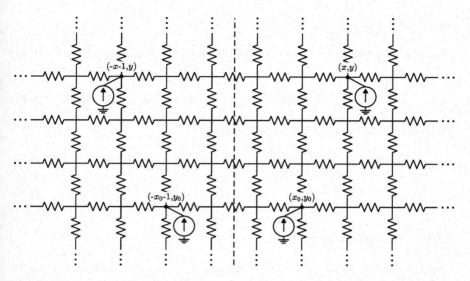

Fig. 6.6 Image method to model truncation in a half-plane mesh. The dashed line illustrates the boundary between the real and image half-planes. Two image sources are introduced in the negative-x plane to model the effect of truncation, ensuring zero effective current across the boundary.

The image technique accomplishes this task. In the following subsections, expressions for the effective resistance in terms of potentials are derived.

6.3.1.1   Half-plane mesh

Consider the circuit topology shown in Fig. 6.6 with ground placed infinitely distant. The positive-x side of the grid remains the same as the truncated grid. The symmetric negative-x side, however, maintains zero voltage between nodes $(0, y)$ and $(-1, y)$, thereby modeling the effect of the grid edge by satisfying the boundary condition (6.9).

A derivation of the effective resistance starts with (6.5). To model the truncation, two image current sources are introduced, as shown in Fig. 6.6. Unlike a fully infinite mesh, the voltage at nodes (x_0, y_0) and (x, y) within a half-plane mesh is the sum of the potential due to four current sources,

$$V_{x_0, y_0} = \phi_{x_0, y_0}(x_0, y_0) - \phi_{x, y}(x_0, y_0) + \phi_{-x_0-1, y_0}(x_0, y_0) - \phi_{-x-1, y}(x_0, y_0),$$
$$\tag{6.12}$$

$$V_{x, y} = \phi_{x_0, y_0}(x, y) - \phi_{x, y}(x, y) + \phi_{-x_0-1, y_0}(x, y) - \phi_{-x-1, y}(x, y). \tag{6.13}$$

Simplifying (6.2) to (6.4),

$$V_{x_0,y_0} = \phi(0,0) - \phi(x-x_0, y-y_0) + \phi(2x_0+1, 0) - \phi(x+x_0+1, y-y_0), \qquad (6.14)$$

$$V_{x,y} = \phi(x-x_0, y-y_0) - \phi(0,0) + \phi(x+x_0+1, y-y_0) - \phi(2x+1, 0). \qquad (6.15)$$

Combining (6.14) and (6.15) with (6.5) yields

$$R_{half} I = 2\phi(0,0) - 2\phi(x - x_0, y - y_0) + \phi(2x_0 + 1, 0) - 2\phi(x + x_0 + 1, y - y_0) + \phi(2x + 1, 0). \tag{6.16}$$

6.3.1.2 Quarter-plane mesh

Consider the case shown in Fig. 6.2, where an infinite mesh is truncated along the
x- and y-axes. Similar to the half-plane case, this topology can be modeled by
introducing six image current sources, as shown in Fig. 6.7, thereby satisfying the
boundary conditions in (6.10)-(6.11). The resulting voltages at (x_0, y_0) and (x, y)
are the sum of the potentials due to eight current sources, which, after simplification,
yields

$$V_{x_0,y_0} = \phi(0,0) + \phi(2x_0 + 1, 0) + \phi(0, 2y_0 + 1) + \phi(2x_0 + 1, 2y_0 + 1) -$$
$$\phi(x - x_0, y - y_0) - \phi(x + x_0 + 1, y - y_0) - \phi(x - x_0, y + y_0 + 1) - \phi(x + x_0 + 1, y + y_0 + 1), \tag{6.17}$$

$$V_{x,y} = \phi(x - x_0, y - y_0) + \phi(x + x_0 + 1, y - y_0) + \phi(x - x_0, y + y_0 + 1) +$$
$$\phi(x + x_0 + 1, y + y_0 + 1) - \phi(0,0) - \phi(2x + 1, 0) - \phi(0, 2y + 1) - \phi(2x + 1, 2y + 1). \tag{6.18}$$

The effective resistance is, therefore,

$$R_{qt.} I = 2\phi(0,0) + \phi(2x_0 + 1, 0) + \phi(0, 2y_0 + 1) + \phi(2x_0 + 1, 2y_0 + 1) +$$
$$\phi(2x + 1, 0) + \phi(0, 2y + 1) + \phi(2x + 1, 2y + 1) - 2\phi(x - x_0, y - y_0) -$$
$$2\phi(x + x_0 + 1, y - y_0) - 2\phi(x - x_0, y + y_0 + 1) - 2\phi(x + x_0 + 1, y + y_0 + 1). \tag{6.19}$$

Expressions (6.16) and (6.19) describe, respectively, the effective resistance in a
half-plane mesh and a quarter plane mesh. By adding the electric potentials at certain
nodes due to the current injected at $(0, 0)$, the effective resistance can be determined.
Derivation of the electric potential is presented in the upcoming subsection.

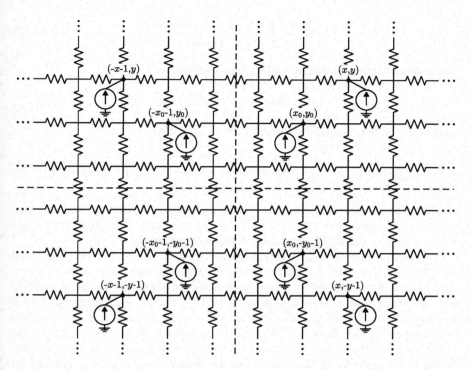

Fig. 6.7 Image method to model truncation in a quarter-plane mesh. The dashed lines illustrate the boundary between the real and image portions of the circuit. Six image sources are introduced in three quadrants of an infinite plane to model the effect of the truncations, ensuring zero effective current across the boundaries.

6.3.2 Integral expressions for effective resistance

The integral expression for the effective resistance in an anisotropic infinite grid is determined in [469, 477] and is

$$R_\infty = \frac{kr}{\pi} \int_0^\pi \frac{1 - e^{-|x-x_0|\alpha} \cos(|y - y_0|\beta)}{\sinh(\alpha)} d\beta, \tag{6.20}$$

where

$$\alpha = \cosh^{-1}(1 + k - k \cos(\beta)). \tag{6.21}$$

$\Omega_k(x, y)$ is defined as

$$\Omega_k \equiv \frac{k}{2\pi} \int_0^\pi \frac{1 - e^{-|x|\alpha}\cos(y\beta)}{\sinh(\alpha)}d\beta. \tag{6.22}$$

The potential within an infinite grid (6.8) is described as

$$\phi(x - x_0, y - y_0) = \phi(0,0) - rI\Omega_k(x - x_0, y - y_0). \tag{6.23}$$

Expression (6.16), therefore, reduces to

$$\frac{R_{half}}{r} = 2\Omega_k(x - x_0, y - y_0) + 2\Omega_k(x + x_0 + 1, y - y_0) - \Omega_k(2x_0 + 1, 0) - \Omega_k(2x + 1, 0). \tag{6.24}$$

Note that due to symmetry along the y-axis, the effective resistance of a half-plane mesh depends upon $(y - y_0)$ and not on y and y_0 separately. In contrast, both x and x_0 are necessary due to the symmetry broken by the truncation.

The exact value of (6.24) for the special case of $x_0 = 0, k = 1$ is listed in Table 6.1. Note that due to truncation, the effective resistance in the x and y directions is not equal, with the resistance along the y-axis increasing at a higher rate. A similar trend is observed for $x_0 > 0$. The effective resistance is evaluated using (6.24) for $x_0 = \{0, 5, 10\}$, $x \in [0, 25]$, $y \in [-25, 25]$, and $k = 1$. The results are shown in Fig. 6.8. Note that the effective resistance to those nodes near the edge of the mesh is higher. Intuitively, this behavior can be explained by the more difficult access to the points along the edges. While the nodes located along the x-axis $(x, 0)$ receive current from all four sides, the nodes located along the y-axis $(0, y)$ are more difficult to reach due to there being only three sides.

Table 6.1 Exact normalized resistance between $(0, y_0)$ and (x, y) in a half-plane resistive grid with $y \in [-3, 3]$, $x \in [0, 3]$, and $r_h = r_y = r$. The numerical values are within the square brackets.

$|y - y_0|$ \ x	0	1	2	3
0	0 [0.000]	$\frac{8}{\pi} - 2$ [0.546]	$\frac{856}{3\pi} - 90$ [0.824]	$\frac{128224}{15\pi} - 2720$ [0.998]
1	$\frac{2}{\pi}$ [0.637]	$\frac{18}{\pi} - 5$ [0.730]	$\frac{998}{3\pi} - 105$ [0.891]	$\frac{131854}{15\pi} - 2797$ [1.029]
2	1 [1.000]	$\frac{56}{3\pi} - 5$ [0.942]	$\frac{952}{3\pi} - 100$ [1.010]	$\frac{130208}{15\pi} - 2762$ [1.100]
3	$4 - \frac{26}{3\pi}$ [1.241]	$\frac{86}{3\pi} - 8$ [1.125]	$\frac{4766}{15\pi} - 100$ [1.138]	$\frac{43514}{5\pi} - 2769$ [1.187]

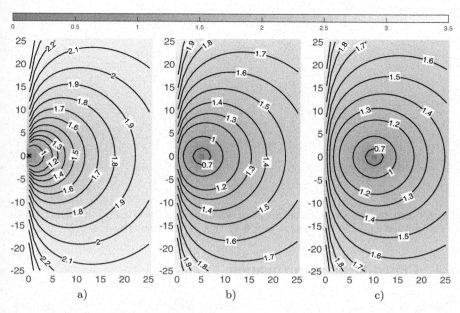

Fig. 6.8 Effective resistance of a half-plane mesh with $k = 1$ between $(x_0, 0)$ and (x, y) for $x \in [0, 25]$ and $y \in [-25, 25]$. a) $x_0 = 0$, b) $x_0 = 5$, and c) $x_0 = 10$

For the quarter-plane mesh, combining (6.22) with (6.19) yields

$$\frac{R_{qt.}}{r} = 2\Omega_k(x-x_0, y-y_0) + 2\Omega_k(x+x_0+1, y-y_0)+2\Omega_k(x-x_0, y+y_0+1)+$$

$$2\Omega_k(x+x_0+1, y+y_0+1)-\Omega_k(2x_0+1, 0) - \Omega_k(0, 2y_0+1)$$

$$-\Omega_k(2x_0+1, 2y_0+1)-\Omega_k(2x+1, 0) - \Omega_k(0, 2y+1) - \Omega_k(2x+1, 2y+1).$$

$$(6.25)$$

Note that due to the broken symmetry in both the x and y directions, the coordinates of both (x_0, y_0) and (x, y) are necessary to determine the effective resistance.

A numerical evaluation of (6.25) for $k = 1$ is shown in Fig. 6.9. As compared to the edges, the effective resistance increases more rapidly near the corner. This trend can be explained using the same intuition: the corner node can be accessed from only two sides unlike the other nodes, which can be accessed from three or four sides. Less current can therefore flow through the node at the same voltage, resulting in a higher effective resistance.

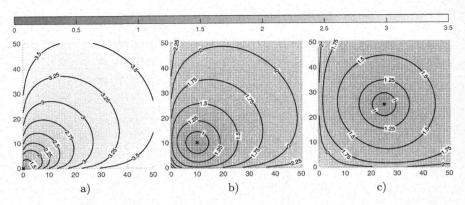

Fig. 6.9 Effective resistance of a quarter-plane mesh with $k = 1$ between (x_0, y_0) and (x, y) for $x \in [0, 50]$ and $y \in [-50, 50]$. a) $x_0 = y_0 = 0$, b) $x_0 = y_0 = 10$, and c) $x_0 = y_0 = 25$

6.4 Closed-form approximation

The exact resistance for a half- and quarter-plane mesh is determined from, respectively, (6.24) and (6.25). For practical purposes, however, an approximate, computationally efficient expression is desirable. A closed-form expression for the integral solution is therefore presented in this section. The derivation is performed in two steps. First, the integral expression for the potential at an arbitrary node within a grid is approximated. The total resistance of a truncated grid is next evaluated using an approximate potential expression.

The derivation of a closed-form expression for the effective resistance is performed in two steps adapted from [476]. First, the integral expression $\Omega_k(x, y)$ is decomposed as

$$\Omega_k(x, y) = J_1 + J_2 + J_3, \qquad (6.26)$$

where

$$J_1 = \frac{\sqrt{k}}{2\pi} \Re \left[E_1 \left(\pi \left(x\sqrt{k} + iy \right) \right) + \ln \left(\pi \left(x\sqrt{k} + iy \right) \right) + \gamma \right], \qquad (6.27)$$

$$J_2 = \frac{k}{2\pi} \int_0^\pi \left(\frac{e^{-x\beta\sqrt{k}}}{\beta\sqrt{k}} - \frac{e^{-x\alpha}}{\sinh(\alpha)} \right) \cos(y\beta) d\beta, \qquad (6.28)$$

$$J_3 = \frac{k}{2\pi} \int_0^\pi \left(\frac{1}{\sinh(\alpha)} - \frac{1}{\beta\sqrt{k}} \right) d\beta, \qquad (6.29)$$

$$E_1(z) = \int_z^\infty \frac{e^{-t}}{t} dt, \tag{6.30}$$

and $\gamma \approx 0.5772$ is the Euler–Mascheroni constant. The first integral J_1 can be determined numerically using the exponential integral function $E_1(z)$, available in most popular engineering packages, including SciPy [488] and MATLAB [489]. For large values of x and y, the integral J_1 reduces to

$$J_1 \approx \frac{\sqrt{k}}{4\pi} \left[\ln\left(x^2 + ky^2\right) + 2\ln(\pi) + 2\gamma \right]. \tag{6.31}$$

To analyze the second integral, note that for small β, $\sinh(\alpha) \approx \beta$ and, for large values of β, the numerator of the integral vanishes for large β with values x and y above 10. This term is, therefore, neglected in the closed-form expression.

The third integral is a function of a single variable k and is approximated as a fourth degree polynomial,

$$J_3 \approx \sum_{i=0}^{4} a_i k^i, \tag{6.32}$$

where the coefficients of the expression are listed in Table 6.2. The final closed-form expression for $\Omega_k(x, y)$ is, therefore,

$$\Omega_k^*(x, y) = \frac{\sqrt{k}}{4\pi} \left[\ln\left(x^2 + ky^2\right) + 2\ln(\pi) + 2\gamma \right] + \sum_{i=0}^{4} a_i k^i. \tag{6.33}$$

With a closed-form expression for $\Omega(x, y)$, the effective resistance of a half-plane resistive mesh is

$$\frac{R_{half}}{r} \approx 2\Omega_k^*(x - x_0, y - y_0) + 2\Omega_k^*(x + x_0 + 1, y - y_0) - \Omega_k^*(2x_0 + 1, 0)$$

$$- \Omega_k^*(2x + 1, 0). \tag{6.34}$$

Table 6.2 Coefficients for the polynomial approximation of J_3 (6.32).

i	$1 \leq k \leq 5$	$5 \leq k \leq 50$
0	4.748×10^{-2}	2.559×10^{-2}
1	-5.989×10^{-2}	-3.356×10^{-2}
2	8.153×10^{-4}	-1.078×10^{-2}
3	-1.274×10^{-5}	2.260×10^{-3}
4	9.092×10^{-8}	-1.669×10^{-4}

Similarly, for the quarter-plane mesh,

$$\frac{R_{qt.}}{r} \approx 2\Omega_k^*(x-x_0, y-y_0) + 2\Omega_k^*(x+x_0+1, y-y_0) + 2\Omega_k^*(x - x_0, y+y_0+1)+$$

$$2\Omega_k^*(x+x_0 + 1, y+y_0+1) - \Omega_k^*(2x_0+1, 0) - \Omega_k^*(0, 2y_0+1)$$

$$-\Omega_k^*(2x_0+1, 2y_0+1) - \Omega_k^*(2x + 1, 0) - \Omega_k^*(0, 2y + 1) - \Omega_k^*(2x + 1, 2y + 1). \tag{6.35}$$

6.5   Model evaluation

The primary contribution of this chapter is the accurate and fast estimation of the IR drop between nodes located close to a grid edge. To evaluate the applicability of the model, in Subsection 6.5.1, the accuracy of the exact expressions (6.24) and (6.25), closed-form expressions (6.34) and (6.35), and nodal analysis is compared. In Subsection 6.5.2, the computational speed of the model is examined.

6.5.1   Accuracy evaluation

The relationship between the relative error and the position of the probed nodes is shown in Figs. 6.10 and 6.11. Note that a larger error is produced when the resistance is evaluated between nearby nodes and with nodes along the y-axis. A peak error of 4.77% is produced when the resistance is evaluated between the adjacent nodes. For a large distance between $(x_0, 0)$ and (x, y), the relative error approaches zero.

Note from Fig. 6.3 that a large error is induced if the expression for a fully infinite grid is used to estimate the voltage drop near the edge of a finite mesh. A drastic increase in accuracy is observed when using (6.24) or (6.34). The error of (6.24) and (6.34) as compared to the resistance evaluated through a nodal analysis on a 101 × 201 mesh is shown in Fig. 6.12. As compared to (6.1), the error in (6.24) and (6.34) is below 3% along the edge. A considerably larger error is produced by (6.34) as compared to (6.24) when one of the nodes is at the grid edge and another node is in close proximity. In addition, the error is significantly increased when the effective resistance is evaluated using closed-form expression (6.34) between nodes within close proximity. Note that the closed-form expression is derived with the assumption of large separation between the target nodes, which leads to larger error. In other cases, the accuracy of (6.24) and (6.34) is approximately equal. Likewise, a significant increase in accuracy is achieved with (6.25) or (6.35), as is evident from Fig. 6.13. Near the corner and edges, the error is below 2%.

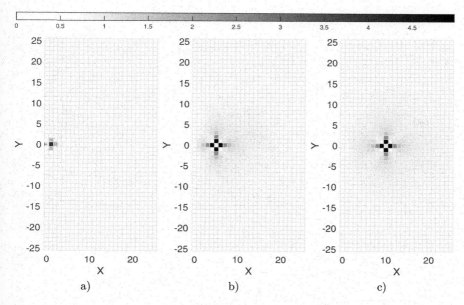

Fig. 6.10 Relative error (in per cent) of (6.34) as compared to (6.24) with respect to x and y for $k = 1$ and a) $x_0 = 0$,b) $x_0 = 5$, and c) $x_0 = 10$.

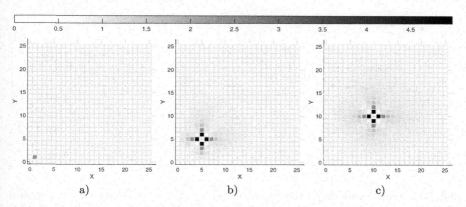

Fig. 6.11 Relative error (in per cent) of (6.35) as compared to (6.25) with respect to x and y for $k = 1$ and a) $x_0 = y_0 = 0$,b) $x_0 = y_0 = 5$, and c) $x_0 = y_0 = 10$.

6.5.2 Computational speed

The speedup of the analysis and simulation process is an important contribution of this chapter. Recall from Chapter 5 that the conventional method for determining the effective resistance is modified nodal analysis, where a linear system of equations is analyzed [65, 490],

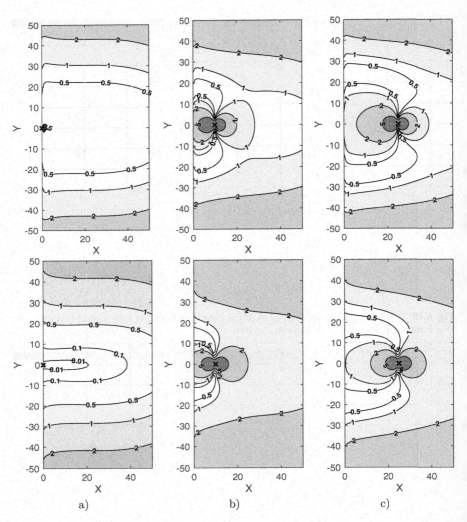

Fig. 6.12 Relative error (in per cent) of (6.24) (top row) and (6.34) (bottom row) as compared to the resistance determined using a nodal analysis between node $(x_0, 0)$ and node (x, y) for a) $x_0 = 0$, b) $x_0 = 10$, and c) $x_0 = 25$. The grid dimensions are 101×201. The point $(x_0, 0)$ is indicated by the $\times$-mark.

$$R_{eff} = \mathbf{1} \operatorname{diag}(H)^T + \operatorname{diag}(H)\mathbf{1}^T - 2H, \qquad (6.36)$$

where $H \in \mathbb{R}^{(MN-1) \times (MN-1)}$ is the inverse of the reduced conductance matrix, $\mathbf{1} \in \mathbb{R}^{(MN-1)}$ is the vector with all entries equal to 1, and $\operatorname{diag}(A)$ is the diagonal of matrix A. The advantage of this method is the effective resistance between any pair of nodes in a circuit is evaluated. If the resistance between only a small subset of

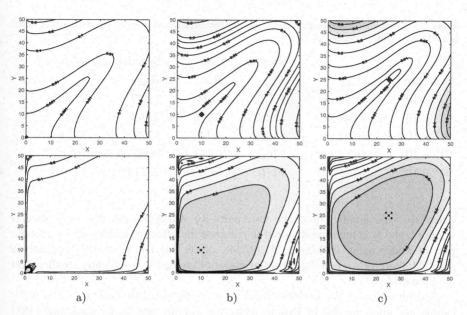

Fig. 6.13 Relative error (in per cent) of (6.25) (top row) and (6.35) (bottom row) as compared to the resistance determined using a nodal analysis between node (x_0, y_0) and node (x, y) for a) $x_0 = y_0 = 0$, b) $x_0 = y_0 = 10$, and c) $x_0 = y_0 = 25$. The grid dimensions are 101×101. The point (x_0, y_0) is indicated by the $\times$-mark.

nodes is needed, however, this approach is highly inefficient. Assuming the pitch of the top metal layer is 2 μm for a 45 nm technology node [491], the grid size of the power delivery network in a 4 cm^2 die size is on the order of $10^4 \times 10^4$. The nodal analysis of this matrix requires the solution of a $10^8 \times 10^8$ linear system, resulting in significant analysis time. The computational time t_{nodal} required to determine the effective resistance in a nonuniform grid using nodal analysis is therefore a superlinear function of the dimensions of the grid,

$$t_{nodal} = t_1 (MN)^c,$$ (6.37)

where M and N are dimensions of the grid, t_1 is a proportionality constant, and c is the degree of the solver complexity, typically larger than one. Importantly, the method allows the effective resistance between all pairs of nodes to be determined.

In contrast, the method proposed in this chapter does not require solving a system of linear expressions. The time required to determine the effective resistance using the proposed method does not depend on the grid dimensions. The method proposed here has constant complexity where the total computational time t_{image} is

$$t_{image} = t_2 n,$$ (6.38)

Table 6.3 Computational speedup for determining the effective resistance between a pair of nodes in an $M \times N$ grid.

Grid Size	t_{nodal}	Speedup (exact)		Speedup (closed-form)	
		R_{half}	R_{qrt}	R_{half}	R_{qrt}
$10^2 \times 10^2$	1.252 ms	1.375	0.296	11.47	8.888
$10^3 \times 10^2$	9.646 ms	11.51	1.936	227.3	80.12
$10^3 \times 10^3$	60.32 ms	53.83	5.739	1278	712.6
$10^4 \times 10^3$	462.4 ms	536.2	43.61	9062	5621
$10^4 \times 10^4$	10.85 s	7216	517.0	232148	135018

where n is the number of target node pairs for which the effective resistance is required, and t_2 is the time required to compute the effective resistance for a single pair of nodes using (6.24), (6.25), (6.34), or (6.35). The proposed approach is justified, therefore, when the subset of nodes of interest is sufficiently smaller than the total grid size.

A comparison of the computational speed is provided in Table 6.3. The algorithms are implemented in Python using the Numpy and Scipy packages [488] on an eight core 3.40 GHz Intel Core i7-6700 machine with 24 GB RAM. The nodal analysis has been performed using the Scipy sparse matrix solver [488]. Note the rapid increase in speedup with grid size. For the exact integral equations, the speedup reaches three to four orders of magnitude in a $10^4 \times 10^4$ grid. Larger speedup is achieved with the closed-form expressions, exhibiting six orders of magnitude improvement in computational time in a $10^4 \times 10^4$ grid. Simulation of grids larger than $10^4 \times 10^4$ is not possible using the Scipy sparse matrix solver due to limited memory.

6.6   Conclusions

Image and superposition methods are utilized to investigate truncated infinite anisotropic mesh structures. Exact integral and closed-form expressions for the effective resistance are presented. A closed-form expression offers a computationally efficient method for evaluating the effective resistance, which can be beneficial in several VLSI circuit applications such as resistive noise analysis, placement of decoupling capacitors, and substrate noise models. Significant speedup is achieved using the proposed expressions, reaching six orders of magnitude with the closed-form expressions. The proposed framework can be utilized in a variety of VLSI oriented applications, including circuit optimization, analysis, and synthesis.

Chapter 7
Effective resistance of finite grids

A rectangular mesh is a common structure in science and engineering. In engineering, a rectangular mesh is used to model on-chip power and ground networks and silicon substrates, as well as electrically and thermally conductive media. Applications specific to very large scale integration (VLSI) circuits include digital logic, memory, and power and ground distribution networks [467]. In modern VLSI systems, large grid sizes are common. Conventional numerical analysis techniques to solve a large system of linear equations result in prohibitive computational time.

The effective resistance is an important characteristic of these grid structures. Applications include static power and ground network analysis [480, 492], power network synthesis [65, 145], decoupling capacitor allocation [493–495], RC delay optimization [386, 490], electrically and thermally conductive media [481, 496], and certain graph characteristics, such as coverage and commute times [497]. From the perspective of circuit analysis, the effective resistance can be utilized to significantly reduce the computational complexity of the grid analysis process [480].

The effective resistance of an infinite resistive lattice is a classical problem in circuit theory [498]. The objective is to determine an equivalent resistance between two arbitrary points within an infinite two-dimensional grid of resistors. The effective resistance between two adjacent points within a two-dimensional isotropic mesh has been determined using symmetry and superposition [482]. In the case of non-adjacent nodes, however, more advanced methods are required. At least six different solutions have been developed since 1940 for this problem, random walk theory [471], elliptic integrals [473], Fourier transforms [476, 477, 485] and Green's function [479]. The problem has been extended to a variety of infinite structures, such as hypercubes [477, 479, 484], triangles [477, 479], hexagons [477, 479], tori and cylinders [487], and anisotropic rectangular lattices [469].

Expressions describing an infinite grid exhibit good agreement with nodal analysis if the effective resistance is measured between nodes located far from the boundary of the grid. Prohibitively large error can however be produced when the resistance is measured between nodes located close to the grid boundaries [71].

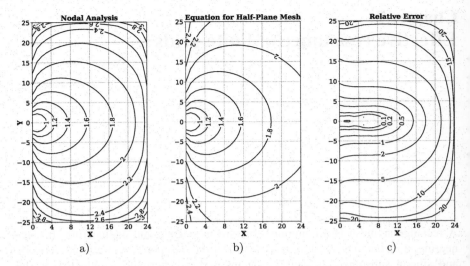

Fig. 7.1 Effective resistance and relative error of a 25 × 51 isotropic grid between node $(0, 0)$ and (x, y). a) Evaluation using nodal analysis, b) evaluation using the half-plane mesh equation [71], and c) error (in per cent) of (b) relative to (a).

Despite the well studied nature of this problem, less attention has been devoted to the analysis of truncated and finite rectangular grids.

Different finite regular structures have been investigated in the literature, including generalized linear chains [499] and circulant graphs [500]. Truncation along one and two dimensions has been analyzed in previous work using the circuit-level image technique [71]. While the expressions described in [71] are in good agreement with nodal analysis if the resistance is within close proximity of a single boundary or corner, these expressions become inaccurate if the resistance is measured between terminals located at opposite boundaries and corners. The double-sum expressions for the effective resistance in the various grid structures have previously been described in [501]. The computational complexity of these expressions, however, increases linearly with the number of nodes. In this chapter, the infinity mirror technique extends the image method to finite structures. With this technique, the effective resistance is determined with high accuracy in a finite rectangular grid of arbitrary size with potential extension to cubic and hypercubic topologies. The computational complexity of the proposed expressions does not depend on the grid size and number of nodes.

To illustrate the relevance of the infinity mirror technique, consider a 25 × 51 uniform resistive grid. The effective resistance is determined between the node on the left boundary and all other nodes using nodal analysis (Fig. 7.1a) and the half-plane mesh equation (Fig. 7.1b) [71]. The relative error is shown in Fig. 7.1c. While the error is low close to the left boundary, the error may exceed 15% if the half-plane equation is used to evaluate the resistance at the opposite boundary of the grid.

This chapter is organized as follows. In Section 7.1, the infinity mirror technique is reviewed, which extends the image method described in [71] to finite structures. The effective resistance in grids with finite dimensions is also presented. In Section 7.2, these expressions are modified to enhance the efficiency while maintaining accuracy below 1%. Application of the infinity mirror technique to practical problems is presented in Section 7.3 using three case studies. Computational speedup of up to five orders of magnitude is demonstrated for certain scenarios. Summary comments are provided in Section 7.4.

7.1 Infinity mirror technique

The effective resistance between two points within a mesh is determined using the method adapted from [476]. Consider a two-dimensional resistive mesh, as illustrated in Fig. 7.2. Pick two nodes, (x_0, y_0) and (x, y), at a finite distance between each other with ground infinitely far. Connect the current source injecting current I into (x_0, y_0). The resulting potential at (x_0, y_0) and (x, y) due to the current source at (x_0, y_0) is, respectively, $\phi^{x_0, y_0}(x_0, y_0)$ and $\phi^{x_0, y_0}(x, y)$. Remove the current source at (x_0, y_0) and inject current $-I$ into (x, y). The resulting potential at (x_0, y_0) and (x, y) is, respectively, $-\phi^{x, y}(x_0, y_0)$ and $-\phi^{x, y}(x, y)$. The effective resistance can be determined by superimposing these solutions,

$$R_{eff} = \frac{V(x_0, y_0) - V(x, y)}{I}, \qquad (7.1)$$

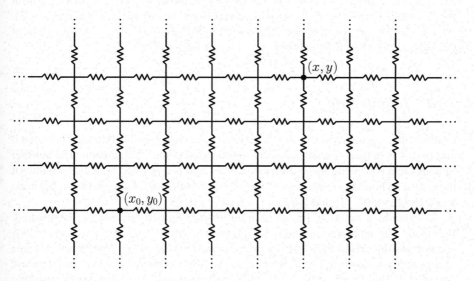

Fig. 7.2 Infinite two-dimensional grid.

where $V(x_0, y_0)$ and $V(x, y)$ are the effective voltage at, respectively, (x_0, y_0) and (x, y) due to all current sources within the grid. $V(x_0, y_0)$ and $V(x, y)$ can be expressed, respectively, as the superposition of potentials due to each individual current source,

$$V(x_0, y_0) = \phi^{x_0, y_0}(x_0, y_0) - \phi^{x, y}(x_0, y_0), \tag{7.2}$$

$$V(x, y) = \phi^{x_0, y_0}(x, y) - \phi^{x, y}(x, y). \tag{7.3}$$

The problem of determining the effective resistance within a grid reduces to finding the electric potential caused by the injected current. A similar approach is applicable to truncated grids. As in the case of a fully infinite mesh, the effective resistance in a truncated mesh structure is determined from (7.1). The voltages, $V(x_0, y_0)$ and $V(x, y)$, however, change to consider the effects of the boundaries modeled as image current sources.

The image method for an infinite grid was introduced in [71] and applied to half- and quarter-plane mesh structures. The resulting effective resistance expressions exhibit good agreement with the resistance of a large grid near a boundary or a corner, where the effects of opposite boundaries can be neglected. If however the effects of the opposite boundaries are significant; for example, if the effective resistance is measured between the opposite corners of a finite rectangular mesh, these expressions are no longer accurate. Efficient methods for determining the effective resistance in a grid where at least one dimension is finite are presented in this section. In Subsection 7.1.1, an expression is presented for an infinite strip, a mesh which is finite in one dimension and unbounded in another dimension ($y \in \mathbb{Z}$). This result is utilized in Subsection 7.1.2 to determine the effective resistance within a semi-infinite strip, an infinite strip truncated along the infinite dimension ($y \in \mathbb{N}_0$). An expression for a finite mesh is presented in Subsection 7.1.3. Generalization of the method to higher dimensions is provided in Subsection 7.1.4

7.1.1 Infinite strip

Consider the circuit shown in Fig. 7.3a, where a resistive grid is bounded between 0 and $(w_x - 1)$ in the x-dimension and is unbounded in the y-dimension. The number of nodes in a row along the x-dimension is w_x and is described here as the width of the grid. The bounds of the strip obstruct the current from flowing between the node pairs, $\{(-1, y), (0, y)\}$ and $\{(w_x - 1, y), (w_x, y)\}$.

To provide a solution for an infinite strip, symmetry needs to be restored. Following the approach outlined in [71], the current through the $\{(-1, y), (0, y)\}$ resistor within an infinite resistive grid can be eliminated by applying the image of the strip, as shown in Fig. 7.3b. The image current sources produce a symmetric potential within the strip that equalizes the potential at $(-1, y)$ and $(0, y)$, resulting

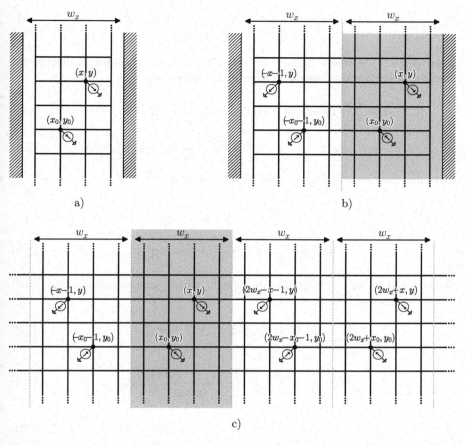

Fig. 7.3 Infinity mirror method applied to an infinite resistive strip of width w_x. a) Original resistive strip, b) first iteration of image method, and c) infinity mirror technique. In each case, the potential distribution is preserved for $0 \leq x \leq w_x$.

in zero current flowing between the pair of nodes. The width of the strip is therefore doubled, while maintaining the potential distribution within the strip. By iteratively repeating the image process for the left and right boundaries, the topology shown in Fig. 7.3c is produced. Intuitively, the topology is similar to placing an object between two parallel mirrors, leading to infinite images of the object.

The resulting voltage at (x_0, y_0) and (x, y) can be described, respectively, as

$$V_{x_0, y_0} = \sum_{i \in \mathbb{Z}} \left(\phi_{i0}(0, 0) + \phi_{i0}(2x_0 + 1, 0) - \phi_{i0}(x - x_0, y - y_0) - \phi_{i0}(x + x_0 + 1, y - y_0) \right),$$

$$(7.4)$$

$$V_{x,y} = \sum_{i \in \mathbb{Z}} \left(-\phi_{i0}(0,0) - \phi_{i0}(2x+1,0) + \phi_{i0}(x-x_0, y-y_0) + \phi_{i0}(x+x_0+1, y-y_0) \right),$$

$$(7.5)$$

where $\phi_{ij}(x,y) \equiv \phi(x+2iw_x, y+2jw_y)$. The effective resistance is determined from the difference between the voltage at (x_0, y_0) and (x, y),

$$R_{w_x,\infty} I = \sum_{i \in \mathbb{Z}} \left(2\phi_{i0}(0,0) + \phi_{i0}(2x_0+1,0) + \phi_{i0}(2x+1,0) \right.$$

$$\left. -2\phi_{i0}(x-x_0, y-y_0) - 2\phi_{i0}(x+x_0+1, y-y_0) \right).$$

$$(7.6)$$

From the effective resistance of a fully infinite mesh [71],

$$\frac{R_{w_x,\infty}}{r} = \sum_{i \in \mathbb{Z}} \left(2\Omega_{i0}^k(x-x_0, y-y_0) + 2\Omega_{i0}^k(x+x_0+1, y-y_0) \right.$$

$$\left. -2\Omega_{i0}^k(0,0) - \Omega_{i0}^k(2x_0+1,0) - \Omega_{i0}^k(2x+1,0) \right),$$

$$(7.7)$$

where r and kr are the resistance of a single resistor in, respectively, the x- and y-dimensions, and

$$\Omega_{ij}^k \equiv \Omega^k(x+2iw_x, y+2jw_y) \tag{7.8}$$

$$\Omega^k(x,y) \equiv \frac{k}{2\pi} \int_0^\pi \frac{1 - e^{-|x|\alpha}\cos(y\beta)}{\sinh(\alpha)} d\beta, \tag{7.9}$$

$$\alpha = \cosh^{-1}(1 + k - k\cos(\beta)). \tag{7.10}$$

The contour of (7.7) is shown in Fig. 7.4. Note that the effective resistance increases close to the boundaries of the strip, similar to the half- and quarter-plane meshes [71] due to the limited accessibility of the nodes near the boundaries. This behavior is consistent with the Monotonicity Law where the effective resistance increases with the removal of branches [484, 502]. Also note that the effective resistance evaluated from the middle of the strip ($x_0 = \frac{w_x-1}{2}$ for $w_x = 2n+1, n \in \mathbb{N}_0$) is symmetric with respect to $x = x_0$ and $y = y_0$ (see Fig. 7.4c).

7.1.2   Semi-infinite strip

Consider the case where the infinite strip is truncated, bounding the strip between 0 and infinity along the y-dimension (see Fig. 7.5a). The effective resistance in this case is determined by applying an image of the infinite strip along $x = 0$, as shown in Fig. 7.5b.

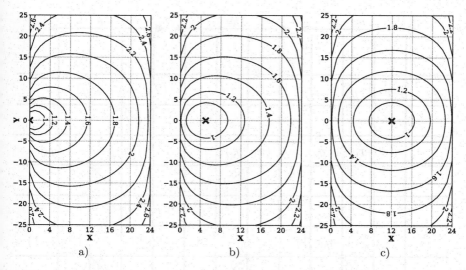

Fig. 7.4 Effective resistance of an infinite strip of width $w_x = 25$ between point (x_0, y_0) and (x, y) for $k = 1$, a) $x_0 = y_0 = 0$, b) $x_0 = y_0 = 5$, and c) $x_0 = y_0 = 12$. The point (x_0, y_0) is indicated by an ×.

$$\frac{R_{w_x,\infty/2}}{r} = \sum_{n\in\mathbb{Z}} \left(2\Omega_{i0}^k(x - x_0, y - y_0) + 2\Omega_{i0}^k(x + x_0 + 1, y + y_0 + 1)\right.$$

$$+ 2\Omega_{i0}^k(x - x_0, y + y_0 + 1) + 2\Omega_{i0}^k(x + x_0 + 1, y - y_0) - 2\Omega_{i0}^k(0, 0)$$

$$- \Omega_{i0}^k(2x_0 + 1, 0) - \Omega_{i0}^k(2x + 1, 0) - \Omega_{i0}^k(0, 2y_0 + 1)$$

$$\left. - \Omega_{i0}^k(0, 2y + 1) - \Omega_{i0}^k(2x_0 + 1, 2y_0 + 1) - \Omega_{i0}^k(2x + 1, 2y + 1)\right). \tag{7.11}$$

A contour of (7.11) is shown in Fig. 7.6. As compared to Fig. 7.4, the effective resistance increases at a higher rate, particularly in the x-direction due to the truncation at $y = 0$. Note that the effective resistance evaluated at the middle of the semi-infinite strip is symmetric along the x-dimension, similar to the infinite strip.

7.1.3   Finite mesh

Consider the case where a semi-infinite strip is truncated at $y = w_y - 1$, resulting in a $w_x \times w_y$ finite mesh. The effective resistance can be determined by applying the infinity mirror technique in two dimensions, as shown in Fig. 7.7. This topology can be modeled using the infinite mirror technique twice, along the x- and y-directions,

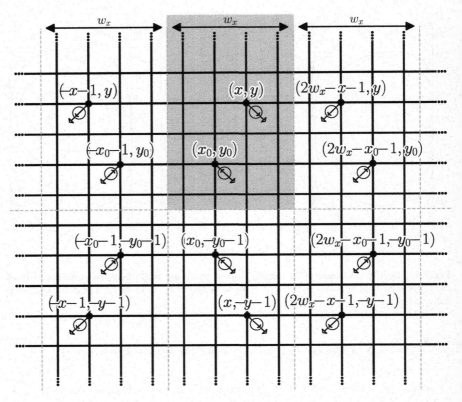

Fig. 7.5 Infinity mirror technique applied to a semi-infinite resistive strip of width w_x. Original semi-infinite strip is shaded. The potential distribution is preserved for $0 \leq x \leq w_x$ and $y \in \mathbb{N}_0$.

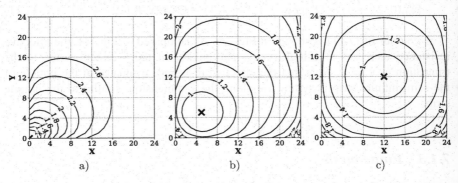

Fig. 7.6 Effective resistance of a semi-infinite strip of width $w_x = 25$ between point (x_0, y_0) and (x, y) for $k = 1$, a) $x_0 = y_0 = 0$, b) $x_0 = y_0 = 5$, and c) $x_0 = y_0 = 12$. The point (x_0, y_0) is indicated by an $\times$.

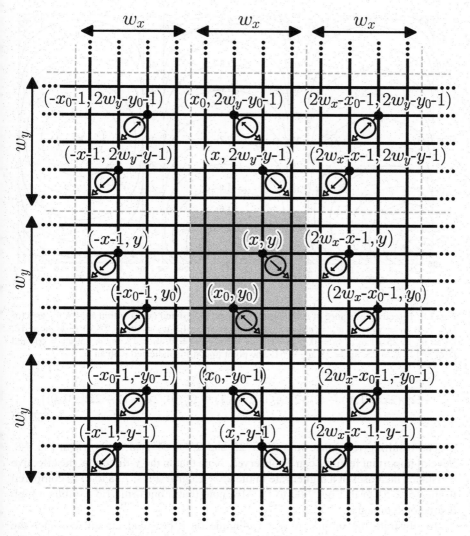

Fig. 7.7 Infinity mirror technique applied to a finite $w_x \times w_y$ resistive mesh. Original mesh is shaded. The potential distribution is preserved for $0 \le x \le w_x$ and $0 \le y \le w_y$.

$$R_{w_x, w_y} = \sum_{i \in \mathbb{Z}} \sum_{j \in \mathbb{Z}} \big(2\Omega_{ij}^k (x - x_0, y - y_0) + 2\Omega_{ij}^k (x - x_0, y + y_0 + 1)$$

$$+ 2\Omega_{ij}^k (x + x_0 + 1, y + y_0 + 1) + 2\Omega_{ij}^k (x + x_0 + 1, y - y_0)$$

$$- \Omega_{ij}^k (2x_0 + 1, 0) - \Omega_{ij}^k (2x_0 + 1, 2y_0 + 1) - \Omega_{ij}^k (0, 2y_0 + 1) \big)$$

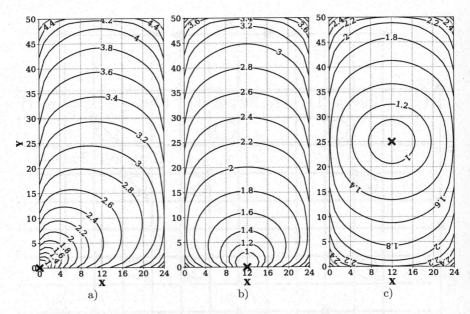

Fig. 7.8 Effective resistance of a finite grid between (x_0, y_0) and (x, y) within a $w_x \times w_y$ grid for $k = 1$, a) $x_0 = y_0 = 0$, $w_x = w_y = 25$, b) $x_0 = 0$, $y_0 = 12$, $w_x = w_y = 25$, c) $x_0 = y_0 = 12$, $w_x = w_y = 25$, d) $x_0 = y_0 = 0$, $w_x = 25$, $w_y = 51$, e) $x_0 = 0$, $y_0 = 12$, $w_x = 25$, $w_y = 51$, and f) $x_0 = 12$, $y_0 = 25$, $w_x = 25$, $w_y = 51$. The point (x_0, y_0) is indicated with an $\times$.

$$-\Omega_{ij}^k(2x + 1, 0) - \Omega_{ij}^k(2x + 1, 2y + 1) - \Omega_{ij}^k(0, 2y + 1) - 2\Omega_{ij}^k(0, 0)).$$
$$(7.12)$$

The resulting resistance in a 25×25 and 25×51 grid is shown in Fig. 7.8. Several important features can be observed. Note that in the y-direction, the effective resistance increases at a higher rate in the 25×25 grid (Figs. 7.8a to c) as compared to the 25×51 grid (Figs. 7.8d to f). Also note that a uniform grid exhibits a high degree of symmetry.

To illustrate the infinity mirror technique on a practical circuit, consider the extreme case of the 2×2 nonuniform resistive network shown in Fig. 7.9. The effective resistance between nodes $(0, 0)$ and $(1, 0)$ is, by Ohm's law, 0.833 ohms. Iteratively evaluating the summands of (7.12) around $n = k = 0$ efficiently converges to the actual resistance, as listed in Table 7.1. Note that the summands of $|n|, |k| \geq 3$ do not exceed 0.001, indicating that only a small number of images around the origin needs to be considered.

Fig. 7.9 2×2 resistive grid.
The resistance is measured
between nodes $(0, 0)$ and
$(1, 0)$

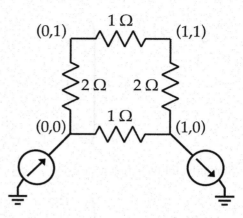

Table 7.1 Summands of (7.12) for $(x_0, y_0) = (0, 0)$ and $(x, y) = (0, 1)$ in a 2×2 resistive grid.

n \ k	-3	-2	-1	0	1	2	3
-3	0.000	0.000	0.000	0.000	0.000	0.000	0.000
-2	0.000	0.000	0.001	0.000	−0.001	−0.001	0.000
-1	0.000	0.000	0.003	0.004	−0.005	−0.001	0.000
0	−0.001	−0.002	−0.013	0.794	0.044	0.005	0.001
1	0.000	0.000	0.003	0.010	−0.009	−0.001	0.000
2	0.000	0.001	0.001	0.000	−0.001	−0.001	0.000
3	0.000	0.000	0.000	0.000	0.000	0.000	0.000

7.1.4   *Generalization to higher dimensions*

The proposed technique can be extended to higher dimensions to evaluate the
resistance of a multidimensional finite grid. Consider an n-dimensional finite grid
with dimensions $\mathbf{w} = [w_1, w_2, \ldots, w_n]$. The resistance is evaluated between source
node $\mathbf{x}_s = \left[x_s^1, x_s^2, \ldots, x_s^n\right]$ and target node $\mathbf{x}_t = \left[x_t^1, x_t^2, \ldots, x_t^n\right]$. Applying the
image technique in each dimension of the grid yields the 2^n source nodes around
the origin,

$$X_s = \begin{pmatrix} x_s^1, & x_s^2, & \cdots & x_s^n \\ x_s^1, & x_s^2, & \cdots -x_s^n - 1 \\ \vdots & \vdots & \ddots & \vdots \\ x_s^1, & -x_s^2 - 1, & \cdots & x_s^n \\ -x_s^1 - 1, & x_s^2, & \cdots & x_s^n \\ \vdots & \vdots & \ddots & \vdots \\ -x_s^1 - 1, & -x_s^2 - 1, & \cdots -x_s^n - 1 \end{pmatrix}. \tag{7.13}$$

Similarly, for the target node,

$$X_t = \begin{pmatrix} x_t^1, & x_t^2, & \cdots & x_t^n \\ x_t^1, & x_t^2, & \cdots -x_t^n - 1 \\ \vdots & \vdots & \ddots & \vdots \\ x_t^1, & -x_t^2 - 1, & \cdots & x_t^n \\ -x_t^1 - 1, & x_t^2, & \cdots & x_t^n \\ \vdots & \vdots & \ddots & \vdots \\ -x_t^1 - 1, & -x_t^2 - 1, & \cdots -x_t^n - 1 \end{pmatrix}. \tag{7.14}$$

Nodes in X_s and X_t are order sources since these nodes are closest to the origin. The higher order sources arise from translating the nodes in sets X_s and X_t along each dimension, yielding countably infinite sets,

$$X_s^{im} = \{\mathbf{x} + 2\mathbf{w} \circ \mathbf{a}; \mathbf{x} \in X_s, \mathbf{a} \in \mathbb{Z}^n\}, \tag{7.15}$$

$$X_t^{im} = \{\mathbf{x} + 2\mathbf{w} \circ \mathbf{a}; \mathbf{x} \in X_t, \mathbf{a} \in \mathbb{Z}^n\}, \tag{7.16}$$

where $\mathbf{w} \circ \mathbf{a}$ denotes the Hadamard (element-wise) vector product of vectors $\mathbf{w}$ and $\mathbf{a}$. The potential difference between nodes $\mathbf{x}_s$ and $\mathbf{x}_t$ is

$$V = \sum_{\mathbf{x} \in X_s^{im}} (\phi(\mathbf{x} - \mathbf{x}_s) - \phi(\mathbf{x} - \mathbf{x}_t)) + \sum_{\mathbf{x} \in X_t^{im}} (\phi(\mathbf{x} - \mathbf{x}_t) - \phi(\mathbf{x} - \mathbf{x}_s)). \tag{7.17}$$

Assume the effective resistance of an n-dimensional infinite grid is $\Omega_{(\mathbf{k},\mathbf{w})}(\mathbf{x})$, where $\mathbf{k}$ is the ratio of the unit resistance along each dimension to the unit resistance along the first dimension. The effective resistance of a finite mesh can be described as

$$\frac{R_{eff}}{r} = \sum_{\mathbf{x} \in X_s^{im}} (\Omega_{(\mathbf{k},\mathbf{w})}(\mathbf{x} - \mathbf{x}_s) - \Omega_{(\mathbf{k},\mathbf{w})}(\mathbf{x} - \mathbf{x}_t)) + \sum_{\mathbf{x} \in X_t^{im}} (\Omega_{(\mathbf{k},\mathbf{w})}(\mathbf{x} - \mathbf{x}_t) - \Omega_{(\mathbf{k},\mathbf{w})}(\mathbf{x} - \mathbf{x}_s)).$$

$$\tag{7.18}$$

7.2   Simplification of the effective resistance expressions

Although the effective resistance is accurately determined with (7.7), (7.11), and (7.12), more computationally efficient equations are desirable. An efficient approximation of (7.9) is described in [71],

$$\hat{\Omega}^k(x, y) = \frac{\sqrt{k}}{4\pi}\Big[\ln\left(x^2 + ky^2\right) + 2\ln(\pi) + \gamma\Big] + \sum_{i=0}^{4} a_i k^i, \qquad (7.19)$$

where $\gamma \approx 0.5772$ is the Euler–Mascheroni constant [471], and the coefficients a_i of the expression are listed in Table 7.2.

The error of (7.19) as compared to (7.9) is shown in Fig. 7.10. Note that the error is reduced to zero for large x and y. At small values of x and y, the error dramatically increases, significantly affecting the accuracy of the effective resistance. To alleviate this issue, the following function is proposed,

$$\Psi_{ij}^k(x, y) = \Psi^k(x + 2iw_x, y + 2jw_y) \qquad (7.20)$$

$$\Psi^k(x, y) = \begin{cases} \Omega^k(x, y), & \text{if } \epsilon(x, y) > 10^{-2} & (7.21a) \\ \hat{\Omega}^k(x, y), & \text{otherwise;} & (7.21b) \end{cases}$$

where $\epsilon(x, y)$ is the relative error of (7.19) as compared to (7.9). Since $\Omega^k(x, y)$ only needs to be evaluated for a small subset of nodes where the error of (7.19) is large, the evaluation of $\Omega^k(x, y)$ can be replaced with a look-up table, providing an effective tradeoff between computational speed and accuracy.

Evaluation of the effective resistance in a finite mesh requires computing a double-infinite sum. The series in (7.12), however, quickly converges to 0. Using additional terms results in higher accuracy while requiring greater computational time. It is of interest to determine the optimal number of terms in series (7.12) to achieve acceptable accuracy in minimum time. Consider the following approximate equation for the effective resistance of a finite mesh,

Table 7.2 Coefficients for the polynomial approximation of J_3 (7.19) [71].

i	$1 \le k \le 5$	$5 \le k \le 50$
0	4.748×10^{-2}	2.559×10^{-2}
1	-5.989×10^{-2}	-3.356×10^{-2}
2	8.153×10^{-4}	-1.078×10^{-2}
3	-1.274×10^{-5}	2.260×10^{-3}
4	9.092×10^{-8}	-1.669×10^{-4}

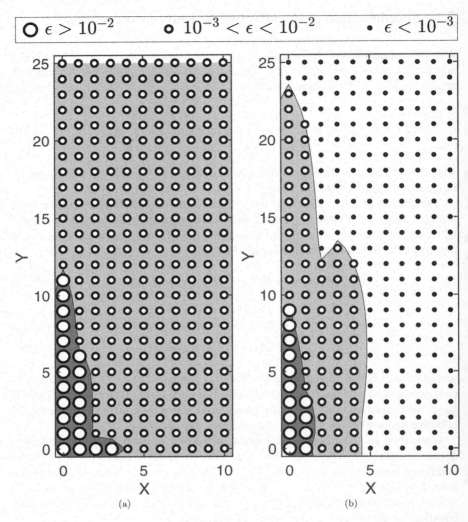

Fig. 7.10 Relative error ϵ between (7.9) and (7.19) for $0 \leq x \leq 10$, $0 \leq y \leq 25$, and $1 \leq k \leq 30$.
a) Maximum error, and b) average error.

$$R_{w_x,w_y}^{N,M} = \sum_{i=-N}^{N} \sum_{j=-M}^{M} \left(2\Omega_{ij}^k(x - x_0, y - y_0) + 2\Omega_{ij}^k(x - x_0, y + y_0 + 1)\right.$$

$$+2\Omega_{ij}^k(x + x_0 + 1, y + y_0 + 1) + 2\Omega_{ij}^k(x + x_0 + 1, y - y_0) - \Omega_{ij}^k(2x_0 + 1, 0) - \Omega_{ij}^k(2x_0 + 1, 2y_0 + 1)$$

$$-\Omega_{ij}^k(0, 2y_0 + 1)) - \Omega_{ij}^k(2x + 1, 0) - \Omega_{ij}^k(2x + 1, 2y + 1) - \Omega_{ij}^k(0, 2y + 1) - 2\Omega_{ij}^k(0, 0)), \quad (7.22)$$

where $N, M \in \mathbb{N}_0$ are the number of iterations required to evaluate the effective resistance of a finite mesh. The accuracy of (7.22) is evaluated for an 11×11,

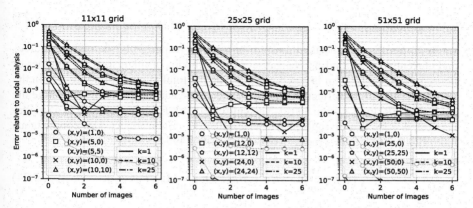

Fig. 7.11 Relative error of normalized effective resistance between $(0, 0)$ and (x, y) determined from (7.22) for $N = M \in [0, 5]$ as compared to a nodal analysis within a mesh with size a) 11×11, b) 25×25, and c) 51×51.

25×25, and 51×51 grid. The relative error of (7.22) is illustrated in Fig. 7.11. Observe that in all cases setting $N = K = 4$ is sufficient to achieve 0.3% accuracy. Note that due to the low error of (7.21) for small x and y, the error is smaller if the effective resistance is evaluated between nearby nodes.

7.3 Case studies

The primary contribution of this chapter is the efficient estimation of the effective resistance of a finite grid of arbitrary size, exhibiting constant complexity. The proposed framework is particularly suitable for circuit analysis techniques based on an effective resistance [480, 492]. In this section, three applications of the proposed framework are presented. In Subsection 7.3.1, the method accelerating the nodal analysis of a grid is presented. In Subsection 7.3.2, the method is applied to the analysis of a capacitive touch screen. In Subsection 7.3.3, a three-dimensional analysis of resistive substrate noise is described.

7.3.1 Mesh reduction based on effective resistance

The nodal analysis can be significantly accelerated by applying these effective resistance techniques if the grid dimensions are large and the number of nodes of interest are small. Consider a large grid Γ with dimensions $N_x \times N_y$. Define the nodes of interest as a set,

$$S \equiv S_v \cup S_i \cup S_o, \tag{7.23}$$

where S_v and S_i are subsets of nodes connected to, respectively, the voltage sources and current sources, and S_o is a subset of other nodes of interest. If the number of nodes of interest $|S| = n$ is much smaller than the total number of nodes within the network $|\Gamma| = N$, the effective resistance technique can significantly accelerate the analysis of IR drops within a grid. The entire network Γ can be reduced to a smaller network Γ_S by preserving the pairwise effective conductance. The conductance matrix $G \in \mathbb{R}^{n \times n}$ of this reduced network is [490, 499]

$$G^\dagger = -\frac{1}{2}\Big(R_S - \frac{1}{n}\big(\mathbf{1}_{n,n} R_S + R_S \mathbf{1}_{n,n}\big) + \frac{1}{n^2}\mathbf{1}_{n,1} R_S \mathbf{1}_{1,n} \Big), \tag{7.24}$$

where $G^\dagger$ denotes the Moore-Penrose pseudoinverse of matrix G, $R_S \in \mathbb{R}^{n \times n}_{\geq 0}$ is the matrix of the effective resistance between each pair of nodes in S, and $\mathbf{1}_{a,b}$ is an $a \times b$ matrix with all entries equal to one. After the conductance matrix is recovered, the reduced network can be evaluated by solving the linear system,

$$\begin{bmatrix} G & B \\ B^T & \mathbf{0} \end{bmatrix} \begin{bmatrix} V \\ I \end{bmatrix} = \begin{bmatrix} J \\ F \end{bmatrix}, \tag{7.25}$$

where V and I are, respectively, the node voltages and currents through the voltage sources. B, J, and F encode the current and voltage sources.

The speed of (7.24) and (7.25) for estimating the effective resistance within a mesh is compared to nodal analysis using the Numpy and Scipy Python packages [488] on an eight core 3.40 GHz Intel Core i7-6700 machine with 24 GB RAM. The comparison is depicted in Fig. 7.12. Nodal analysis in circuits larger than 10^7 nodes could not be performed due to insufficient memory. Note that while the computational time of the nodal analysis process scales with grid size N, the computational time of the infinity mirror technique scales with the number of nodes of interest n. The bottom-right corner of the plot in Fig. 7.12a is the area where the grid size is large and the number of nodes of interest is small. The infinity mirror technique provides the largest speedup in this situation. In Fig. 7.12b, the relationship between the speedup due to (7.24) and (7.25) and the fraction of nodes of interest is presented. The results suggest that the framework provides significant computational speedup if finding the voltage at only 0.23% of nodes is required (i.e., one in 430 nodes). For example, in a $10^3 \times 10^4$ grid, determining the voltage at $1,000$ nodes using nodal analysis would require $3,430$ seconds. Applying (7.24) and (7.25) results in a 17 fold speedup, requiring only 196 seconds to complete. If the number of nodes of interest is reduced to 100, the speedup reaches $1,400$, completing in 2.37 seconds.

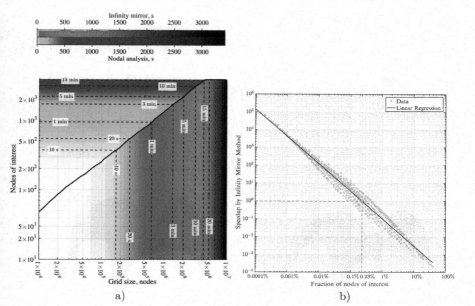

Fig. 7.12 Comparison of (7.24) and (7.25) to nodal analysis for $10 < n < 200$ and $10^4 < N < 10^7$. (a) Computational time (in seconds) to calculate the voltage at n nodes of interest in a grid with N nodes. The black line indicates n and N for which both techniques exhibit approximately equal time. (b) Speedup due to the use of (7.24) as compared to a pure nodal analysis as a function of $\frac{n}{N}$.

7.3.2   Resistive noise in capacitive touch screen

A possible application of the infinity mirror technique is the analysis of conductive media. An example of a conductive medium is a capacitive screen. The typical structure of a capacitive touch screen is shown in Fig. 7.13a [503]. An important component of the touch screen panel is the display cathode electrode providing a reference voltage for the screen. Resistive noise in the electrode layer of the display cathode may affect the accuracy of the touch recognition process. The accuracy of the touch sensor can therefore be enhanced by considering resistive noise during the sensor design process. An accurate estimate of the resistance typically requires significant computational time due to the finite element method extraction process often utilized for this task. The analysis can however be vastly accelerated by applying the infinity mirror technique to the equivalent model of the panel shown in Fig. 7.13b [503]. The method of mesh reduction presented in Subsection 7.3.1 is utilized to accelerate this analysis process.

The results for the effective resistance evaluation for trace resistances of 0.1 Ω and 100 Ω are shown, respectively, in Figs. 7.13c and 7.13d. The results are consistent with the Q3D extraction described in [503], significantly reducing the

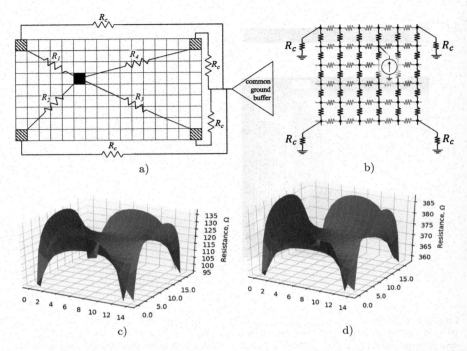

Fig. 7.13 Estimation of the effective resistance in a touch screen panel. a) Structure of the panel, b) equivalent circuit model, c) effective resistance with a 0.1 ohm trace resistance, and d) effective resistance with a 100 ohm trace resistance

analysis time while maintaining the high accuracy of the effective resistance estimation.

7.3.3 Resistive substrate noise

A three-dimensional mesh is widely utilized to model conductive media, including thermal paths and substrate noise. Substrate noise is a common issue in mixed-signal VLSI circuits [504]. While several advanced techniques for mitigation of substrate coupling exist, including guard rings and silicon-on-insulator technology [505], these techniques may significantly complicate the fabrication process. It is therefore necessary to estimate the magnitude of the substrate noise. The application of a three-dimensional network to substrate noise analysis is presented in this case study.

A frequent scenario in mixed-signal circuits is noise coupling between a digital aggressor and an analog victim. An equivalent circuit model of a mixed-signal circuit is shown in Fig. 7.14. The current I in the digital circuit would ideally flow

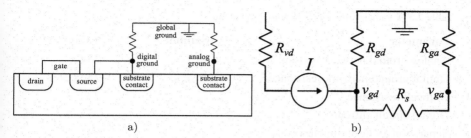

Fig. 7.14 Resistive substrate coupling mechanism in a mixed-signal complementary metal-oxide-semiconductor (CMOS) circuit. The substrate ground contacts for the analog and digital grounds are connected to the global ground through, respectively, the analog and digital ground distribution networks. (a) Side view of the substrate, and (b) equivalent circuit model of the noise injection process.

into the digital ground. With substrate coupling, however, a sizable current flows into the analog ground, affecting the performance of the sensitive analog circuits.

The voltage v_{ga} at the analog ground terminal is

$$v_{ga} = \frac{I R_{ga} R_{gd}}{R_{gd} + R_s + R_{ga}}, \tag{7.26}$$

where R_{gd} and R_{ga} are, respectively, the resistance of the digital and analog ground distribution networks, and R_s is the substrate resistance. Note that if the substrate resistance is large, the analog ground voltage converges to zero while reducing the substrate resistance, increasing the analog ground voltage.

The infinity mirror technique can be used to evaluate the effective resistance between substrate contacts. Consider a uniform three-dimensional grid with unit resistance r_s, infinite x-y dimensions, and finite z-dimension w_z. The analog and digital substrate contacts are represented by two terminals on the top surface of the grid separated by an $(x, y, 0)$ vector. Applying (7.18) yields

$$R_s = 2r_s \sum_{p \in \mathbb{Z}} \Omega_{00p}(x, y, 0) + \Omega_{00p}(x, y, 1) - \Omega_{00p}(0, 0, 0) - \Omega_{00p}(0, 0, 1),$$

$$\tag{7.27}$$

where

$$\Omega_{ijp}(x, y, z) = \Omega(x + 2w_x i, y + 2w_y j, z + 2w_z p). \tag{7.28}$$

Expression (7.27) is applied to (7.26) to determine the minimum distance between analog ground terminals. The parameters are listed in Table 7.3. The resulting ground voltage is shown in Fig. 7.15. If the spacing is small, the substrate noise is significantly lower with increasing separation. After 20 μm, however, the space does not have a significant effect on the coupling noise. Note again that the analysis time is significantly reduced by avoiding a costly nodal analysis process [232].

Table 7.3 Parameters for substrate noise evaluation.

Parameter	Symbol	Value
Analog ground network resistance	R_{ga}	25 Ω
Digital ground network resistance	R_{gd}	25 Ω
Unit cell resistance	r_s	1 Ω
Digital circuit current	I	25 mA
Separation along y-dimension	y	0
Grid z-dimension parameter	w_z	10
Cell dimensions	dx,dy,dz	1 μm

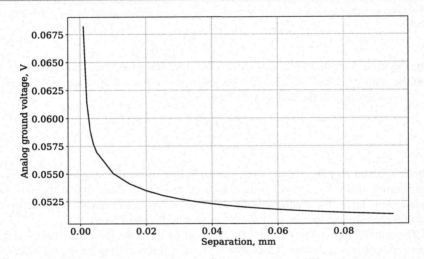

Fig. 7.15 Analog ground voltage as a function of the distance between the digital and analog ground terminals

The resistance measurement for each separation is completed on average in 2.76 seconds. A nodal analysis of a three-dimensional substrate with a size of 200 μm × 200 μm × 10 μm requires approximately 30.2 seconds, consistent with Fig. 7.12, indicating an approximate tenfold speedup.

7.4   Conclusions

An infinity mirror technique is proposed here that maps a rectangular resistive grid structure with finite dimensions into an infinite grid. Extending the contributions in [71], where semi-infinite structures are considered, the methodology described here is applicable to those structures where one or both dimensions are finite. In addition, the framework is extended to higher dimensional topologies, evaluating the effective resistance in finite structures with three and more dimensions. The proposed expressions exhibit high accuracy and outperform the nodal analysis method in terms of computational speed. Using the infinity mirror technique, the effective resistance between two points in an anisotropic finite mesh can be determined within 1% accuracy. Several orders of magnitude speedup in IR drop analysis in large grids is achieved in case studies by utilizing closed-form expressions for the effective resistance. The most significant reduction in computational time is achieved in those cases where only a small fraction of nodes needs to be evaluated. These results can be beneficial to a variety of applications, including power grid and substrate analysis in VLSI circuits, estimation of commute times in random walks, and the analysis of isotropic and anisotropic conductive media [499–501, 506–509].

Chapter 8
Placement of on-chip distributed voltage regulators

The primary objective of a VLSI power delivery system is to supply and maintain a nearly constant (i.e., low ripple) voltage across the load circuitry. Additional objectives include dissipating less power while limiting the current density to reduce the likelihood of electromigration. Different techniques have been proposed to accomplish these tasks, including multiple voltage domains [361], on-chip decoupling capacitors [493], and on-chip voltage regulation [381].

In a conventional VLSI system, a power management IC (PMIC), also known as a voltage regulator module (VRM), is placed at the board level and supplies multiple voltages to different on-chip voltage domains [145, 232], as illustrated in Fig. 8.1a. The primary limitation of this approach is the large physical distance between the off-chip regulator and the many billions of on-chip loads. The interconnect and I/O pins connecting the off-chip voltage converter with the load circuitry exhibit a high parasitic resistance and inductance, producing significant power noise [65]. The supply voltage is often increased to compensate for the voltage drop caused by the parasitic impedance of the power network [510] (see Fig. 8.1a), degrading the overall energy efficiency of the system. Furthermore, the parasitic impedance between the converter and load circuitry slows the load regulation process. Considerable variations in supply voltage can be experienced by the load circuitry, potentially violating the noise margins of the many signals.

Heterogeneous voltage regulation [511] is a recent advancement in power delivery systems. The power efficient voltage converters within a PMIC are supplemented by area efficient on-chip regulators, as shown in Fig. 8.1b. The on-chip converters are placed in close proximity to the load devices. Since the physical distance and impedance between the on-chip regulator and device are small, this configuration provides superior power quality despite load dependent current fluctuations.

Increasing the number and enhancing the placement of the on-chip voltage regulators may greatly improve overall power integrity as compared to a single regulator, since the distance between the regulator and the load is much smaller.

© The Author(s), under exclusive license to Springer Nature Switzerland AG 2023   217
R. Bairamkulov, E. G. Friedman, *Graphs in VLSI*,
https://doi.org/10.1007/978-3-031-11047-4_8

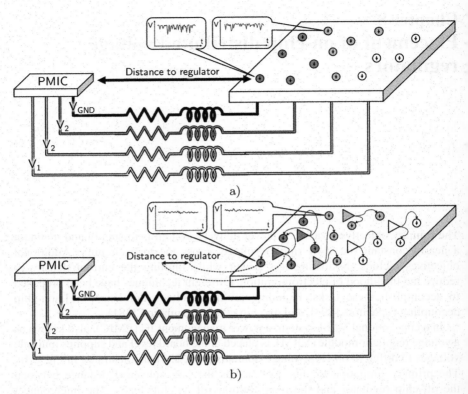

Fig. 8.1 Overview of power delivery systems. a) Conventional power delivery system. The voltage converter within a power management IC (PMIC) provides multiple supply voltages to several power delivery systems. These networks are connected to the functional circuitry via dedicated power networks. Due to the significant distance to the regulators, fluctuations in the load current degrade the quality of the power supply. b) Heterogeneous power delivery system with on-chip voltage regulators. The on-chip regulators are placed near the load devices. A stable voltage is more effectively supplied to the functional circuits.

Multiple regulators however may occupy significant on-chip area. The number of voltage regulators is therefore limited. The regulators should therefore be judiciously distributed within an IC to enhance the power quality.

Most works discussing regulator distribution approach this problem from a purely electrical perspective, focusing on the stability, power efficiency, and thermal behavior of the on-chip voltage regulation system [512–516]. Placement in the context of power delivery is however discussed in several works. The distribution of the power supply input/output (I/O) pads using mixed integer linear programming is discussed in [517]. Based on a predefined set of I/O pad locations, a subset of locations is chosen to minimize the voltage drop within a power network. A framework for power supply and decoupling capacitor distribution is proposed in [495]. Based on a closed-form expression for the effective resistance within a two layer mesh [469], the location of the decoupling capacitors is chosen to reduce the response time while lowering the voltage drop within the network.

Based on [469], a novel voltage regulator distribution algorithm is presented in this chapter. Similar to the optimization process described in Chapter 9 [75], the distribution is formulated as an optimization problem, where the voltage drops due to the parasitic impedances within the power network are minimized. With the infinity mirror technique (see Chapter 7 [72]), a several orders of magnitude improvement in the speed of the power network analysis process is achieved while maintaining high accuracy. The physical and electrical constraints of the voltage regulators, such as current capacity and electromigration, are supported by the algorithm. Based on this algorithm, the position of the voltage regulators is efficiently determined using particle swarm optimization [518]. In case studies, regulators are distributed in less than two minutes on a Linux workstation powered by a dual core 2.3 GHz Intel Core i5 processor with 16 GB of RAM, achieving a significant reduction in voltage drop as compared to a uniform distribution of regulators.

The rest of the chapter is organized as follows. In Section 8.1, the basic principles of on-chip voltage regulation are described. A computationally efficient model of an on-chip power network is discussed in Section 8.2. To improve the runtime of the optimization process, load clustering is performed, as described in Section 8.3. The problem setup for register placement optimization is reviewed in Section 8.4. Several case studies are described in Section 8.5, followed by the conclusions in Section 8.6.

8.1 On-chip voltage regulation

Unlike traditional voltage regulation schemes, where the voltage converter is placed far from the load devices, the proposed on-chip voltage regulation methodology places the voltage regulators closer to the load [381, 519]. Two major advantages of distributed on-chip power regulation exist. Multiple voltage domains can be supported using distributed on-chip voltage regulators [520]. Localized control of the power flow within a VLSI system is therefore possible, enabling fine grain dynamic voltage scaling [521] and power gating [522]. Furthermore, the shorter distance between the regulator and the load circuitry greatly enhances the communication speed, allowing the regulator to quickly react to changes in the workload. Similar to decoupling capacitors that exhibit high efficiency when placed near the load circuitry [493], placing the regulators near the point-of-load (POL) drastically improves the power quality. Due to input regulation and load regulation, the voltage regulators distributed within an IC separate the power network into two loosely dependent parts, as illustrated in Fig. 8.2. The load circuitry is effectively shielded from fluctuations in the input voltage, while fluctuations in current demand are quickly accommodated by the regulator. The power noise is therefore significantly suppressed, improving the overall power integrity of the system.

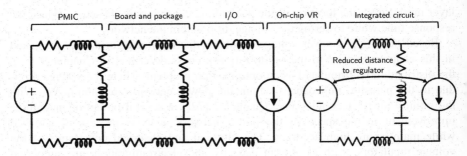

Fig. 8.2 On-chip voltage regulators effectively split a power network into independent parts. The off-chip part connects the PMIC to the regulators, while the on-chip part connects the regulators to the load circuitry.

Table 8.1 Comparison of major types of on-chip converters

Converter type	SMPS	SC	Linear
Power efficiency	High	Medium	Low
Regulation quality	High	Low	High
Physical area	Large	Medium	Small

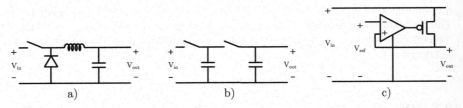

Fig. 8.3 Common on-chip voltage regulators. a) Switching mode power supply (SMPS), b) switched capacitor (SC), and c) low dropout (LDO) linear voltage regulator.

Three major classes of integrated voltage regulators exist; namely, switching mode power supply (SMPS), switched capacitor, and linear [381]. Essential properties of these regulators are summarized in Table 8.1. An SMPS converter is a power efficient converter that operates by energizing the LC branch during the charging phase, transferring the stored energy to the load during the discharge phase [358, 385]. A SMPS buck converter is shown in Fig. 8.3a. The output voltage is efficiently controlled by adjusting the duty cycle, i.e., the relative duration of the charge and discharge phases. SMPS converters exhibit superior power efficiency and voltage regulation but require large area to accommodate the inductor and capacitor.

Switched capacitor (SC) converters do not require inductors, as shown in Fig. 8.3b. SC converters are therefore more area efficient than SMPS converters. Similar to SMPS converters, the voltage conversion process is accomplished by switching. Since the primary mechanism of conversion is to transfer charge between capacitors, the power efficiency is degraded due to the phenomenon of charge sharing [232]. Furthermore, the output voltage is highly sensitive to the load current, reducing the regulation quality of the SC converters.

Superior regulation is achieved by linear regulators which regulate the voltage by the principle of voltage division, as illustrated in Fig. 8.3c. The output voltage is produced by inducing a resistive voltage drop within a variable resistance. A low dropout (LDO) regulator is the most common type of linear regulator, where the resistance of a pass transistor is controlled by an error amplifier. Due to the power dissipated by the variable resistor, linear regulators exhibit poor power efficiency bounded by the ratio of the output voltage to the input voltage [232]. LDO regulators however have gained significant popularity in modern high performance systems due to the small area and fast voltage regulation.

On-chip regulators typically exhibit nonlinear behavior and require significant computational time for analysis. For the purpose of optimization, however, the accurate behavior of a regulator is less important. The regulator model should rather exhibit high *fidelity*, i.e., track the general behavior of the target metric rather than accurately estimate the metric. Assuming the input regulation and load regulation are sufficiently fast, a regulator can be modeled as a constant voltage source. Despite the poor accuracy of this model, the model exhibits high fidelity, appropriate for the optimization process (high computational efficiency rather than accuracy). To demonstrate the fidelity of modeling a regulator with a voltage source, consider the relationship between the position of a regulator and the maximum voltage drop within a grid, as shown in Fig. 8.4. Two models of a voltage regulator are considered; namely, a SPICE-level transient LDO model and a constant voltage source. Observe the poor accuracy of the voltage source-based model shown in Fig. 8.4c, exhibiting significant deviation from the SPICE-level model. Both of the functions however exhibit similar behavior (see Figs. 8.4a and 8.4b), increasing and decreasing within the same regions, and achieving the minimum at position (6, 3). The constant voltage source model of a regulator therefore exhibits high fidelity, supporting the use of this model within the algorithm which distributes the on-chip regulators within a grid.

8.2 Model of power network

Many VLSI systems utilize global power grids spanning large portions of the physical area of an IC. These grids consist of two or more layers of orthogonal interconnects connected by vias, as illustrated in Fig. 8.5a. The advantages of this topology include ease of design, robustness, and low impedance, as compared to routed power networks [232]. Due to the regularity and symmetry of a power grid, the power network can be modeled as a resistive mesh, as depicted in Fig. 8.5b. Due to the size of the mesh, an infinite two-dimensional model of the grid can be used to analyze this network. This approach supports the use of closed-form expressions for the effective resistance between two nodes within an infinite grid [469],

$$R(\mathbf{x}) = 2r\,\Omega_k(\mathbf{x}), \tag{8.1}$$

where $\mathbf{x} = (x, y)$,

$$\Omega_k(\mathbf{x}) = \frac{\sqrt{k}}{4\pi}\left[\ln\left(x^2 + ky^2\right) + 2\ln(\pi) + 2\gamma\right] + J(k). \tag{8.2}$$

r and x (kr and y) are, respectively, the resistance and physical distance between the nodes in the horizontal (vertical) dimension, and $J(k)$ is a polynomial function of k. Due to the finite size, however, this model exhibits a significant error near the boundaries of the grid. This issue is overcome in [71, 72] where the boundaries of the grid are modeled with image current sources. With this approach, the effective

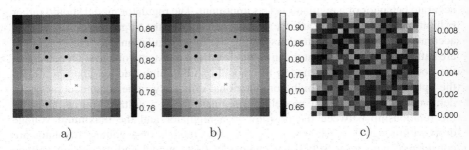

Fig. 8.4 Relationship between minimum voltage within a 20×20 grid and the location of the voltage regulators. The minimum voltage is obtained from an analysis of a power grid. The LDO is modeled as a) an operational amplifier driving a pass transistor, and b) a constant voltage source. Note that the same optimum location is obtained for each model. c) The relative error of the constant voltage source model of a regulator based on normalized voltages.

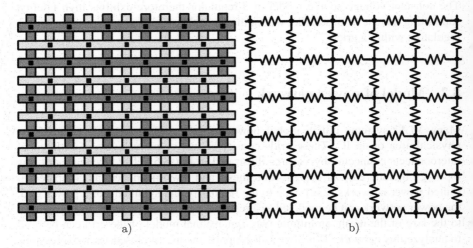

Fig. 8.5 On-chip power grid. a) Layout of power (dark grey) and ground (light grey) distribution networks, and b) power distribution network modeled as a resistive mesh.

resistance can be determined in $O(N_x N_y)$ time, where N_x and N_y denote the number of images, respectively, in the x and y dimensions. To maintain the error of this method below 1% for a 100×100 grid, only three images are sufficient [72]. Fewer images are required in practical power networks due to the significantly larger grid size. Observe that the analysis runtime does not depend upon the size of the mesh. Based on this feature, an efficient power grid analysis algorithm is presented in Subsection 8.2.1 to efficiently determine the minimum voltage within a grid.

Although practical power networks are typically grid structured, significant deviations, such as missing vias or variable interconnect pitch, do exist. Furthermore, a global mesh may span more than two layers, complicating the two layer model. To analyze practical grids, a power network should be converted into an equivalent resistive mesh while preventing excessive deviations from the original grid.

To simplify the structure of the network, a 3-D to 2-D grid regularization technique is proposed in [523]. By ignoring the via impedance, multiple grid layers are initially collapsed into a single layer based on location information. The 2-D network is mapped into a two-dimensional grid with a fixed pitch, yielding a resistive mesh with a fixed pitch. An analysis of the resulting grid exhibits an error of less than 1% as compared to SPICE.

A similar approach is followed in this chapter. By examining each benchmark circuit, a dominant wire pitch and resistance are observed. Consider, for example, the ibmpg4 power network [524]. The dominant resistivity and pitch of all of the interconnects in the x dimension are, respectively, 48 units and 35 milliohms per unit length, as depicted in Figs. 8.6a and 8.6b. Similarly, the dominant pitch in the y direction is 24 units with a resistivity of 32.5 microohms per unit length, as shown in Figs. 8.6c and 8.6d. The resulting simplified grid has dimensions $\mathbf{w} = (284, 571)$ and $k = 2.15$. The parameters of a simplified grid for each benchmark circuit are listed in Table 8.2.

8.2.1 Fast grid analysis

The primary objective of the voltage regulator distribution process is to deliver a stable voltage to the functional circuitry. To minimize the maximum voltage drop within a power grid, minimizing the voltage drop at the loads is sufficient. Standard circuit analysis techniques based on MNA typically analyze the entire network, even if the voltage at only a single node is of interest. An efficient grid analysis algorithm based on the infinity mirror technique is a faster alternative where the voltage is only determined at specific locations. This analysis technique is based on the approach described in [480]; however, the effective resistance is not explicitly evaluated. Rather, the potential induced by the supply and load currents is determined.

Let $\ell = (\mathbf{x}(\ell), I(\ell))$ be a load located at position $\mathbf{x}(\ell)$ and drawing current $I(\ell)$ from a resistive grid of size $\mathbf{w} = (w_x, w_y)$. The set $\mathcal{L} = \ell_p | p \in [1, \ldots, n]$ is a set

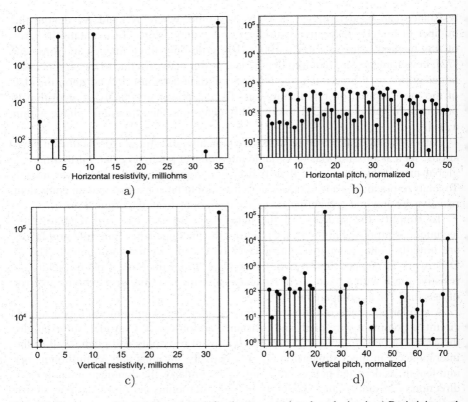

Fig. 8.6 Frequency of resistivity and pitch for the ibmpg1 benchmark circuit. a) Resistivity, and b) pitch along the x dimension. c) Resistivity, and d) pitch along the y dimension. The equivalent grid is constructed based on the dominant resistivity and pitch within the network.

Table 8.2 Parameters of the equivalent grids used to model the ibmpg benchmark circuits [524]

	Pitch		Resistivity, mΩ		Dimensions	
	x	y	x	y	x	y
ibmpg1	2,062	33	5.714	0.635	10	629
ibmpg2	48	72	4.000	16.25	169	113
ibmpg3	864	1,296	0.714	2.407	354	236
ibmpg4	48	24	35.00	32.50	284	571
ibmpg5	82	12	10.00	21.67	129	882
ibmpg6	280	280	0.286	0.464	3,630	3,644

of all loads within the network. Based on the infinity mirror technique, the finite grid is converted into an infinite two-dimensional resistive lattice, as illustrated in Fig. 7.7. The images of each load $\ell_p \in \mathcal{L}$ are described by a set of loads,

$$\ell_p^* = \{ \left(\mathbf{x}_p^{(i,j)}, I_p \right) \mid i \in [-N_x, \dots, N_x] , j \in \left[-N_y, \dots, N_y \right] \}, \tag{8.3}$$

where, for brevity, $\mathbf{x}_p^{(i,j)} = \mathbf{x} \left(\ell_p^{(i,j)} \right) = \left(x_p^i, y_p^j \right)$, $I_p = I \left(\ell_p \right)$, and

$$x_p^i = \begin{cases} w_x i + x_p, & \text{if } i \text{ is even}, & (8.4a) \\ w_x(i+1) - x_p - 1, & \text{if } i \text{ is odd}, & (8.4b) \end{cases}$$

$$y_p^j = \begin{cases} w_y j + y_p, & \text{if } j \text{ is even}, & (8.5a) \\ w_y(j+1) - y_p - 1, & \text{if } j \text{ is odd}. & (8.5b) \end{cases}$$

$\mathcal{L}^*$ is the set of all loads within an infinite grid, including the mirrored loads,

$$\mathcal{L}^* = \bigcup_{p=1}^{n} \ell_p^*. \tag{8.6}$$

The electric potential at node $\mathbf{u} = (x_{\mathbf{u}}, y_{\mathbf{u}})$ in response to a unit load $\hat{\ell} = (\mathbf{x}, 1))$ with respect to a ground node at infinity is

$$\phi(\mathbf{u}, \mathbf{x}) = \sum_{\hat{\ell} \in \hat{\ell}^*} \Omega_k(\mathbf{u} - \mathbf{x}(\hat{\ell})). \tag{8.7}$$

By selecting arbitrary ground node $\mathbf{g}$, the voltage at node $\mathbf{u}$ becomes

$$v^{\mathbf{g}}(\mathbf{u}, \mathbf{x}) = \phi(\mathbf{u}, \mathbf{x}) - \phi(\mathbf{g}, \mathbf{x}). \tag{8.8}$$

Due to the principle of superposition, the voltage at node $\mathbf{u}$ is the weighted sum of the potentials caused by each current source within a grid,

$$V^{\mathbf{g}}(\mathbf{u}) = \sum_{\ell_p \in \mathcal{L}} I_p v^{\mathbf{g}}(\mathbf{u}, \mathbf{x}_p). \tag{8.9}$$

If a grid contains only current sources, the voltage at any node within a grid can be determined using (8.9). The power network however includes voltage regulators that maintain a constant voltage by changing the current supplied to the network. Any voltage source can therefore be transformed into a current source supplying equivalent current into a network.

Finding the current injected by each voltage source requires additional processing. Suppose m voltage regulators are connected to a network. The set of voltage regulators within the network is

$$\mathcal{S} = \{s_q \mid q \in [1, \dots, m]\}, \tag{8.10}$$

where

$$s_q = \left(\mathbf{x}_q, I_q\right). \tag{8.11}$$

The target voltage at each node $\mathbf{x}_q, q \in [1, \dots, m]$ is known *a priori*, producing a vector $\mathbf{v}(\mathcal{S}) \in \mathbb{R}^m$ of target voltages,

$$\mathbf{v}(\mathcal{S}) = [V_1, \dots, V_m]^T. \tag{8.12}$$

To determine the current injected by each voltage regulator, an arbitrary node $\mathbf{g}$ is initially designated as ground. Without loss of generality, suppose $\mathbf{g} = \mathbf{x}_m$, producing set $\mathcal{S}^{\mathbf{g}} = \mathcal{S} \setminus s_m$. The target voltages are therefore adjusted, yielding a vector $\mathbf{v}^{\mathbf{g}}(\mathcal{S}) \in \mathbb{R}^{m-1}$,

$$\mathbf{v}^{\mathbf{g}}(\mathcal{S}) = \left[V_1^{\mathbf{g}}, \dots, V_{m-1}^{\mathbf{g}}\right]^T, \tag{8.13}$$

where

$$V_q^{\mathbf{g}} = V_q - V_m. \tag{8.14}$$

The voltage $V_r^{\mathbf{g}}$ is determined by superimposing the effect of the supply and load currents,

$$V_r^{\mathbf{g}} = \sum_{q=1}^{m} I(s_q) v^{\mathbf{g}}(s_r, s_q) + \sum_{p=1}^{n} I(\ell_p) v^{\mathbf{g}}(s_r, \ell_p), \tag{8.15}$$

where, for brevity,

$$v^{\mathbf{g}}\left(s_r, s_q\right) = v^{\mathbf{g}}\left(\mathbf{x}\left(s_r\right), \mathbf{x}\left(s_q\right)\right), \tag{8.16}$$

$$v^{\mathbf{g}}\left(s_r, \ell_p\right) = v^{\mathbf{g}}\left(\mathbf{x}(s_r), \mathbf{x}\left(\ell_p\right)\right). \tag{8.17}$$

Reformulating (8.15) in matrix form yields

$$\begin{bmatrix} v^{\mathbf{g}}(s_1, s_1) & \cdots & v^{\mathbf{g}}(s_m, s_1) \\ \vdots & \ddots & \vdots \\ v^{\mathbf{g}}(s_1, s_{m-1}) & \cdots & v^{\mathbf{g}}(s_m, s_{m-1}) \end{bmatrix} \begin{bmatrix} I(s_1) \\ \vdots \\ I(s_{m-1}) \end{bmatrix}$$

$$= \mathbf{v}^{\mathbf{g}}(\mathcal{S}) - \begin{bmatrix} v^{\mathbf{g}}(\ell_1, s_1) & \cdots & v^{\mathbf{g}}(\ell_n, s_1) \\ \vdots & \ddots & \vdots \\ v^{\mathbf{g}}(\ell_1, s_{m-1}) & \cdots & v^{\mathbf{g}}(\ell_n, s_{m-1}) \end{bmatrix} \begin{bmatrix} I(\ell_1) \\ \vdots \\ I(\ell_n) \end{bmatrix} \tag{8.18}$$

or, equivalently,

$$\Phi^g(\mathcal{S}, \mathcal{S}^g)\mathbf{i}(\mathcal{S}) = \mathbf{v}^g(\mathcal{S}) - \Phi^g(\mathcal{L}, \mathcal{S}^g)\mathbf{i}(\mathcal{L}). \tag{8.19}$$

The system described by (8.19) is underdetermined with $m - 1$ equations and m unknowns. To obtain the remaining equation, note that the total current drawn by the loads is equal to the total current injected by the voltage regulators,

$$\mathbf{1}_{1,m}\mathbf{i}(\mathcal{S}) + \mathbf{1}_{1,n}\mathbf{i}(\mathcal{L}) = 0, \tag{8.20}$$

where $\mathbf{1}_{a,b}$ is an $a \times b$ matrix with all entries equal to 1. The current $\mathbf{i}(\mathcal{S})$ supplied by the voltage regulators can therefore be determined by solving a system of linear equations,

$$\begin{bmatrix} \Phi^g(\mathcal{S}, \mathcal{S}^g) \\ \mathbf{1}_{1,m} \end{bmatrix} \mathbf{i}(\mathcal{S}) = \begin{bmatrix} \mathbf{v}^g(\mathcal{S}) \\ 0 \end{bmatrix} - \begin{bmatrix} \Phi^g(\mathcal{L}, \mathcal{S}^g) \\ \mathbf{1}_{1,n} \end{bmatrix} \mathbf{i}(\mathcal{L}). \tag{8.21}$$

By combining $\mathcal{L}$ and $\mathcal{S}$, the set of current injections $\mathcal{I} = \mathcal{L} \cup \mathcal{S}$ is obtained. The voltage at each load is therefore

$$\mathbf{v}^g(\mathcal{L}) = \Phi^g(\mathcal{I}, \mathcal{L})\mathbf{i}(\mathcal{I}) + V_m \mathbf{1}_{||\mathcal{I}||,1}. \tag{8.22}$$

8.2.2 Limited regulator current

The amount of current reliably delivered by a linear regulator is a strong function of the regulator area [511]. LDO regulators with wider power transistors can supply more current to the loads. Since on-chip regulators occupy silicon layer, other circuitry may constrain the placement and size of the regulator. The maximum current supplied by an LDO is therefore limited by the size of the regulator. Furthermore, even if the size of the regulator is unlimited, electromigration [525] limits the maximum current density that can be produced by a regulator. The current capacity of a regulator is therefore limited.

To consider this limitation during the optimization process, the fast grid analysis algorithm described in Subsection 8.2.1 is extended to support the limited current capacity of a regulator. Let $I_{max} : \mathcal{S} \to \mathbb{R}$ be a function mapping each regulator s to the maximum current $I_{max}(s)$ that s can supply. Note that the total capacity of the regulators should be equal to or exceed the current demand of the circuit,

$$\mathbf{1}_{1,m}\mathbf{i}_{max} \geq \mathbf{i}(\mathcal{L}), \tag{8.23}$$

where $\mathbf{i}_{max}$ is a vector describing the current capacity of each regulator,

$$\mathbf{i}_{max} = [I_{max}(s_1), \ldots, I_{max}(s_m)]^T. \tag{8.24}$$

Suppose, after solving (8.21), the estimated current of subset $\mathcal{S}^* \subset \mathcal{S}$ exceeds the corresponding maximum current. Vector $\mathbf{i}(\mathcal{S})$ therefore does not realistically represent the current supplied by each regulator. This result however indicates that the regulators in $\mathcal{S}^*$ operate at maximum capacity, i.e., $I(s_i) = I_{max}(s_i) \forall s_i \in \mathcal{S}^*$. Since the current supplied by these regulators is known, these nodes can be treated as loads. Transferring $\mathcal{S}^*$ into $\mathcal{L}$ yields

$$\mathcal{S}_1 \leftarrow \mathcal{S} \setminus \mathcal{S}^* \tag{8.25}$$

and

$$\mathcal{L}_1 \leftarrow \mathcal{L} \cup \mathcal{S}^*. \tag{8.26}$$

Note that a different ground node $\mathbf{g}$ should be selected if $\mathbf{g} \in \mathcal{S}^*$. The system of (8.21) is transformed into

$$\begin{bmatrix} \Phi^{\mathbf{g}}(\mathcal{S}_1, \mathcal{S}_1^{\mathbf{g}}) \\ \mathbf{1}_{1,\|\mathcal{S}_1\|} \end{bmatrix} \mathbf{i}(\mathcal{S}_1) = \begin{bmatrix} \mathbf{v}^{\mathbf{g}}(\mathcal{S}_1) \\ 0 \end{bmatrix} - \begin{bmatrix} \Phi^{\mathbf{g}}(\mathcal{L}_1, \mathcal{S}_1^{\mathbf{g}}) \\ \mathbf{1}_{1,\|\mathcal{L}_1\|} \end{bmatrix} \mathbf{i}(\mathcal{L}_1). \tag{8.27}$$

If no current $\mathbf{i}(\mathcal{S}_1)$ exceeds the current limit, the process is completed and the voltage at any node can be determined. Otherwise, the process is repeated until all of the regulator currents satisfy the constraints.

8.3 Load clustering

The functional circuits within an IC are typically distributed across the entire area of the power network. A large number of load currents within each functional block is therefore connected to the power grid. Since the runtime of the proposed method increases with the number of loads, individually considering each load incurs a significant computational penalty. Recall however from Chapter 5 that a power grid is a smooth system, i.e., a small variation in position correlates with a small variation in voltage [438]. Multiple loads can therefore be merged into a single load if located sufficiently close to each other.

At the global level, this procedure is accomplished by clustering, where the number of loads can be reduced by several orders of magnitude while exhibiting minimal effect on the voltage within the grid. To illustrate this effect, consider the example shown in Fig. 8.7 where the current sources are randomly distributed within a network. The voltage within the grid is depicted in Fig. 8.7a. After clustering the current sources, the number of loads is reduced by tenfold. The voltage within a network with clustered loads is shown in Figs. 8.7b to 8.7h. Observe that the average voltage within a grid is relatively unchanged since the total current within the grid remains constant. The maximum voltage drop however increases since the load is concentrated within a smaller area. The relationship between the number of clusters

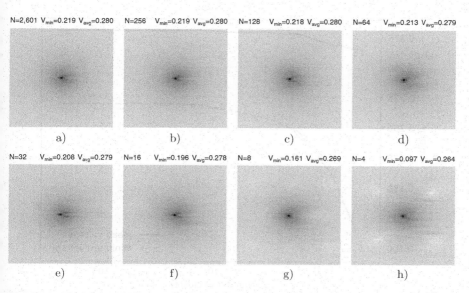

N=2,601 V_{min}=0.219 V_{avg}=0.280   N=256 V_{min}=0.219 V_{avg}=0.280   N=128 V_{min}=0.218 V_{avg}=0.280   N=64 V_{min}=0.213 V_{avg}=0.279

a)   b)   c)   d)

N=32 V_{min}=0.208 V_{avg}=0.279   N=16 V_{min}=0.196 V_{avg}=0.278   N=8 V_{min}=0.161 V_{avg}=0.269   N=4 V_{min}=0.097 V_{avg}=0.264

e)   f)   g)   h)

Fig. 8.7 Effect of clustering the load current on the accuracy of the power grid analysis process. a) Original 51 × 51 power grid with 2, 601 loads. Each node is connected to a random load current. Loads within the grid are split into b) 256, c) 128, d) 64, e) 32, f) 16, g) 8, and h) 4 clusters. N denotes the number of clusters, and Vmin and Vavg denote, respectively, the minimum and average voltage within the grid. The minimum voltage is not significantly affected until the number of loads is reduced below 16, i.e., 0.6% of the total number of loads. Observe that the average voltage does not significantly change.

and the minimum voltage is shown in Fig. 8.8. The number of clusters is therefore a tradeoff between accuracy and runtime. Note however that the goal of the proposed framework is to optimize the position of the voltage regulators. A minor penalty in accuracy can therefore be tolerated if the effect on the optimization result is small.

8.4   Optimization setup

Constrained global optimization is applied in this chapter to determine the optimal location of the regulators. The voltage drop caused by the parasitic impedances within the power network is expressed as a function of the position of the voltage regulators,

$$v_{drop}(\mathcal{S}) = -\min\left(\mathbf{v}^{\mathbf{g}}(\mathcal{L})\right)|_{\mathcal{S}}, \qquad (8.28)$$

where the right hand side denotes the maximum voltage drop as a function of $\mathcal{S}$. During the optimization process, however, an objective function is evaluated hundreds of times before achieving convergence. To use (8.28) as the objective

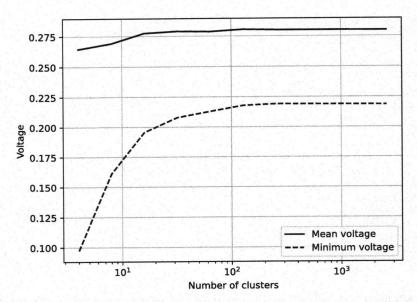

Fig. 8.8 Minimum and average voltage within a 51 × 51 grid during clustering. A two orders of magnitude reduction in the number of loads is possible with only a minor effect on the accuracy of the estimates of the minimum voltage.

function, the runtime $v_{drop}(\mathcal{S})$ should be small. Function (8.28) can be evaluated using modified nodal analysis [64]. Due to the size of modern integrated systems, power network models are extremely large, containing many millions to billions of nodes. Conventional MNA-based analysis of power grids is therefore not suitable for optimizing the position of the regulators. In contrast, using the proposed fast grid analysis method, the voltage within a grid can be determined in $O\,(n\,(m+n))$ time, where m and n denote the number of, respectively, voltage regulators and loads. Note that the proposed method does not depend upon the size of the mesh. Arbitrarily large grids can therefore be analyzed with this method. The voltage for a subset of nodes is determined in milliseconds, many orders of magnitude faster than state-of-the-art MNA-based algorithms.

The optimization problem is therefore described as

$$\textbf{Minimize} : v_{drop}(\mathcal{S}) \tag{8.29}$$

$$\textbf{subject to} :$$

$$\mathbf{x}(s) \in A \forall s \in \mathcal{S}, \tag{8.30}$$

$$\mathbf{i}(\mathcal{S}) \leq \mathbf{i}_{max}, \tag{8.31}$$

where A is the set of whitespace nodes, i.e., unoccupied positions available for placing voltage regulators. Since the regulators occupy silicon layer, congested regions cannot be used to place the regulators. Constraint (8.30) restricts the position of the voltage regulators within a grid to those regions capable of accommodating the regulators. Due to physical limitations, the regulators cannot provide arbitrarily large currents. An upper bound on the current therefore exists for each regulator and is expressed by constraint (8.31).

Since the convexity of objective function (8.29) is unknown, a global optimization algorithm is required, such as basin hopping [526], evolutionary [527], or swarm intelligence algorithms [518]. The discrete particle swarm optimization algorithm [518] is used in the case studies described in Section 8.5.

8.5 Case studies

The analysis and optimization algorithms are implemented in Python and applied to IBM power grid benchmarks [524]. The algorithms are run on a Linux workstation powered by a dual core 2.3 GHz Intel Core i5 processor with 16 GB of RAM. Three optimization scenarios are considered. In the first case study, the voltage regulators are distributed within the entire grid without restricting the placement and maximum current supplied by the regulators, as described in Subsection 8.5.1. In Subsection 8.5.2, a second case study is considered, where the regulators are placed within specific whitespace regions. In the final case study, as described in Subsection 8.5.3, the maximum current of the regulators is restricted.

8.5.1 Unrestricted placement – case study one

In this case study, no constraints on the maximum current are placed on the location of the voltage regulators. The number of clusters is set to 100 while the number of voltage regulators is varied from 10 to 50. The results are summarized in Table 8.3. Consistent with expectations, more regulators provide superior regulation, raising the minimum voltage of the system. Observe that the runtime of the optimization process does not increase with grid size, but increases with the number of regulators. The voltage within the ibmpg4 benchmark is depicted in Fig. 8.9. Since the load current is uniformly spread throughout the grid, the regulators are also uniformly spread throughout the integrated circuit. Observe that the benefit of additional regulators diminishes with increasing number of regulators.

Table 8.3 Summary of the distribution of voltage regulators within the ibmpg benchmark circuits for case studies one and three. The results of case two greatly depend upon the particular geometry of the restricted placement and are therefore omitted from this table.

	Number of regulators	Case one		Case three	
		Voltage drop, V	Runtime, s	Voltage drop, V	Runtime, s
ibmpg1	10	9.25	37.2	11.02	76.2
	20	4.52	135.2	8.53	119.1
	30	3.59	136.5	7.04	167.1
	40	4.42	242.0	7.22	251.2
	50	1.82	587.2	5.26	621.8
ibmpg2	10	10.35	35.1	10.33	59.9
	20	8.07	414.9	7.86	180.0
	30	6.93	575.4	6.39	154.7
	40	6.46	337.5	7.16	409.8
	50	6.59	420.8	5.93	468.2
ibmpg3	10	9.48	34.1	9.56	30.9
	20	6.17	242.2	5.94	183.4
	30	4.70	203.4	5.27	160.1
	40	4.15	360.6	3.79	460.2
	50	4.22	233.2	3.48	308.4
ibmpg4	10	0.23	50.4	0.33	45.8
	20	0.14	123.1	0.12	253.2
	30	0.10	484.2	0.09	338.0
	40	0.08	304.9	0.11	414.7
	50	0.06	438.1	0.05	465.0
ibmpg5	10	0.41	73.5	0.36	68.4
	20	0.20	209.8	0.33	162.5
	30	0.10	274.3	0.17	592.7
	40	0.08	224.6	0.16	989.9
	50	0.10	427.6	0.14	811.6
ibmpg6	10	2.63	107.0	2.36	60.9
	20	1.36	451.0	1.37	318.8
	30	0.99	187.7	0.95	313.6
	40	0.72	239.8	0.77	640.4
	50	0.61	309.6	0.73	913.6

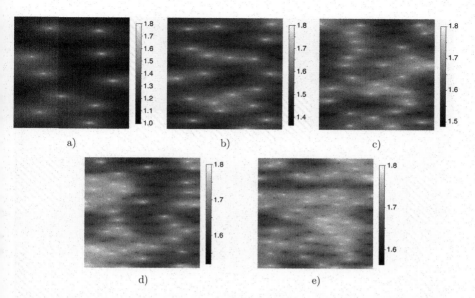

Fig. 8.9 Voltage after distributing a) 10, b) 20, c) 30, d) 40, and e) 50 voltage regulators within the ibmpg4 benchmark circuit. Note that the voltage drop decreases with additional voltage regulators.

8.5.2 Restricted placement – case study two

The congested areas within each benchmark circuit are manually selected. The restricted regions within the ibmpg4 benchmark circuit are depicted in Fig. 8.10. The voltage regulators are prevented from being placed into these areas by adding a penalty (or cost) to the objective function for each regulator placed within a restricted area. The resulting distribution of voltage regulators is shown in Fig. 8.11. This restriction produces larger voltage drops within the network, since the regulators cannot be necessarily placed near the hot spots, i.e., regions exhibiting high current demand. The degradation of the voltage drops however depends upon the location and size of the blockages and loads. A system with spatial constraints located near large load currents will likely exhibit greater power noise.

8.5.3 Restricted current – case study three

In this case study, the maximum current supplied by a voltage regulator s is set as

$$I_{max}(s) = 1.2 \times \frac{\mathbf{1}_{1,n}\mathbf{i}(\mathcal{L})}{m}. \tag{8.32}$$

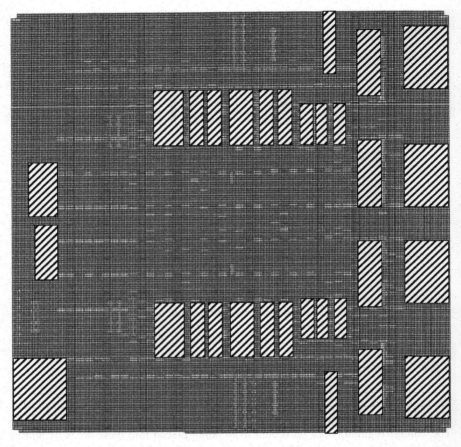

Fig. 8.10 Layout of wires within the ibmpg4 benchmark circuit. Those regions where the placement of the voltage regulators is prohibited are denoted by the shaded rectangles.

The total current supplied by the regulators is therefore 20% higher than the total current demand of the circuit. The results of the placement process are shown in Fig. 8.12. Observe that the regulators are more uniformly spread within the layout. Multiple voltage regulators placed in close proximity provide less current since each regulator sources current to fewer loads. A higher current demand is therefore experienced by the remaining regulators, potentially causing these regulators to operate at maximum capacity. Inadequate current supplied by these regulators degrades the voltage within the grid, incentivizing the optimization algorithm to spread the clustered regulators throughout the IC. The voltage drop and runtime are listed in Table 8.3. Note that the computational runtime of the algorithm is greater due to the additional processing to consider the limited current of the regulators.

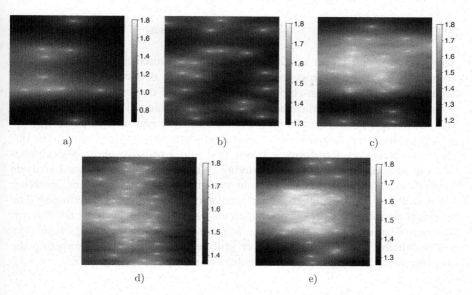

Fig. 8.11 Voltage after distributing a) 10, b) 20, c) 30, d) 40, and e) 50 voltage regulators within the ibmpg4 benchmark circuit. The regulators are not placed outside the restricted zones depicted in Fig. 8.10. The voltage drop is approximately 10% greater than in the unrestricted case study.

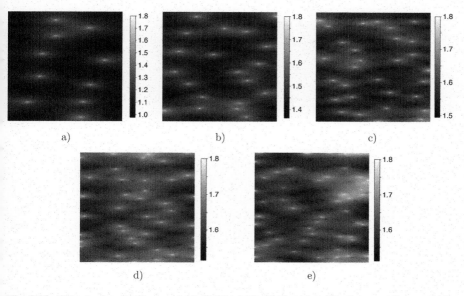

Fig. 8.12 Voltage after distributing a) 10, b) 20, c) 30, d) 40, and e) 50 voltage regulators within the ibmpg4 benchmark circuit. The regulator current is restricted according to (8.32). The voltage drop is approximately 24% greater than in the unrestricted case study.

8.6   Conclusions

To tackle stringent power quality and efficiency requirements of modern VLSI complexity systems, heterogeneous power regulation is necessary, incorporating both off-chip as well as on-chip point-of-load voltage converters. The number of voltage regulators is however limited due to area constraints. The on-chip regulators should therefore be strategically placed to maximize the system-wide power quality. A novel voltage regulator allocation algorithm is presented in this chapter. The on-chip power network transforms a power grid into a resistive mesh. The number of loads is reduced by applying clustering. With the proposed fast grid analysis method, the on-chip power distribution system is efficiently analyzed, enabling the position of the distributed voltage regulators to be optimally determined. The proposed algorithm does not depend upon the size of the grid, enabling the efficient analysis of large scale power networks. In three case studies, the voltage regulators are distributed within industrial power grid benchmark circuits, minimizing the parasitic voltage drop given a fixed number of on-chip voltage regulators.

Chapter 9
Exploratory methodology for power delivery

Power delivery is pivotal to the performance of modern integrated systems [381]. Violating limitations in power delivery such as load voltage drop, thermal characteristics, and power dissipation, may cause a variety of issues, such as circuit malfunction or overheating. Due to the high level of complexity in modern systems, it is difficult to monitor power delivery characteristics throughout the system development process [65]. This approach adds risks to the entire development flow. Unsatisfied power quality constraints at later stages of the design process may require unacceptable time and resources.

One strategy for reducing the burden of modifying the power network is overdesign, such as using additional interconnections and pins for power or larger and more numerous decoupling capacitors. This strategy increases cost and allocates less metal and pin resources for signaling, and less area for the functional circuitry [528]. In addition, external factors, such as cooling power or cost, shift the resulting system even farther from the optimal objective.

Numerous works on power delivery optimization at varying levels of abstraction exist in the literature. On-chip voltage regulation is discussed in [359, 495, 511, 529]. In [511], a framework for combining switching and linear regulators within a single system is presented that combines high efficiency linear regulators with superior regulation characteristics in switching converters. Power management has been deeply investigated from an architectural perspective. The work of [530] presents a framework for system-wide dynamic voltage scaling with thermal considerations that improves overconstrained circuits based on worst case scenarios. In [531], the GradualSleep strategy has been proposed to minimize on-chip static energy dissipation. More recent works describe paradigms suitable for modern circuit-level power management solutions. A system-level framework for optimizing decoupling capacitor and parasitic inductance is proposed in [145, 532]. A system-level power management system is described in [533], where the electrical and thermal characteristics are monitored to make appropriate adaptations, such as dynamic voltage and frequency scaling (DVFS) based on system temperature and workload.

Despite the maturity of the field, power delivery in VLSI systems is rarely approached from a constrained optimization perspective. In [534], quadratic programming methods are exploited to reduce the impedance profile of the power delivery network at frequencies of interest by sacrificing the impedance at less relevant frequencies. More recent work [535] utilizes differential evolutionary optimization to suggest the impedance profile of a physical structure. A significant omission in the literature is the almost exclusive focus on optimizing the electrical parameters, only indirectly addressing external metrics such as power and cost. Constrained global optimization provides a natural framework for design exploration of power delivery systems. The primary strength of the proposed technique is flexibility, allowing different design objectives and constraints to be considered including thermal and cost parameters. The subsequent sections provide a deeper insight into this proposed methodology. In Section 9.1, the necessary components of the proposed framework are described. Two case studies are presented in Section 9.2 to demonstrate the validity and discuss the strengths and limitations of the proposed approach The chapter is concluded by a summary in Section 9.3.

9.1   Optimization framework

The standard design process in the absence of power network design exploration is shown in Fig. 9.1 [536]. Due to the lack of preliminary information, power delivery network analysis is performed during the placement and routing stage [536]. If the circuit does not comply with power quality and voltage drop objectives, the power network is changed or resynthesized. The verification and redesign processes repeat until the resulting power network satisfies the required specifications. Due to the significant time required to evaluate and refine the power delivery network at the system level, multiple design iterations at later stages of the development process are highly undesirable, as these changes may cause delays on the order of days.

To mitigate potential losses, the number of power network redesigns needs to be minimized, preferably to zero. Power delivery exploration can provide valuable guidelines for power network synthesis, bringing the resulting system close to the optimal state. Two important characteristics of the early design stages are worth noting. First, the lack of accurate electrical data creates a high degree of uncertainty in the power network development process. The assumptions made at this stage are crucial. Second, before the primary design parameters are fixed, a high degree of flexibility exists. For example, the number of voltage domains may significantly affect the efficiency of the system at the expense of additional metal resources or increased power noise. Exploiting these tradeoffs is crucial to unlocking the full potential of the overall power delivery system.

Fig. 9.1 Conventional IC
development process [536]

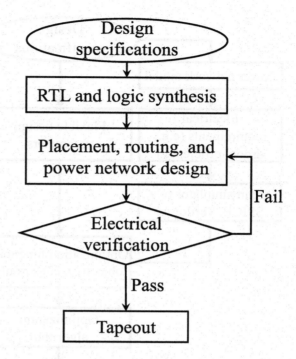

The proposed power delivery exploration process is illustrated in Fig. 9.2. The process is general and varies greatly with different inputs. The process starts with the analysis of the design specifications. A model of the power network is used to estimate the electrical metrics. Non-electrical metrics of interest are also identified and certain design flexibilities are identified. After the required components are characterized, the functions are passed to optimization algorithms. The result of the optimization process is a set of design guidelines that ensure proper operation without excessive overdesign. A more detailed explanation of the proposed exploration process is provided in the following subsections.

9.1.1 Specification of the electrical design requirements

A model of the power delivery network consists of four components: topology, voltage sources, load currents, and impedances. The topology reflects the relative placement of the elements within the netlist, supporting a comprehensive circuit analysis process. Technology information, such as the number of metal layers or interconnect conductivity, and design specifications, such as the interconnect dimensions, determine the parameters of the power network model [384]. One of the simplest and most widespread power network models is the hierarchical

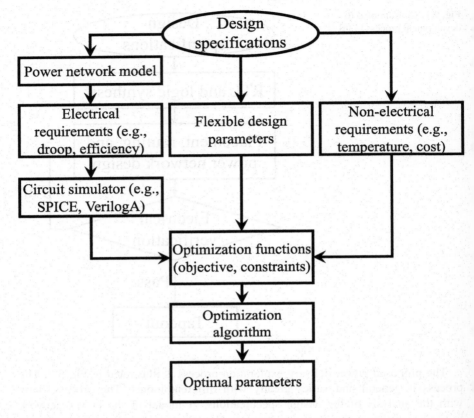

Fig. 9.2 Proposed power network optimization process

model shown in Fig. 9.3 [381], composed of cascaded lumped sections consisting of series RL segments, representing the interconnects and solder bumps, interleaved with parallel RLC segments, representing the decoupling capacitors, with an equivalent series resistance and inductance. More advanced topologies are necessary to evaluate the information from lower abstraction levels, such as the on-chip mesh [71, 72]. However, due to the lack of topology information during the early design phase, the development of a more accurate circuit model of a power network is a complex task.

The voltage source represents an idealized on-board regulator. For simplicity, a constant voltage supply is assumed. The main source of power consumption is modeled as a current source, representing the current delivered to the functional blocks, on-chip regulators, and leakage current. A current profile is necessary to evaluate the reliability of the network. Functional block information is used to model the profile of the load current [537]. Alternatively, the current profile may be modeled as a constant average current with a worst case current pulse [536].

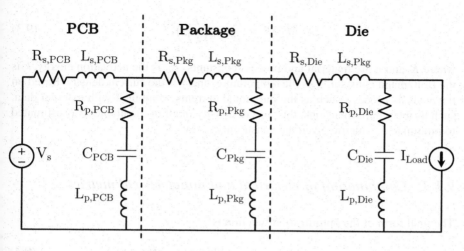

Fig. 9.3 Simplified model of power delivery network for optimization purposes

Once the power network model is determined, the design goals and technology limitations are converted into a functional form. For example, any limitations on voltage drop can be represented as

$$V_{drop} = \frac{\min(V_{Load}(t))}{V_s}, \tag{9.1}$$

where $V_{Load}(t)$ is the load voltage and V_s is the supply voltage. The power distribution efficiency, in turn, is

$$\eta = \frac{P_{Load}}{P_{in}}, \tag{9.2}$$

where P_{Load} and P_{in} are, respectively, the power dissipated by the current source and the total dissipated power. These specifications are necessary to convert the metrics of interest into the optimization functions.

9.1.2   Specification of non-electrical design requirements

In this chapter, non-electrical parameters are described as the system characteristics that are not directly inferred from the circuit model of the power network. These nonelectrical parameters include the on-chip temperature, manufacturing cost of the components, and area of the circuit elements. An externally supplied model is required to link the nonelectrical metrics and electric performance of the system. For example, if the mean time to failure (MTTF) is of concern, optimizing MTTF would place an upper limit on the current density and temperature, as shown in [538],

$$MTTF = \frac{K}{j^n}\exp\left(\frac{E_a}{kT}\right),\tag{9.3}$$

where K and n are material and process constants, E_a is the activation energy, k is the Boltzmann constant, T is the temperature, and j is the current density. Based on the analysis process, such as the individual currents, combined with external data, such as the wire dimensions, the current density in all of the elements is estimated to minimize this metric given the constraints.

9.1.3 Combination of electrical and nonelectrical metrics

The final form of the optimization function is

$$\mathbf{x}_{opt} = \min\left(f\left(\mathbf{x}\right)\right), \quad \text{subject to} \quad c(\mathbf{x}) \le 0,\tag{9.4}$$

where $\mathbf{x}$ and $\mathbf{x}_{opt}$ are variables and correspond to the optimal parameter vectors, $f(\mathbf{x})$ is the function being optimized, and $c(\mathbf{x})$ is a set of constraint functions. The power delivery exploration process is formulated as in (9.4) to allow the application of constrained optimization algorithms.

The electrical analysis process needs to provide sufficient information to allow the nonelectrical metrics to be evaluated. The comprehensive optimization function requires an expression of the external metrics in terms of the variable parameters, electrical metrics, or both. For example, with adaption of [539], the MTTF of the interconnect segment can be approximated in terms of the interconnect dimensions and current,

$$MTTF = \frac{K_1 W^n H^n}{I_{rms}^n}\exp\left(\frac{K_2 W^2 H^2}{I_{rms}^2}\right),\tag{9.5}$$

where W and H are, respectively, the interconnect width and thickness, I_{rms} is the RMS current through the segment, and K_1, K_2, and n are process and material related constants. Electrical metrics, such as the RMS current through the segment, are evaluated from simulations of the power network. The variable parameters determine the characteristics of the power network model. For example, the dimensional parameters can be used to determine the impedance of the circuit elements. The formulated metrics are combined to create the objective function and set of constraints.

If multiple design objectives exist, a weighted sum of each objective is used to minimize each objective. The resulting formulation is shown in (9.6) to (9.9), where V_s is the supply voltage, W and H are, respectively, the top level interconnect width and thickness, w_1 and w_2 are weight parameters, $A_{int}(\mathbf{x})$ is the total area of the metal expended for the interconnect, and $V_{drop,max}$ and η_{min} are design constraints on, respectively, the voltage droop and efficiency. The objective function is the weighted sum of the MTTF and cost, minimizing both metrics. To be satisfied, both $c_1(\mathbf{x})$ and $c_2(\mathbf{x})$ need to be greater than or equal to 0, ensuring that the droop is not larger than $V_{drop,max}$ and the efficiency is not less than η_{min}.

$$\mathbf{x} = [V_s, W, H], \tag{9.6}$$

$$f(\mathbf{x}) = \frac{w_1}{MTTF(\mathbf{x})} + w_2 A_{int}(\mathbf{x}), \tag{9.7}$$

$$c_1(\mathbf{x}) = V_{drop}(\mathbf{x}) - V_{drop,max}, \tag{9.8}$$

$$c_2(\mathbf{x}) = \eta_{min} - \eta(\mathbf{x}). \tag{9.9}$$

9.1.4 Circuit simulation procedure

During the optimization process, the circuit parameters are varied and the corresponding electrical parameters are evaluated. An efficient circuit simulator is the cornerstone of this procedure as the quality and timeliness depend upon the speed and accuracy of the simulator. Two simulation methods are utilized. The first method is commercial HSPICE [540] which requires a special interface with the programming language. The primary advantage of this approach is the versatility of the simulator. With the variety of available models, a wide range of circuits can be simulated and, therefore, optimized. The disadvantage of this approach is the communication overhead between the programming language and HSPICE which dramatically increases the simulation time.

Another approach is a custom Laplace transform-based simulator, requiring no interface with the programming language. The Laplace transform is widely used for simulation and optimization of linear circuits and systems [541, 542]. The primary advantage of this approach is the higher speed of the simulation due to the lack of communication with an external language and application-specific code optimization. A significant limitation is the narrow applicability of the method - only linear systems can be simulated using this approach due to the Laplace transform. A variety of methods exist, however, to extend the Laplace transform to nonlinear circuits. In [541], the switching transistors are replaced with lumped RC elements. A piecewise-linear model is another common approach for applying Laplace transforms to nonlinear systems. This method is particularly compatible with sequential switching [543, 544]. A modification of the Laplace Transform applicable to a certain class of nonlinear systems is introduced in [545]. Incorporating this method into the proposed framework may significantly extend the applicability of the proposed tool.

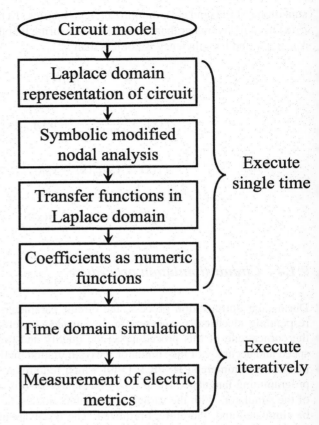

The proposed optimizer is applied to a model of a power network, which typically consists of passive RL-RLC branches [381]. The active devices, such as a voltage regulator or load transistors, are replaced with equivalent linear models to offset the error due to the assumption of linearity, which enables the use of a Laplace transform-based optimizer. In cases where the power network model is nonlinear (e.g., a power gated network), typically slower, numerical simulation tools can be utilized, such as HSPICE [540] or Verilog-AMS [546]. The choice between an active and passive power network model, therefore, becomes a tradeoff between accuracy and computational speed.

The Laplace transform-based process is shown in Fig. 9.4. The circuit elements are represented in the s domain. The fixed parameters are expressed numerically, while the variables are represented as symbolic variables. For instance, the impedance of a capacitor with a variable capacitance, fixed equivalent series resistance of 1 mΩ, and fixed equivalent series inductance of 10 pH can be presented as

$$Z_c = 1\text{m}\Omega + 10\text{pH} \times s + \frac{1}{Cs}, \tag{9.10}$$

where the capacitance C is shown as a symbolic variable, Z_c is the equivalent impedance of the capacitor, and s is the Laplace domain parameter.

After the circuit elements are expressed in the Laplace domain, a modified nodal analysis is applied. The circuit is modeled in terms of six input matrices, representing connections and parameter values, as shown in [64]

$$\begin{bmatrix} Y & B \\ C & D \end{bmatrix} \begin{bmatrix} V \\ I \end{bmatrix} = \begin{bmatrix} J \\ F \end{bmatrix}, \tag{9.11}$$

where V and I are, respectively, the node voltages and currents through the voltage sources, Y is the matrix of nodal admittances, while B, C, D, J, and F encode current and voltage sources, including controlled sources. The constructed matrix equation is solved for $[V, I]^T$.

The resulting vector represents the node voltages and source currents in terms of symbolic parameters in the Laplace domain. Dividing the resulting vectors by the source produces the transfer function, as shown in

$$H(s) = \frac{b_n s^n + \ldots + b_0}{a_m s^m + \ldots + a_0}. \tag{9.12}$$

The coefficients of the transfer function are expressed as a function of the variable parameters,

$$b_i = f_{i,num}(\mathbf{x}), \tag{9.13}$$

$$a_i = f_{i,den}(\mathbf{x}). \tag{9.14}$$

While the aforementioned procedure is computationally expensive, requiring a solution of the symbolic matrix system, the process only needs to be performed once for a particular circuit topology. Modifications of the variable parameters only change the value of the coefficients, $b_n \ldots b_0 \, a_n \ldots a_0$, while the symbolic representation remains intact. The speedup due to the proposed simulator is, therefore, largely dependent upon the number of iterations N during the optimization process. The speedup is estimated as

$$Speedup = \frac{t_n}{\frac{t_{setup}}{N} + t_{\mathcal{L}}}, \tag{9.15}$$

where t_n and $t_{\mathcal{L}}$ are the time per iteration using, respectively, numerical analysis and the Laplace transform-based simulator, and t_{setup} is the time required to determine the transfer function (9.12). Note that typically $t_{setup} > t_n > t_{\mathcal{L}}$, thus the speedup converges to a positive value with large N, while approaching zero with small N. Since most optimization procedures require a large number of iterations to determine the global minimum, the creation of a symbolic transfer function represents a negligible fraction of the total computational time.

To simulate the transfer functions and extract the numeric data, the coefficients of the transfer functions of interest are calculated and converted into a state space model. A variety of efficient state space model simulation packages are available, such as LAPACK [547] and LTITR [548]. The input waveform and state space model are passed to the simulators to calculate the output waveform. This approach achieves significant speedup as compared to conventional, purely numerical algorithms. Applying a state-space model, the output waveform can be determined without solving the matrix equation during each time step. Conversion of a circuit into a matrix form is performed only once, greatly reducing the computational overhead. With the large number of circuit simulations during the optimization process, significant optimization speedup is achieved, as described in Section 9.2.

9.2 Case studies

Two practical case studies are presented in this section. Allocation of area for decoupling capacitors within a single rail system is analyzed in Subsection 9.2.1. The cost of decoupling capacitor placement is minimized while satisfying power consumption and the voltage droop constraints. The framework is then applied to a multi-rail system to determine the optimal number of voltage domains as described in Subsection 9.2.2.

9.2.1 Single rail system

A typical power network represented by serially cascaded RL branches and parallel RLC branches is shown in Fig. 9.5. A three-level power network including the PCB, package, and die levels is considered here. The series resistance and inductance of the power network are assumed fixed. The on-die parallel inductance is neglected assuming point-of-load on-die decoupling capacitors with small inductance [511]. The profile of the load current has been adapted from [536] and shown in Fig. 9.6(a). The load current profile models the fluctuations of the workload during system operation. The supply voltage is used as a design variable to explore the effects of supply voltage on system performance. Other controllable parameters are the number and magnitude of the decoupling capacitors within the PCB, package, and die levels. Minimization of the decoupling capacitor placement cost is the primary objective of this case study, subject to power consumption, power quality, and frequency requirements.

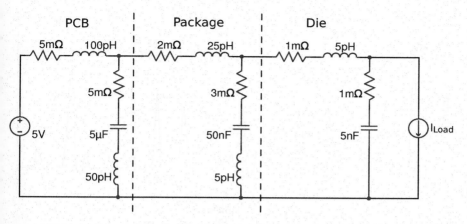

Fig. 9.5 Model of 1-D power delivery network with initial parameters

9.2.1.1 Optimization setup

The cost of each system level (PCB, package, die) is assumed to be a function of the physical area which is affected by the area of the decoupling capacitors. The decoupling capacitor placement cost Q_{die} is

$$Q_{die} = w_{die} A_{die}, \qquad (9.16)$$

where A_{die} is the area of the on-chip decoupling capacitor and w_{die} is the cost of the unit on-die area. The total cost of the decoupling capacitors is therefore

$$Q = \frac{1}{\varepsilon_0} \sum_{i \in S} \frac{w_i C_i d_i}{\varepsilon_i}, \qquad (9.17)$$

where S is the set of levels in the system (e.g., PCB, package, and die), ε_0 is the permittivity of free space, C_i is the parallel plate capacitance at level i, and d_i and ε_i are, respectively, the insulator thickness and relative permittivity at level i.

The oxide thickness and dielectric constant are described in [549–551]; however, the cost per area is not as clear. Based on the review of publicly available cost information [552–556], the cost per unit area of a package is approximately 3 to 6 times greater than the cost of unit PCB area, and approximately 3 to 10 times lower than the cost of unit die area. To simplify the cost estimate, the cost per unit area of a PCB is normalized to 1, the package area cost is assumed to be 4.5, and the cost per unit on-die area is assumed to be 20.25, 4.5 times greater than the cost per unit area of the package. The normalized cost estimates used in this case study are listed in Table 9.1.

Note the important tradeoffs that affect the optimization process [381]. A higher supply voltage enhances the speed but significantly increases the power

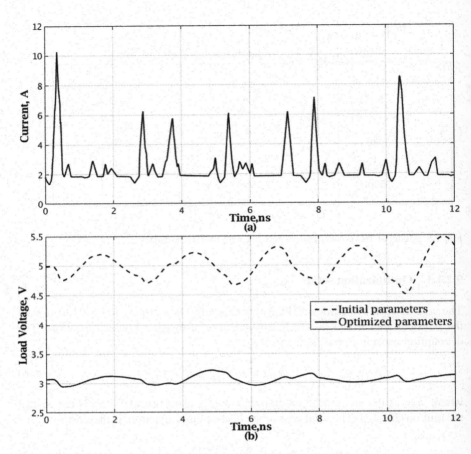

Fig. 9.6 Waveform of power network, a) load current adapted from [536], and b) load voltage with initial and optimized parameters

Table 9.1 Parameters of decoupling capacitor cost

Parameter	Die	Package	PCB
Cost per unit area, normalized	20.25	4.5	1
Insulator thickness	0.9 nm [549]	12 μm [550]	250 μm [551]
Insulator permittivity	3.9 [549]	4.6 [550]	4.5 [551]

consumption. Insertion of parallel decoupling capacitances is a powerful technique for reducing ripple currents since the high frequency components of the current bypass the load. Larger decoupling capacitors, however, require significant on-chip area, leading to greater system cost.

The target constraint metrics are power consumption, power quality, and speed. The power consumption is directly measured through simulation, and the corresponding constraint function is

$$c_1(\mathbf{x}) = P - P_{max}, \tag{9.18}$$

where $c_1(\mathbf{x})$ is the initial constraint function, P is the measured power, and P_{max} is the upper bound on the power consumption. Since the constraint function is negative, (9.18) ensures that the power dissipation does not exceed the maximum allowable power level.

For frequency, the constraint is

$$t_{p,CP} \leq T_{min}, \tag{9.19}$$

where $t_{p,CP}$ is the propagation delay of the critical path and T_{min} is the lower bound on the clock period. Evaluation of this metric, however, is computationally expensive and requires identification of the critical paths and extensive parameter extraction. This level of precision is typically not available during the early stages of the design process. In this case, accuracy is sacrificed for higher computational efficiency. The load voltage is, therefore, used as the speed metric,

$$c_2(\mathbf{x}) = V_{min} - \min(V_{load}(t)), \tag{9.20}$$

where $V_L(t)$ is the instantaneous voltage at the load, and V_{min} is the minimum voltage to maintain reliable high speed operation.

The third design constraint is power quality, described as voltage fluctuations, and is formulated as

$$c_3(\mathbf{x}) = \frac{\max(V_{load}(t)) - \min(V_{load}(t))}{V_{rail}} - \Delta V_{max}, \tag{9.21}$$

where V_{rail} is the supply voltage, and ΔV_{max} is the maximum allowed fluctuation. The optimization constraints are listed in columns two and three of Table 9.2.

9.2.1.2 Optimization results

The Interior Point Algorithm, part of MATLAB Optimization Toolbox [557] and HSPICE [540], is used in this case study. The optimization functions, circuit parameters, and external parameters are inputs to the optimization algorithm. The optimization procedure has been run on an Intel Core i7-6700 3.40 GHz 8-core computer using different initial conditions to avoid any local minima. The initial parameters that produce the lowest cost under specified constraints are listed in column four of Table 9.2.

Table 9.2 Optimization constraints, with initial and optimal parameters

	Lower bound	Upper bound	Initial value	Optimized value
Supply voltage	1.4 volts	10.0 volts	5.0 volts	3.09 volts
PCB decap	25.0 nF	10.0 μF	5.00 μF	2.71 μF
Package decap	50.0 pF	100 nF	50.0 nF	9.77 nF
Die decap	2.00 pF	10.0 nF	5.00 nF	9.32 nF
Minimum load voltage	1.40 volts	—	2.96 volts	2.94 volts
Power dissipation	—	10.0 watts	10.6 watts	6.51 watts
Load voltage	—	10.0%	19.3%	9.07%
Normalized cost	—	—	0.317	0.270

The optimization process is completed in 28 seconds, requiring 66 function evaluations to converge. The load voltage waveforms are shown in Fig. 9.6(b). The power network initially exhibits an underdamped response, resulting in relatively large droops and overshoots. After optimization, the voltage fluctuations are reduced in the optimized power network by choosing an appropriate decoupling capacitor. The reduction in the load voltage fluctuations allows the supply voltage to be scaled since fluctuations are less likely to drop below the minimum allowed level. Reducing the supply voltage, in turn, leads to lower power dissipation.

The optimization results are listed in column five of Table 9.2. As compared to the initial suboptimal parameters, the cost has decreased by almost 15% from 0.317 to 0.270. The initial parameters do not satisfy the power dissipation and load voltage constraints. A 38.6% reduction in power consumption is achieved, from 10.6 watts to 6.51 watts. Most of the reduction in power originates from the reduced supply voltage, from 5 volts to 3.09 volts. In addition, a 53% decrease in fluctuations is achieved, from 19.3% to 9.07%. As a result, the optimized parameters satisfy the target requirements, including the power and voltage constraints.

9.2.2   Multiple rail system

The problem of choosing the optimal number of rails is an important power delivery exploration issue. Utilizing several voltage domains may bring considerable savings in terms of power, while achieving performance goals [361]. At early stages of the design process, planning the circuit topology is problematic since the resulting power delivery characteristics are difficult to estimate in advance. In particular, it is unclear whether the power network is sufficiently conductive to satisfy voltage droop requirements. Separation of the low voltage circuitry from the rest of the IC is an attractive option to reduce power consumption due to the quadratic

Table 9.3 Voltage domain specifications of power delivery network adapted from [558]

Power network	Rail #	Voltage, V		Current, mA		Peak slew rate, A/μs	Function
		max	min	max	min		
A	A1-4	1.42	0.97	5,830	416	1,000	CPU core
	A5	1.20	0.99	3,150	225	500	GPU
	A6	1.33	1.00	10	1	500	USB
	A7	1.93	1.67	10	1	500	GPS
	A8	1.93	1.72	30	1	500	DSP
	A9	1.93	1.67	10	1	500	Camera
	A10	1.93	1.67	10	1	500	Audio
	A11	1.93	1.67	1,500	58	500	LTE+WiFi
	A12	1.55	1.00	3,150	225	500	Memory
B	B1-4	1.42	0.97	5,830	416	1,000	CPU core
	B5	1.20	1.00	3,160	226	*	GPU+USB
	B6	1.93	1.67	1,500	58	500	LTE+WiFi
	B7	1.93	1.72	60	4	*	GPS+DSP+ Camera+Audio
	B8	1.55	1.00	3,150	225	500	Memory
C	C1	1.42	1.00	26,470	1,889	*	CPU+Memory
	C2	1.20	1.00	3,160	226	*	GPU+USB
	C3	1.93	1.72	1,560	62	*	GPS+DSP+Camera +Audio+LTE+WiFi

relationship between power consumption and operating voltage. The scaled voltage is, however, less robust to sudden load current fluctuations, possibly violating droop requirements, allowing the device to malfunction. Moreover, utilizing separate power networks requires less metal resources for each rail, resulting in a power delivery network exhibiting higher impedance.

To investigate this problem, three power networks are considered, twelve rail (A), eight rail (B), and three rail (C) systems. The impedance characteristics of these networks are based on [558] and assume the power network topology shown in Fig. 9.3. The rail specifications are listed in Table 9.3. The maximum and minimum voltages represent the range of allowed values of the voltage. The model of the load current is a worst case triangular current waveform [532].

In system B, the rails with the closest voltage levels are merged to minimize energy losses due to the voltage conversion process. Rail A5 is merged with rail A6 to produce rail B5, and rails A7 through A10 are merged into rail B7, resulting in the eight rail system B. Further, rails B1 to B4 and B8 are merged, while rail B6 is merged with rail B7 to produce the three rail system C. The variables are the voltage supply of each rail, as well as the decoupling capacitance at each level of each rail.

For simplicity, the power rails are assumed to be mutually isolated, allowing each rail to be evaluated separately.

The objective of the design exploration process is to determine the set of rails that delivers the lowest possible cost of decoupling capacitance area. The objective function of the multiple rail system is adapted from (9.17),

$$Q = \frac{1}{\varepsilon_0} \sum_{j \in D} \sum_{i \in S_j} \frac{w_i C_i d_i}{\varepsilon_i}, \tag{9.22}$$

where D is the set of rails (voltage domains), and S_j is the set of layers of the power network (printed circuit board (PCB), package, or die) within the rail j.

Moving the decoupling capacitance farther from the load makes the system more vulnerable to inductive noise [493], limiting the cost benefits of a small on-chip capacitance. The greater fluctuations in the load voltage result in a need for a higher voltage supply to offset the potential voltage droops, resulting in higher power consumption. In addition, the inductive system response may result in significant overshoots [354] that may damage the transistors. For each rail in D, the aforementioned tradeoffs are expressed as constraint functions, as shown in (9.23) to (9.25),

$$c_1(V_s, C_{PCB}, C_{Pkg}, C_{Die}) = V_{load,min} - \min(V_{load}(t)), \tag{9.23}$$

$$c_2(V_s, C_{PCB}, C_{Pkg}, C_{Die}) = \max(V_{load}(t)) - V_{load,max}, \tag{9.24}$$

$$c_3(V_s, C_{PCB}, C_{Pkg}, C_{Die}) = P_{total} - P_{max}, \tag{9.25}$$

where $V_{load}(t)$ is the waveform of the load voltage, $V_{load,min}$ and $V_{load,max}$ are, respectively, the minimum and maximum bounds on the load voltage, and $Power_{total}$ and $Power_{max}$ are, respectively, the total power consumption and upper limit on the consumed power. The constraint functions place strict requirements on the quality of the power rails. If the voltage waveform violates the constraint functions, the objective function (or cost) is severely penalized, invalidating the result.

The power network model used in this case study does not include any nonlinear elements. A Laplace transform-based simulator has therefore been chosen. Particle swarm optimization is chosen as the optimization algorithm due to the robustness and efficiency characteristics of this algorithm. The optimization procedure is run on an eight core 3.40 GHz Intel Core i7-6700 machine. The results for 23 separate rail configurations are obtained in 26 minutes, with an average time of 67 seconds per rail. The results of the optimization are shown in Fig. 9.7. Note that the lowest value of the objective function is achieved with eight rails. In the eight rail and twelve rail scenarios, certain rails (e.g., rails seven to eleven in the twelve rail scenario) do

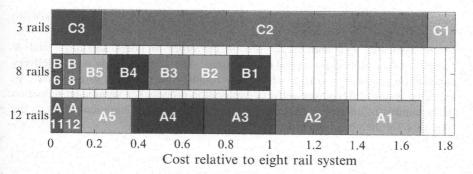

Fig. 9.7 Decoupling capacitor placement for three power delivery networks

not require decoupling capacitors due to the low load currents and high tolerance to variations.

To evaluate the benefits of the Laplace Transform optimization process, a similar optimization is performed using HSPICE [540]. The optimization results are identical to those results obtained from the Laplace transform optimization process due to the absence of nonlinear elements in the model. The total computational time, however, is 265 minutes, ten times greater than the Laplace simulator.

Distribution of the decoupling capacitor costs across the voltage domains normalized to the least expensive system is shown in Fig. 9.7. Certain patterns can be inferred. Comparing the eight and twelve rail systems, allocation of metal resources for separate power rails is unjustified from a cost perspective. The higher contribution of the CPU cores (A1 to A4) in the twelve rail network indicates that voltage fluctuations in this network are greater due to less metal resources allocated to each CPU rail, as compared to the eight rail system. The combination of rails A5 and A6 allocates more metal resources for both networks, resulting in reduced decoupling capacitor cost in combined rail B5.

As compared to the three rail system, where rails B1 to B4 and B8 (CPU cores and memory) are merged into a single voltage domain, the three rail system requires a large decoupling capacitance for the combined rail C2. The reason for the increased decoupling capacitance is the poor compatibility between voltage ranges. While rails B1 to B4 require a range of 0.97 to 1.42 volts, rail B8 has a range of 1.00 to 1.55 volts. The combined rail, therefore, needs to satisfy both ranges and is effectively shrunk to 1.00 to 1.42 volts, placing greater limitations on the voltage fluctuations. The narrow voltage range is compensated by placing a larger on-chip decoupling capacitance, increasing the overall cost of the power network.

A conventional power network design process may require a series of late design backtracking iterations to satisfy target noise performance requirements [559, 560]. Assuming that the post-floorplan power network model requires time t_{sim} for simulation and $t_{correct}$ for hotspot correction, and N iterations are required to reach the acceptable characteristics, the total time for the power integrity analysis process without early exploration is

$$t_{noEE} = (N - 1)t_{sim} + Nt_{correct}, \tag{9.26}$$

where, typically, t_{sim} and $t_{correct}$ are on the order of hours and days, and N typically ranges between two and ten iterations. Alternatively, early power delivery exploration requires time t_{exp}, which may require several hours to complete. An expected result of the power delivery exploration process is a significant reduction in the number of iterations. Assuming the updated number of iterations is N_{new}, the total time for the power integrity analysis process is

$$t_{EE} = t_{exp} + (N_{new} - 1)t_{sim} + N_{new}t_{correct}. \tag{9.27}$$

The savings in time due to the early power integrity analysis process is

$$t_{noEE} - t_{EE} = (N - N_{new})(t_{sim} + t_{correct}) - t_{exp}, \tag{9.28}$$

therefore, to ensure that the power delivery exploration is justified from the perspective of computational time, the following condition must be satisfied:

$$(N - N_{new})(t_{sim} + t_{correct}) > t_{exp}. \tag{9.29}$$

Note that typically $t_{sim} + t_{correct} > t_{exp}$, therefore, to justify early design exploration, it is sufficient to reduce the number of post-floorplan backtracking iterations, i.e., $N_{new} < N$.

The proposed early power delivery exploration framework may reduce the number of costly iterations by providing an estimate of the optimal parameters at an earlier phase of the development process, shrinking both time and labor. The non-electrical parameters, such as area and cost, are combined with the electrical parameters to produce a system with minimum cost while satisfying target performance metrics. This approach provides useful information for early system exploration, allowing more effective design decisions to be made.

Several limitations of the proposed framework exist. First, the computational time largely depends upon the circuit simulator. Therefore, optimization of more complex circuits with a larger number of nodes may require significant computational time. A Laplace transform-based simulator is proposed for optimization of linear circuits. The speedup due to the Laplace transform-based simulator, however, largely depends upon the number of iterations during the optimization process. Second, a function for the metrics of interest needs to be determined to conduct the power delivery exploration process. Practical assumptions, therefore, need to be made to achieve useful results. An issue of premature convergence exists, resulting in the optimization converging to a local minimum rather than a global minimum [527]. It is, therefore, necessary to ensure that the design space is thoroughly explored, for example, by increasing population sizes (evolutionary algorithms), mutation and migration rates (genetic algorithm), swarm velocities and inertia (particle swarm), and the initial temperature and frequency of reheating (simulated annealing).

9.3 Conclusions

A versatile methodology for power delivery design exploration is described in this chapter. The primary strength of the framework is applicability to a wide range of objectives and constraints, including external, non-electrical parameters. The procedure supports the application of robust, general purpose algorithms to solve power delivery problems. A fast, optimization oriented Laplace transform-based simulator is described. Limitations of the proposed framework include the dependence on the computational time of the circuit simulator, the need for optimization functions during the preliminary design stages, and careful tuning of the optimization algorithms. The effectiveness of the framework is demonstrated by a case study, where the appropriate power delivery network is chosen among existing options.

Chapter 10
SPROUT - Smart Power ROUting Tool for board-level exploration and prototyping

Modern high performance VLSI systems require stable power [384]. Voltage scaling combined with shrinking interconnect dimensions and increasing current consumption result in significant power noise, degrading power integrity [561]. Fast transition times significantly broaden the spectrum of the power noise [66]. Different strategies are employed at the die, package, and board levels to mitigate this power noise. The board-level power delivery network is a crucial component of the power delivery system, connecting the power management integrated circuit (PMIC) with the die or package. Careful design of the board level power delivery system is crucial for connecting the power management IC with the package or die as well as the on-board decoupling capacitors.

The flow of the power delivery design process for printed circuit boards (PCB) is illustrated in Fig. 10.1. The quality and cost of the PCB is governed by a set of system-level parameters, such as the location and model of the components, and the number and thickness of the metal layers. These parameters affect the floorplan and placement of the components. After the location of the components is known, the power management IC is connected to the target ball grid array and decoupling capacitors. If the impedance profile of the resulting layout does not satisfy the target requirements, the layout is iteratively adjusted. These adjustments range from minor changes to the routed shape to altering the entire floorplan. Several iterations are often necessary to comply with the target impedance requirements [562].

The influence of the system parameters on power integrity and cost is qualitatively well understood. For example, adding decoupling capacitors would likely reduce the inductive noise while adding cost [75, 510]. Quantifying these effects prior to floorplanning and routing is however difficult. Due to the lack of information at early stages of the system design process, the system level parameters are often arbitrarily chosen. These power delivery systems may fail to satisfy target impedance requirements, leading to a costly redesign process. Early exploration of the design space may eliminate or decrease the number of layout adjustments at later stages of the design process.

© The Author(s), under exclusive license to Springer Nature Switzerland AG 2023   257
R. Bairamkulov, E. G. Friedman, *Graphs in VLSI*,
https://doi.org/10.1007/978-3-031-11047-4_10

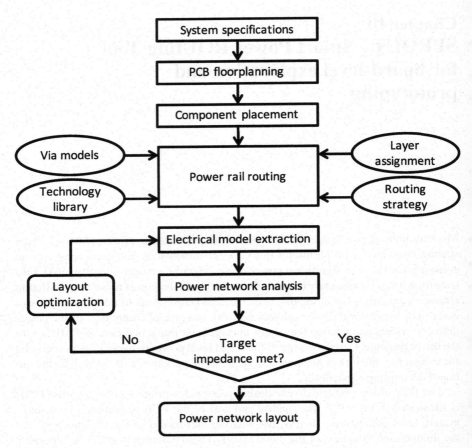

Fig. 10.1 Conventional design flow for power delivery networks for printed circuit boards.

The objective of the proposed Smart Power ROUTing algorithm for printed circuit boards (SPROUT) is to produce a prototype of the power network based on a target set of design parameters (see Fig. 10.2). The resulting layout is suitable for impedance extraction. Therefore, the impedance of the layout based on the target set of design parameters may be efficiently and automatically evaluated. This capability supports a more rigorous evaluation of the design space and better exploration of design tradeoffs, such as cost and performance. An informed choice of design parameters early in the development process reduces the likelihood of not satisfying the target impedance. In addition, the layout prototype may guide the final layout, further accelerating the development process.

Unlike automated signal routing, which is extensively studied in the literature, the automated *synthesis* of board-level power nets has received minimal attention [65]. Most works in the literature focus on the analysis of existing power delivery networks. For example, in [468, 563–565], fast methods for estimating the

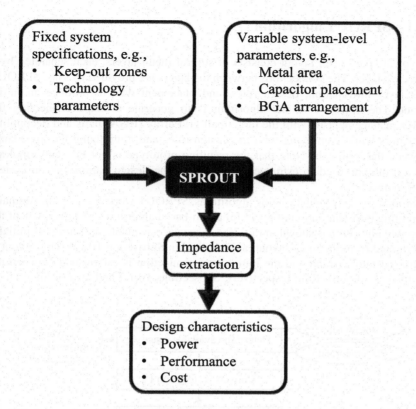

Fig. 10.2 Proposed prototyping flow for printed circuit boards using SPROUT. The PCB layout parameters are the inputs to SPROUT that produce a prototype of the power network. The parasitic impedance of the prototype is estimated. This process is repeated for different sets of system-level parameters. The power, performance, and cost of each prototype is evaluated and compared to other prototypes to determine the most favorable system parameters.

impedance of board-level power networks are described. In [566], a simplified circuit model is described to evaluate inductive power noise. An accurate PCB analysis methodology is proposed in [567] where the finite difference model is integrated with SPICE. Methods for enhancing electromagnetic compatibility and power integrity are discussed in [493, 568–571].

SPROUT is the first automated power network prototyping algorithm for PCBs. The remaining portion of this chapter is organized as follows. In Section 10.1, the power routing algorithm is described. The algorithm is validated using industrial case studies in Section 10.2. Some conclusions are provided in Section 10.3. A modification of SPROUT to support multilayer routing is presented in the Appendix C.

10.1 SPROUT algorithm

A typical board-level layout consists of several metal layers, each separated by a dielectric layer. The connections between the layers are provided by vias. SPROUT uses layer information, design rules, and placement data to produce an initial layout. The objective of the algorithm is to generate a shape connecting the power management IC with the target ball grid array (BGA) balls and decoupling capacitors while complying with the design rules and minimizing the impedance between the terminals. Note that the resulting prototype is not the final topology but a prototype used to estimate the effects of the design parameters on system performance.

Similar to many signal routing algorithms, SPROUT works in the graph domain, permitting the exploitation of powerful graph-based algorithms. An overview of the proposed algorithm is shown in Fig. 10.3. The space available for routing is initially determined from the input layout, as described in Subsection 10.1.1. This layout is converted into a graph and the initial seed connection is established between the terminals as described in, respectively, Subsections 10.1.2 and 10.1.3. SmartGrow

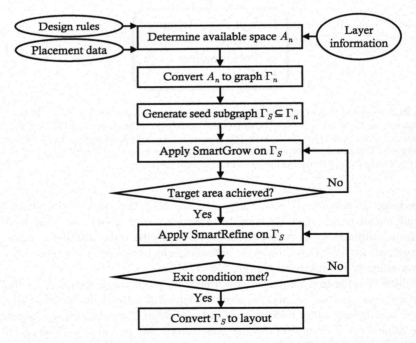

Fig. 10.3 Overview of SPROUT algorithm. The available space A_n is converted into an equivalent graph Γ_n. The subgraph seed Γ_n^s is generated by SPROUT and expanded using the SmartGrow algorithm described in Subsection 10.1.4. After achieving the target area, the nodes in Γ_n^s are rearranged using the SmartRefine algorithm to enhance the electrical characteristics. The final subgraph is converted into a physical layout.

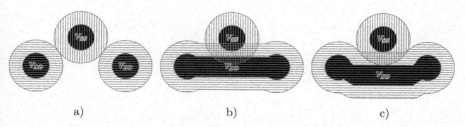

Fig. 10.4 One V_{SS} (vertical hatch), two V_{DD} (horizontal hatch) via pads (dark), and buffers (light).
a) Initial layout. b) The connection to the V_{DD} vias is invalid since the buffer around the V_{DD} connection overlaps the V_{SS} via, and the V_{DD} connection overlaps the V_{SS} via buffer. c) Example of valid routing. Neither the V_{DD} nor the V_{SS} buffer intersects the vias or connections of a different net. Note that the V_{DD} connection can be placed in the buffer around the V_{DD} vias because both the via and connection belong to the same net.

and SmartRefine algorithms are introduced in, respectively, Subsections 10.1.4 and 10.1.5. Using these algorithms, the impedance between the terminals is iteratively reduced by adding and rearranging the nodes. A subgraph reheating technique, inspired by simulated annealing, is proposed in Subsection 10.1.6 where the size of the graph is temporarily increased to reduce the probability of a suboptimal graph impedance. In Subsection 10.1.7, the placement of the resulting graph into the original layout is described. The complexity of SPROUT is discussed in Subsection 10.1.8.

10.1.1 Available routing space

An assessment of the available space commences with processing the input information. Each element of the layout is converted into a polygon with four parameters, layer, net, geometry, and buffer. To understand each component, consider three vias placed on the top layer of a PCB (see Fig. 10.4a). The via pads are converted into polygons. Assuming the vias are placed on layer 1, the layer parameter of the corresponding polygons is 1. Each capacitor pad is assigned a net, namely V_{DD} and V_{SS}. The geometry of each pad is expressed as an ordered set of coordinates. To decrease the likelihood or minimize the effect of manufacturing defects such as unintended shorts, spurs, underetches, and electromagnetic interference [572], each geometry is assigned a buffer. This buffer ensures polygons from different nets are properly spaced. To illustrate the buffering process, consider the example shown in Figs. 10.4b and 10.4c. Contact between the two V_{DD} vias is not possible using a straight interconnect segment because this segment intersects the buffer of the V_{SS} via, and the via intersects the buffer of the interconnect. The bent interconnect segment shown in Fig. 10.4c produces a valid connection since the geometries do not intersect the buffers of the other nets. Note that it is legal for a V_{DD} polygon to cross a V_{DD} buffer because these polygons belong to the same net.

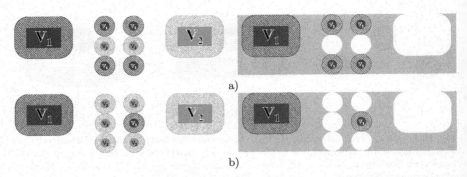

Fig. 10.5 Available space (shaded) for V_1 in two layouts. a) Layout (left) where routing from the pad on the left to four vias is possible, as evident from the connected available space (right). b) Layout (left) where connecting a pad with a via is not possible within a single layer due to the disjoint available space (right).

The entire design space U is initially viewed as available for routing. The available space A_n for a particular net n is determined by removing buffers of the other nets from the design space.

$$A_n = U \setminus \bigcup_{n_j \neq n} b_j. \tag{10.1}$$

Polygon removal is achieved by utilizing efficient polygon clipping algorithms [573, 574] that require negligible time, as discussed in Subsection 10.1.8. After removal, the available space on each layer may become disjoint, leaving no valid path between terminals on the same layer, as illustrated in Fig. 10.5. In this case, routing is accomplished using multiple layers. Based on the algorithm described in Appendix C, the multilayer routing problem is decomposed into several single layer routing problems.

10.1.2 Equivalent graph

Once the available space of the layout is determined, it is converted into an equivalent graph Γ_n. The available space A_n is divided into tiles a_n. Using a bijective map,

$$f : A_n \leftrightarrow \Gamma_n, \tag{10.2}$$

each tile a_n becomes a node γ_n within the graph. This mapping is recorded and used in the last stage of the algorithm to convert each node back into a tile. The

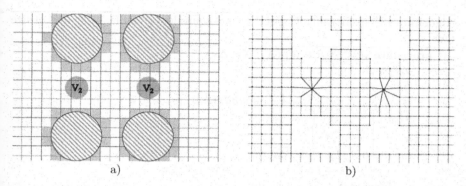

Fig. 10.6 Conversion of the available space for net V_2 into an equivalent graph. a) The available space is split into unit cells. Cells with irregular shapes are shaded. b) Equivalent graph. The tiles overlapping vias are treated as a single node. Nodes are not generated in prohibited areas.

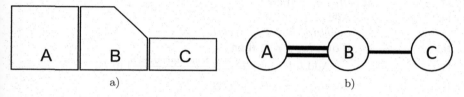

Fig. 10.7 Conversion of irregularly shaped tiles into equivalent graph. a) Tiles A and B have a twice wider contact than tiles B and C, and b) nodes A and B have double conductance as compared to nodes B and C.

dimensions Δ_x and Δ_y of the tiles are set in advance and affect the performance of the algorithm, as described in Subsection 10.1.8. Finer tiling produces smoother shapes and a smaller resistance at the cost of additional runtime. Due to the irregular shape of the available space, tiles near the boundaries may be irregular in shape, as shown in Fig. 10.6.

The adjacent vertices in the equivalent graph are connected with edges, producing a mesh structure [71, 72]. To mimic the electrical behavior of the rail, the weight of the edges is proportional to the conductance between adjacent tiles. An accurate estimate of the resistance between arbitrary shapes requires computationally expensive methods, such as the finite element method [575]. For routing, however, a more efficient heuristic is proposed. The conductance of each edge is proportional to the width of the contact between two corresponding tiles. For example, the conductance between tiles A and B in Fig. 10.7 is twice larger than the conductance between tiles B and C due to the wider contact.

Algorithm 1 Convert available space A_n into equivalent graph Γ_n using tiles of size $(\Delta x, \Delta y)$

1: **procedure** SPACETOGRAPH(A_n, Δx, Δy)
2:     $V_n \leftarrow \varnothing$
3:     $E_n \leftarrow \varnothing$
4:     $[x_{min}, x_{max}, y_{min}, y_{max}] \leftarrow bounds(A_n)$
5:     $n_x \leftarrow \left\lfloor \frac{x_{max} - x_{min}}{\Delta x} \right\rfloor$
6:     $n_y \leftarrow \left\lfloor \frac{y_{max} - y_{min}}{\Delta y} \right\rfloor$
7:     **for** $i = 0, 1, 2, \ldots, n_x$ **do**
8:         $x_{min}^i \leftarrow x + i\Delta x, x_{max}^i \leftarrow x + (i+1)\Delta x$
9:         **for** $j = 0, 1, 2, \ldots, n_y$ **do**
10:            $y_{min}^j \leftarrow y + j\Delta y, y_{max}^j \leftarrow y + (j+1)\Delta y$
11:            $box_{i,j} \leftarrow rectangle(\{x_{min}^i, y_{min}^j\}, \{x_{max}^i, y_{max}^j\})$
12:            $cell_{i,j} \leftarrow box_{i,j} \cap A_n$
13:            **if** $cell_{i,j} \neq \varnothing$ **then**
14:                Add $cell_{i,j}$ to V_n
15:                $overlap_y = cell_{i,j} \cap cell_{i,j-1}$
16:                **if** $overlap_y \neq \varnothing$ **then**
17:                    Add $\{cell_{i,j}, cell_{i,j-1}, \frac{length(overlap_y)}{\Delta x}\}$ to E_n
18:                $overlap_x = cell_{i,j} \cap cell_{i-1,j}$
19:                **if** $cell_{i,j} \cap cell_{i-1,j} \neq \varnothing$ **then**
20:                    Add $\{cell_{i,j}, cell_{i-1,j}, \frac{length(overlap_x)}{\Delta y}\}$ to E_n
21:     **return** $\Gamma_n = (V_n, E_n)$

10.1.3   Seed subgraph

Once the available space is converted into the equivalent graph Γ_n, the power routing problem is transformed into finding the subgraph $\Gamma_n^s \in \Gamma_n$ connecting the terminal nodes such that the resistance between terminals is minimized. The order of the subgraph $|V_n^s|$ is limited by the preset area constraint A_{max}. In SPROUT, the routing process commences with determining the initial connection between the source and target terminals. The location of the source and target terminals is supplied externally as a set $T_n = \{t_n^1, \ldots, t_n^k\}$. Efficient routing algorithms exist to determine the shortest path, such as Dijkstra [576] and Bellman-Ford [195]. This seed subgraph is iteratively improved using SmartGrow and SmartRefine algorithms, as described in, respectively, Subsections 10.1.4 and 10.1.5.

To generate the seed subgraph, the shortest path is determined for each pair of nodes, as shown in Fig. 10.8a. The resulting subgraph can be directly passed to the SmartGrow algorithm. To accelerate convergence, however, the nodes located within the boundary of the seed are added to the subgraph, producing a subgraph without voids, as illustrated in Fig. 10.8b.

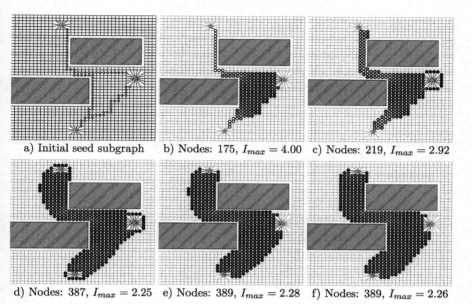

a) Initial seed subgraph   b) Nodes: 175, $I_{max} = 4.00$ c) Nodes: 219, $I_{max} = 2.92$

d) Nodes: 387, $I_{max} = 2.25$ e) Nodes: 389, $I_{max} = 2.28$ f) Nodes: 389, $I_{max} = 2.26$

Fig. 10.8 Example of graph-based routing process among three terminals. a) Initial seed subgraph, b) voidless subgraph after filling the internal voids, c) initial stage of subgraph growth, and d) final stage of subgraph growth. Areas with large current are reinforced with new nodes. e) Initial stage of the refinement process. Areas with small current, specifically those nodes near the terminals, are replaced by nodes in the areas of current crowding, *i.e.*, closer to the obstacles. f) Final stage of the refinement process. The reduction in impedance is negligible, triggering termination of the algorithm.

Algorithm 2 Set of terminal polygons T_n to identify the corresponding nodes in Γ_n and the adjacent terminal nodes

---

1: **procedure** IDENTIFYTERMINALS(Γ_n, $T_n = t_1, \ldots, t_k$)
2:      $\Theta = \{\theta_1, \ldots, \theta_k\}$
3:      **for** $i = 0, 1, 2, \ldots, k$ **do**
4:          **for** each vertex $v \in V_n$ **do**
5:              **if** $v \cap t_i \neq \varnothing$ **then**
6:                  Add v to θ_i
7:          identify nodes in θ_i
8:      **return** $\Gamma_n = (V_n, E_n)$, Θ

10.1.4 Growth stage

The seed subgraph typically exhibits high resistance. The impedance of the subgraph can be improved by increasing the order $|V_n^s|$ of the subgraph. To identify those parts of the subgraph that benefit most from reinforcement, a node current metric is introduced here. Those regions within the subgraph with the highest node current metric indicate a high current density. Additional nodes would likely produce a significant reduction in the impedance. In contrast, those regions with a smaller node current would produce a negligible reduction in impedance, resulting in suboptimal allocation of metal resources. These low current density regions are therefore left unchanged.

Algorithm 3 Generate voidless seed subgraph $\Gamma_n^s = (V_n^s \in V_n, E_n^s \in E_n)$ such that terminals in set $\Theta_n \in V_n$ are connected.

1: **procedure** SEED($\Gamma_n, \Theta_n = \{\theta_1, \ldots, \theta_k\}$)
2:      $V_n^s \leftarrow \varnothing$
3:      **for** each node θ_i in Θ_n **do**
4:          $paths \leftarrow shortestpath(\Gamma_n, \theta_i, \{\theta_i + 1, \ldots, \theta_k\})$
5:          Add $paths$ to V_n^s and E_n^s
6:      $poly \leftarrow exterior(\bigcup(V_n^s))$
7:      **for** each node v in V_n **do**
8:          **if** $v \cap poly \neq \varnothing$ **then**
9:              Add v to V_n^s
10:             Add edges adjacent to v to E_n^s
11:     **return** $\Gamma_n^s = (V_n^s, E_n^s)$

The node current metric is evaluated in three stages. The current is initially injected into each pair of terminals. The magnitude of the current is proportional to the expected current carried by the connection. For example, those pairs of terminals with large current, e.g., between the PMIC and the BGA balls, are injected with larger current as opposed to those pairs requiring relatively smaller current, such as connections between BGA balls. This current injection process is expressed as a current injection matrix, $E \in R^{(|\Gamma_s|-1)\times n_{pairs}}$. Each column of E corresponds to a node within the subgraph. All entries in E are zero except the two nodes where current i is injected. The value of these currents is, respectively, $+i$ and $-i$. The voltage distribution for each current injection is determined using nodal analysis,

$$V = L^{-1}E, \tag{10.3}$$

where L is a grounded Laplacian matrix. The current J within each edge is determined by multiplying the voltage matrix V by the weighted directed incidence matrix B of subgraph Γ_n^s,

$$J = BV = BL^{-1}E. \tag{10.4}$$

The total current carried by an edge is the sum of the absolute value of the current for each pair of terminals. The node current is the sum of the total current in the adjacent edges. Thus, the nodes adjacent to the edges carrying large current exhibit a large node current.

Algorithm 4 Evaluate the current metric for each node in subgraph Γ_n^s and set of terminals $\Theta \in \Gamma_n^s$.

1: **procedure** NODECURRENT(Γ_n^s, Θ)
2:      $N = |\Gamma_n^s|$
3:      $[\Theta]^2 = \{\theta' \subseteq \Theta \mid |\theta'| = 2\}$
4:      $N_{pairs} \leftarrow |[\Theta]^2|$
5:      $L \leftarrow$ Laplacian matrix of Γ_n^s
6:      $E \in \mathbb{R}^{(N-1) \times N_{pairs}}$
7:      **for** each pair (s, t) in $[\Theta]^2$, $i = 1, 2, \ldots, N_{pairs}$ **do**
8:          $E_{s,i} \leftarrow 1$
9:          $E_{t,i} \leftarrow -1$
10:      solve($LV = E$)
11:      $J \in \mathbb{R}^N$
12:      **for** $p \in \Gamma_n^s$ **do**
13:          $J_p \leftarrow \sum_{i=1}^{N_{pairs}} \sum_{j \in N(\Gamma_n^s, i)} g_{pj} |V_i - V_j|$
14:      **return** J

Algorithm 5 Given available space graph Γ_n, seed subgraph Γ_n^s, and set of terminals $\Theta \in \Gamma_n^s$, add k nodes from Γ_n to Γ_n^s to reduce the impedance of the subgraph.

1: **procedure** SMARTGROW(Γ_n, Γ_n^s, Θ, k)
2:      $V_n^c \leftarrow V_n \setminus V_n^s$
3:      $[\Theta]^2 = \{\theta' \subseteq \Theta \mid |\theta'| = 2\}$
4:      $N_{pairs} \leftarrow |[\Theta]^2|$
5:      $J \leftarrow$ NODECURRENT(Γ_n^s, Θ)
6:      $J^c \in \mathbb{R}^{|V_n^c|}$
7:      **for** $p \in V_n^c$ **do**
8:          $J_p^c \leftarrow \sum_{j \in N(\Gamma_n, p), j \in \Gamma_n^s} J_j|$
9:      **for** $i = 1, 2, \ldots, k$ **do**
10:          $m \leftarrow \{c \mid J_c = max(J)\}$
11:          $V_n^s \leftarrow V_n^s \cup m$
12:          $J \leftarrow J \setminus m$
13:      $\Gamma_n^s \leftarrow G_n[V_n^s]$
14:      **return** Γ_n^s

The boundary of subgraph Γ_n^s is defined as C, a set of nodes in Γ_n adjacent but not belonging to Γ_n^s. The nodes in C adjacent to the nodes in Γ_n^s with the highest current are added to the subgraph along with the corresponding edges.

Algorithm 6 Given available space graph Γ_n, subgraph Γ_n^s, and set of terminals $\Theta \in \Gamma_n^s$, replace k nodes in Γ_n^s by k nodes from Γ_n to reduce the impedance of the subgraph.

1: **procedure** SMARTREFINE($\Gamma_n, \Gamma_n^s, \Theta, k$)
2:     $J \leftarrow$ NODECURRENT(Γ_n^s, Θ)
3:     **for** $i = 1, 2, \ldots, k$ **do**
4:         $m \leftarrow \{c \mid J_c = min(J)\}$
5:         $V_n^s \leftarrow V_n^s \setminus m$
6:         $J \leftarrow J \setminus m$
7:     $\Gamma_n^s \leftarrow$ SMARTGROW($\Gamma_n, \Gamma_n[V_n^s], \Theta, k$)
8:     **return** Γ_n^s

This process is iteratively repeated until the area limit A_{max} is reached. Therefore, regions with high current are reinforced whereas those areas with smaller current are left unchanged, maximizing the reduction in resistance per unit of added metal. To illustrate this process, an example seed subgraph is shown in Fig. 10.8b. Brighter nodes correspond to nodes with high current, whereas the darker nodes represent nodes with small current. In the next iteration, brighter zones are reinforced, leading to a reduction in the impedance in that region (see Fig. 10.8c). Further iterations reinforce the brightest zones, increasing the conductance until the target area is reached (see Fig. 10.8d).

10.1.5   Refinement stage

Due to the area constraint, the growth process cannot continue indefinitely. Further lowering of the subgraph impedance is however possible without increasing the area using the SmartRefine procedure described in algorithm 6. The areas with the largest and smallest current are identified using the node current metric described in the previous section. The nodes conducting the smallest current are removed without exhibiting a significant effect on the impedance. Using the vacated metal, those regions carrying large current are reinforced, further reducing the subgraph impedance. This process is illustrated in Figs. 10.8d to 10.8f. The nodes behind the terminals in Fig. 10.8d carry smaller current than the rest of the subgraph. These nodes are removed and replaced by the nodes near the blockages with greater node current.

The SmartRefine process can be viewed as moving nodes from quiescent zones to hot spots. The number of nodes removed per iteration is a design variable. Removing additional nodes each iteration would initially converge faster. At later stages of the refinement process, however, the subgraph is close to being locally optimal; excessive movement would possibly increase the impedance. Moving fewer nodes at later stages of the refinement process would therefore yield a lower impedance.

10.1.6   Subgraph reheating

The graph-based power routing problem can be viewed as an optimization problem,

$$\textbf{Minimize} : R(\Gamma_n^s, \Theta_n) \textbf{ s.t. } : A(\Gamma_n) \le A_{max}. \tag{10.5}$$

From an optimization perspective, the SmartGrow and SmartRefine procedures are a form of gradient descent. The resistance of the subgraph is the objective function and the node current metric is a proxy metric for the gradient of the objective function. These algorithms are, therefore, a form of local optimization where the result is not guaranteed to be a global minimum. To mitigate this issue, the subgraph reheating technique is presented in this section, inspired by the simulated annealing algorithm [577] where the objective function can temporarily increase to explore the design space.

The reheating process consists of two operations, dilation and erosion, inspired by image processing operations. Initially, the subgraph is dilated beyond the area constraint by adding nodes adjacent to the subgraph. After completing the dilation operation, the erosion process commences. Using the node current metric, those nodes with the smallest current are removed from the subgraph, eliminating the redundant nodes while reinforcing the hot spots. The number of dilation iterations determine the extent to which the search space is explored. Additional iterations would explore a wider space while requiring greater runtime for the subsequent erosion process.

10.1.7   Back conversion

Once the reheating process is complete, the resulting subgraph is converted back into a polygon. Recall that each node within the graph Γ_n is associated with a tile within the available space. The subgraph Γ_n^s therefore corresponds to a polygon comprised of multiple merged tiles. A typical PCB consists of several nets. Thus, it is crucial to remove the routed polygon from the available space of other nets.

10.1.8   Algorithm runtime analysis

The runtime of the algorithm depends upon a multitude of parameters including the number of terminals, grain size, and size of the available physical space. The first stage of the algorithm is the available space. Modern polygon clipping algorithms exhibit linear complexity with the number of vertices [578]. The PCB layout may contain more than many hundred thousands of vertices [579]. An early PCB prototype, however, contains much fewer vertices, due to the fewer polygons and simpler geometry. In the case studies presented in Section 10.2, fewer than 10,000 vertices are processed, requiring up to 50 seconds for six power rails.

The complexity of the Dijkstra shortest path algorithm is $O((|V_n| + |E_n|) \log |V_n|)$, where V_n and E_n are sets of, respectively, nodes and edges of Γ_n. Due to the rectangular tiling of the available space, the number of edges is approximately twice larger than the number of vertices, yielding

$$O((|V_n| + 2|V_n|) \log |V_n|) = O(|V_n| \log |V_n|). \tag{10.6}$$

The complexity can be improved by employing alternative algorithms such as A-star [580], which utilizes the location of the nodes to accelerate the search process. The complexity of the Dijkstra algorithm, however, is smaller than the complexity of subsequent stages, namely SmartGrow and SmartRefine. In the case studies, finding the shortest path between all pairs of nodes requires negligible time. Thus, accelerating the shortest path algorithm yields only a marginal improvement in computational performance.

The SmartGrow and SmartRefine algorithms both require computation of the voltages within the graph. These processes require the node current metric to be iteratively computed, requiring a solution of the matrix equation. This step is the main bottleneck of the algorithm, requiring up to 90% of the total runtime. Using sparse linear equation solvers, the complexity of solving a linear equation is $O(|V|^q)$ where $q \in [1.5, 3]$ is the scaling exponent which equals 1.5 in the best case and 3.0 in the worst case [581]. Both SmartGrow and SmartRefine solve a single linear equation per iteration. Thus, the runtime for SmartGrow stage T_g is

$$T_g = c_g \sum_{i=0}^{k_g-1} \left(|V_n^s| - i \Delta V \right)^q, \tag{10.7}$$

where k_g is the number of growth iterations, ΔV is the number of nodes added per iteration, and c_g is the proportionality coefficient. The number of iterations k_g during the growth stage is approximately

$$k_g \approx \frac{A_{max}}{\Delta A}, \tag{10.8}$$

where A_{max} is the area of the resulting polygon, and ΔA is the area added to subgraph during each iteration of SmartGrow. Similarly, the runtime for SmartRefine stage T_r is

$$T_r = c_r k_r |V_n^s|^q, \tag{10.9}$$

where c_r is the proportionality coefficient.

The reheating process exhibits a complexity similar to SmartGrow and SmartRefine. The dilation process requires negligible time as compared to the erosion process which requires the node current metric to be evaluated. The runtime T_e required to apply erosion to a dilated subgraph is

$$T_e = c_e \sum_{i=0}^{k_e-1} \left(c_d|V_n^s| - i\Delta V\right)^q,$$
(10.10)

where $c_d|V_n^s|$ is the number of nodes after the dilation process, c_e is the proportionality coefficient, ΔV is the reduction in order of the subgraph per iteration, and k_e is the number of erosion iterations,

$$k_e = \left\lceil \frac{|V|_d - |V_n^s|}{\Delta V} \right\rceil.$$
(10.11)

The back conversion process reconstructs a set of polygons from the resulting subgraph. The polygons corresponding to each node are iteratively merged using the union operation, exhibiting $O(N \log N)$ complexity for N vertices. In the worst case, the number of vertices grows linearly with each converted node, yielding a worst case complexity $O(|V_n^s|(|V_n^s| - 1)) = O(|V_n^s|^2)$. Practically, however, the union of multiple tiles often yields the same number of vertices. For example, the union of tiles A and B, shown in Fig. 10.7, has the same number of vertices as tile B. The complexity of the back conversion process is therefore between $O(|V_n^s|)$ and $O(|V_n^s|^2)$.

Greater complexity occurs when the node current metric is evaluated, namely, during the SmartGrow, SmartRefine, and erosion procedures. Combining (10.7), (10.9), and (10.10) yields a complexity of approximately

$$O((\frac{A_{max}}{\Delta A} + k_r + k_e)|V_n^s|^q).$$
(10.12)

The number of nodes $|V_n^s|$ is approximately

$$|V_n^s| \approx \frac{A_{max}}{\Delta x \Delta y}.$$
(10.13)

The complexity is therefore

$$O\left(\frac{A_{max}}{\Delta A} + k_r + k_e\right)\left(\frac{A_{max}}{\Delta x \Delta y}\right)^q.$$
(10.14)

Therefore, to reduce the computational time, the tile size and incremental increase in area during the growth stage should be increased, while the number of refinement and erosion iterations should be reduced.

10.2 Validation of case study

Three practical case studies are presented in this section to demonstrate the validity of the proposed tool. In the first case, as described in Subsection 10.2.1, the layout for a portion of the PCB between the PMIC and the two groups of vias is synthesized. In the second case, as described in Subsection 10.2.2, the connections among the PMIC, capacitor, and a congested group of vias are established for the six nets. An example of PCB resource planning using SPROUT is described in Subsection 10.2.3

10.2.1 Two rail system

A part of an eight layer PCB for an industrial wireless application is shown in Fig. 10.9a. The PMIC is placed at the bottom layer and provides power to the two power rails, V_{DD1} and V_{DD2}, and the corresponding BGA balls at the top layer. The power rails connect the PMIC inductor to the group of BGA vias on the penultimate (seventh) layer. Dedicated ground planes are placed in layers two, six, and eight.

The manually generated layout is shown in Fig. 10.9b, and the synthesized layout using SPROUT is shown in Fig. 10.9c. Note the regular geometries utilized primarily in the manual layout whereas the automatically generated layout exhibits greater diversity in the shape of the geometries. The impedance of the layouts is extracted using a commercial parasitic extraction tool and compared in Table 10.1. The two layouts (manual and synthesized) exhibit similar impedance characteristics. The difference in resistance does not exceed 3.1%. The inductance of the V_{DD1} rail is reduced by 12% by using SPROUT, whereas the inductance of the V_{DD2} rail is increased by 1.47%.

10.2.2 Six rail system

In this case study, SPROUT is applied to a congested BGA arrangement, as shown in Fig. 10.10a. 612 BGA (six power supply nets and 306 BGA for ground) are located at the top layer, and two PMICs are located in the bottom layer of a ten layer PCB. Each PMIC regulates the current for the three voltage domains. Layers four, six, and eight are used for ground routing and the power rails are routed on the ninth layer.

The power supply rails are routed and compared to the manual layout. The resulting topologies are shown in Figs. 10.10b and 10.10c. Note the visual similarity between the layouts. The DC resistance and loop inductance of each rail are listed in Table 10.2. The loop inductance of the rails generated by SPROUT are 1 to 4% smaller than the manual layout while the difference in DC resistance is below 11%.

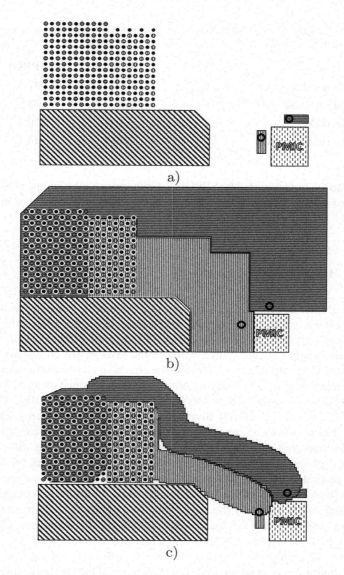

Fig. 10.9 Automated power routing using SPROUT and manual routing. a) Initial layout with blockage (diagonal hatch), and two rails, V_{DD1} (dark horizontal hatch) and V_{DD2} (light vertical hatch). A single PMIC supplies power to the rails using two inductors at bottom layer 8. The inductors are connected to routing layer 7 using a via. Any blockage is shaded with a diagonal pattern. b) Manually routed layout. c) Layout synthesized using SPROUT

The six rail PCB layout is synthesized in approximately 11 minutes using an Intel Core i7-67003.40 GHz eight core computer. Although the manual layout time varies with expertise and software, the typical time for manual layout is significantly

Table 10.1 Comparison of normalized impedance between SPROUT and manual routing for the two rail system shown in Fig. 10.9

	Net	Manual	SPROUT
Normalized inductance	V_{DD1}	100	87.5
@ 25 MHz (picohenrys)	V_{DD2}	136	138
Normalized DC	V_{DD1}	10.0	10.1
resistance (milliohms)	V_{DD2}	12.7	13.1

greater than the time required by SPROUT. Furthermore, after setup, SPROUT does not require active human involvement, providing additional reduction in time and labor.

10.2.3   Area/impedance tradeoff

With the ability to efficiently prototype and evaluate a power network, design tradeoffs can be extensively explored. In this case study, the relationship between the area and impedance is investigated in an industrial PCB. Modem, CPU, and DSP power supply nets are routed within a ten layer board containing 86 BGA, as illustrated in Fig. 10.11a. To determine the effects of additional metal area on the parasitic impedance, nine PCB layout prototypes are generated using SPROUT. The area of the power rails in each prototype is summarized in Table 10.3. The current demand of each rail is uniformly distributed within the ball grid array. The modem and CPU are provided with, respectively, two and five decoupling capacitors.

With greater area, the impedance is reduced while increasing the cost of the PCB. To explore this tradeoff, nine layouts with different area for the power rails are generated using SPROUT. The examples of these layouts are shown in Figs. 10.11b to 10.11d. Note that with smaller area, the BGA are connected while leaving large voids to satisfy the target area. In contrast, the larger area produces congestion due to a lack of space. The relationship between area allocated to each rail and the impedance is shown in Figs. 10.12a and 10.12b. The resistance of the rails is significantly reduced with additional area. The rate of reduction, however, diminishes with larger area. The inductance of the DSP rail exhibits similar behavior. The inductance of the modem and CPU rails is, however, not significantly reduced due to the decoupling capacitors.

The peak voltage drop is shown in Fig. 10.12c. Despite the greater inductance, the voltage drop in the DSP power rail is significantly smaller due to the smaller load current. In contrast, the voltage drop in the modem and CPU rails is significantly larger due to the greater load current and current slew rate. Note that the voltage drop in the modem does not significantly decrease with an area of 27.5 units. The blockages likely prevent adding metal to those regions with a high current density,

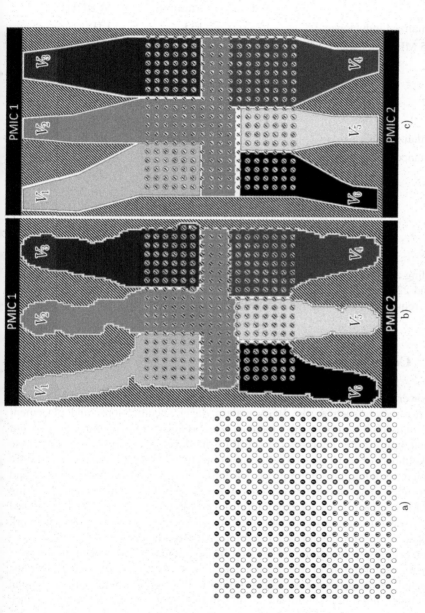

Fig. 10.10 Comparison between the automated power routed layout using SPROUT and manually routed layout. a) BGA placement. The numbers indicate the net of the vias; vias without number are ground vias. b) Layout synthesized using SPROUT and c) manual layout. The routing layer is filled with ground metal shown with diagonal hatch.

Table 10.2 Comparison of normalized impedance between SPROUT and manual routing for the six rail system shown in Fig. 10.10

	Net	Manual	SPROUT
Normalized inductance	V_1	133	131
@ 25 MHz (picohenrys)	V_2	103	99
	V_3	131	127
	V_4	161	155
	V_5	152	150
	V_6	116	114
Normalized DC	V_1	15.0	16.8
resistance (milliohms)	V_2	8.4	9.1
	V_3	13.0	14.2
	V_4	18.4	18.2
	V_5	18.5	18.9
	V_6	9.2	9.2

impeding any reduction in voltage drop. A similar trend is observed in the CPU rail. Beyond 22.5 units, the linear reduction in the voltage drop with area significantly slows, requiring additional metal to produce a similar gain in conductance.

10.3   Conclusions

The power network design process at the board level is highly influenced by system-level parameters such as the BGA pattern, layer specifications, and placement of the individual components. Changing a floorplan if a target impedance is not satisfied significantly degrades the speed of the development process. To increase the likelihood of satisfying target design objectives, system-level parameters are evaluated to determine appropriate tradeoffs among power, performance, and design time. To accelerate this evaluation process, SPROUT, an automated routing algorithm for power network exploration and prototyping, is introduced here. Based on the node current metric introduced in this chapter, a layout of a power network suitable for impedance extraction is automatically synthesized.

The primary contribution of SPROUT is automation of layout prototypes, enhancing exploration of the design space. As compared to manual layouts, automated synthesis requires similar time for PCB prototyping without human involvement, providing significant savings in both time and labor. The impedance of the generated layout is similar to a manual layout, achieving less than a 4% difference in the two case studies. Due to automation, a large number of layout prototypes can be analyzed. By providing greater insight into the layout at early

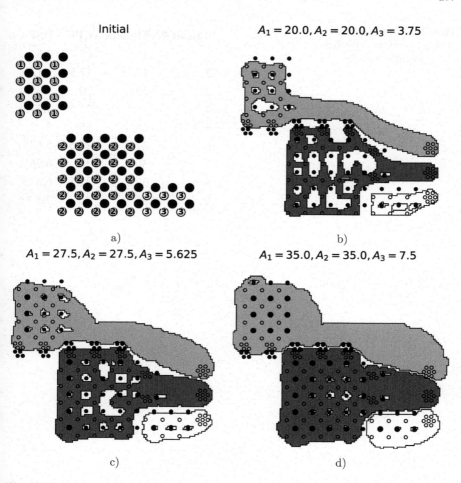

Fig. 10.11 Layout generated using SPROUT for three rails, modem (top left), CPU (center), and DSP module (bottom right), for varying metal area. a) Initial BGA arrangement. The numbers within circles indicate the nets. Vias for ground net are solid black. The size of vias is intentionally exaggerated to show nets b) $a_{modem} = 17.5$, $a_{CPU} = 17.5$, $a_{DSP} = 3.12$, c) $a_{modem} = 25.0$, $a_{CPU} = 25.0$, $a_{DSP} = 5.00$, and d) $a_{modem} = 32.5$, $a_{CPU} = 32.5$, $a_{DSP} = 6.88$. Area is normalized.

stages of the design process, system parameters can be accurately determined, reducing the likelihood of not satisfying target impedance objectives. The tool is demonstrated on two industrial applications.

In addition, the area/impedance tradeoff is explored for a three rail PCB layout. The trends revealed in this case study reveal the potential of automated exploration. SPROUT enables fast PCB prototyping and provides valuable information on design tradeoffs. For example, increasing the area of the modem rail beyond 27.5 units is not likely to yield a lower impedance.

Table 10.3 Target area of the test layouts for exploring area impedance tradeoffs

Layout #	Modem	CPU	DSP
1	15	15	2.5
2	17.5	17.5	3.125
3	20	20	3.75
4	22.5	22.5	4.375
5	25	25	5
6	27.5	27.5	5.625
7	30	30	6.25
8	32.5	32.5	6.875
9	35	35	7.5

Fig. 10.12 Parasitic impedance of PCB rails as a function of area. a) Effective resistance, b) effective inductance, and c) maximum transient voltage drop.

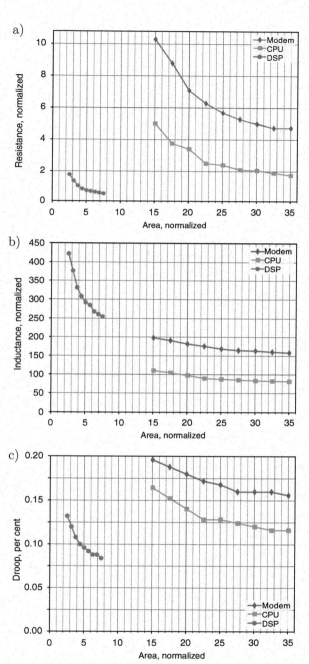

Chapter 11
QuCTS – single flux Quantum Clock Tree Synthesis

Rapid single flux quantum (RSFQ) technology offers a range of advantages as compared to CMOS. Several orders of magnitude greater operating frequency and three orders of magnitude lower power are among the most prominent advantages of RSFQ. Substantial progress has been made in the field of superconductive electronics in the past decades. SFQ manufacturing technology is capable of accommodating over 6,000 Josephson junctions (JJ) per mm^2 [582]. An 8 bit superconductive microprocessor operating at a frequency of 80 gigahertz has been successfully fabricated [583]. Ongoing advancements in electronic design automation for RSFQ circuits are expected to enable the large scale integration of superconductive digital systems [584, 585].

Beyond the necessity for cryogenic operation below approximately 4K and the relatively low density on-chip integration as compared to CMOS, the design of a robust on-chip clock distribution network remains a significant challenge in RSFQ systems [586]. The fundamental properties of RSFQ technology are described in the seminal work of Likharev and Semenov [587]. Unlike traditional CMOS, where the information is represented with a high or low DC voltage level, short quantized voltage pulses are utilized in RSFQ. A logical high or low is represented by, respectively, the presence or absence of a single flux quantum (SFQ) pulse within a certain time interval. Most logic gates in RSFQ are therefore sequential, such as AND and OR gates that are combinatorial in CMOS. This structure drastically increases the pipeline depth as compared to CMOS, complicating the clock network design process. The complexity of the clock distribution network is further exacerbated by the interconnect structures in RSFQ systems [588]. Unlike CMOS, where the connections are established with a simple wire [72], RSFQ interconnect is either a passive transmission line (PTL) requiring a driver, receiver, and impedance matching [589, 590], or an active Josephson transmission line (JTL) requiring bias current for each Josephson junction. Finally, most RSFQ gates have a fanout of one. A splitter gate is used to generate two (or more) SFQ pulses from an input signal [588, 591].

R. Bairamkulov, E. G. Friedman, *Graphs in VLSI*,
https://doi.org/10.1007/978-3-031-11047-4_11

Different approaches to clocking in RSFQ circuits have been reported in the literature. Clockless self-timed systems have been proposed [592–595]. An effective operating frequency of 20 gigahertz has been demonstrated while eliminating the overhead of the clock distribution network. Self-timed circuits, however, remain vulnerable to timing violations, exhibit unpredictable performance due to sensitivity to logic delays, and require handshaking circuitry that requires significant area [596].

Hierarchical chains of homogeneous clover-leaves clocking $(HC)^2LC$ are described in [597]. The primary advantage of this structure is robustness since the clock period of the system adapts to the slowest hierarchical chain. Another advantage is the elimination of race condition hazards due to forced counter-clocking [586, 597]. The primary drawback is reduced clock speed since the worst case path determines the clock period of the entire system. Another drawback of this method is underutilization of clock skew as an additional degree of design freedom. Requiring counter-clocking increases the setup time constraints which limit the minimum clock period [586].

A minimum skew clock tree synthesis algorithm for SFQ circuits is proposed in [598]. The algorithm incorporates the CMOS-based deferred merge embedding (DME) algorithm [392] to generate a zero skew clock tree. Due to the non-negligible dimensions of the splitters, the clock tree generated by DME typically violates RSFQ design rules. A legalization step is therefore proposed [598] to correct the layout at the cost of introducing small skew into the clock tree. Minimizing the clock skew, however, results in a suboptimal clock frequency [231], and does not guarantee correct functionality [599]. Furthermore, nonzero clock skew in data paths can improve the performance and robustness of the synchronous system [239]. With clock skew scheduling, the delay slack in fast data paths is exploited to decrease the effective delay of the critical paths, thereby increasing the maximum attainable operating frequency [230, 231, 233, 238, 600].

While clock skew may provide significant gains in performance and robustness, it is often overlooked in existing RSFQ clocking approaches. To bridge this gap, QuCTS, a single flux Quantum Clock Tree Synthesis algorithm, is introduced. In the clock skew scheduling stage, the arrival time of each clocked gate is based on the algorithm adapted from [231, 241]. In the clock tree synthesis stage, a clock tree layout is generated based on the gate placement information, design rules, and schedule of clock arrival times from the clock scheduling stage. With QuCTS, the number of delay elements and the total wirelength are reduced while satisfying the timing requirements of each clocked gate.

This chapter is organized as follows. In Section 11.1, the clock skew scheduling algorithm is presented. The binary clock tree synthesis process is described in Section 11.2, followed by the delay equilibration process presented in Section 11.3. The performance of the algorithm is evaluated in the case study and benchmark circuits presented in Section 11.4, followed by the conclusions in Section 11.5.

11.1   Clock skew scheduling

Clock skew scheduling is a powerful technique to maximize the speed and robustness of a synchronous system [231, 233, 238]. Despite the potential benefits of useful clock skew, it is however often viewed as a parasitic effect requiring minimization [601, 602]. In addition, achieving zero clock skew is quite difficult due to process and environmental variations as well as electromagnetic interference that permeate not only CMOS but also RSFQ circuits [586, 597, 603].

The first stage of QuCTS, presented in this section, mitigates this issue by adapting clock skew scheduling within the RSFQ circuit design process. QuCTS operates in four stages. The sequential circuit topology, described in Verilog, is initially converted into a timing graph. The minimum clock period is determined by evaluating the delay and delay uncertainty of each data path. The permissible range (PR) of each data path is a function of the clock skew in sequentially-adjacent registers [230, 239, 599]. The clock skew schedule is generated using a quadratic programming algorithm that maximizes the robustness of the circuit to parameter variations [231, 241]. The clock skew schedule is converted into a schedule of clock arrival times that is passed to the clock tree synthesis algorithm.

11.1.1   Timing graph

The first step in the clock skew scheduling process is conversion of the circuit topology into a directed timing graph $G = (V, E, d_{min} : E \rightarrow \mathbb{R}, d_{max} : E \rightarrow \mathbb{R})$, where V is the set of nodes, E is the set of edges, and d^{min} and D^{max} are, respectively, the minimum and maximum delay of an edge in E. A typical sequential circuit consists of inputs, outputs, clocked gates, non-clocked gates, and interconnects. For brevity, the clocked and non-clocked gates are referred to as, respectively, registers and gates. Each edge $(i, j) \in E$ represents a combinatorial data path p_{ij} from a source to target register. The range of delays $d_{ij} = [d_{ij}^{min}, D_{ij}^{max}]$ of an edge (i, j) within a graph is the sum of the delays along a data path,

$$d_{ij} = \sum_{k \in p_{ij}} (d_k^{gate} + d_k^{int}), \qquad (11.1)$$

where d_k^{int} denotes the range of delay of the interconnect between gate k and the next gate, and d_k^{gate} denotes the range of the input-to-output delay of gate k or a clock-to-output delay of register k. The gate and register delays are supplied externally as input data.

The inputs and outputs of a sequential circuit are often described in Verilog as floating signal nets. This structure is not supported in a graph where the edges require both source and target nodes. Furthermore, it is often desired that the clock skew between the input and output nodes of a circuit is zero [231]. A dummy I/O

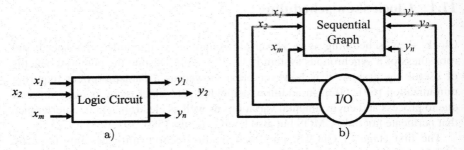

Fig. 11.1 Processing of inputs and outputs of a logic circuit in a timing graph. a) An initial system with inputs $x_1, \ldots, x_m$ and outputs $y_1, \ldots, y_n$. Note that the input and output edges (signal nets in Verilog) typically have one floating terminal. b) Timing graph representation of the input and output edges in QuCTS. The floating terminals of the input and output edges are connected to a dummy I/O node. This node acts as a tail (source) of all input edges and a head (target) of all output edges. The I/O node eliminates any clock skew between the circuit terminals.

node is therefore added to the timing graph, as illustrated in Fig. 11.1. The I/O node is the tail (source) of each input edge and the head (target) of each output edge. The dummy node is treated as a standard node during the clock skew scheduling process. Since a node cannot have a non-zero clock skew with itself, zero clock skew is ensured among the circuit inputs and outputs.

11.1.2   *Minimum clock period*

In the zero clock skew approach, the minimum clock period is determined by the delay of the critical paths. In a non-zero clock skew system, however, finding the minimum clock period requires a significantly more sophisticated process. The minimum clock period is determined by the cycles and reconvergent paths within the timing graph [231], as shown in, respectively, Figs. 11.2a and 11.2b.

An example of a sequential circuit containing cycle p_{ii} with n nodes is shown in Fig. 11.2a. To ensure correct operation of the circuit including this cycle, the clock period cannot be smaller than

$$T^i = \frac{1}{n} \sum_{(j) \in p_{ii}} (D^{max}_{j,j+1} + \delta^s_{j+1}), \tag{11.2}$$

where δ^s_{j+1} is the setup time of the gate following gate j. The clock skew within the cycle is fixed at zero, since, as described in Section 4.1.1.2, a register cannot have a non-zero clock skew with itself [230]. Equation (11.2) therefore requires the average propagation delay of a data path within a cycle to not be greater than the clock period. Finding the minimum clock period requires determining the cycles within

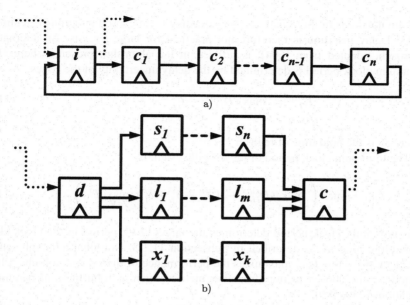

Fig. 11.2 Constraints of the minimum clock period within a sequential circuit. a) Cycle path with n registers starting with node i. The dotted arrows represent the connection to external circuitry. b) Reconvergent path between registers d and c.

the timing graph G. The computational complexity of finding all cycles within a graph is $O((|V| + |E|)(n_c + 1))$, where n_c is the number of cycles within a graph.

The reconvergent paths are distinct sequential paths that begin at the same divergent register d and end at the same convergent register c. Optimization of these reconvergent paths includes delay insertion, *i.e.*, intentionally adding delay to specific data paths to align the arrival time of the signals, thereby reducing the minimum clock period. Consider the example illustrated in Fig. 11.2b. The short path $(s_1, \ldots, s_n)$ with n nodes has the smallest propagation delay, and the long path $(l_1, \ldots, l_m)$ with m nodes has the largest propagation delay. The minimum clock period T^{dc} due to the reconvergent paths between nodes d and c is

$$T^{dc} = \frac{D_l - D_s + \delta_c^s + \delta_c^h}{|m - n + 1|}, \tag{11.3}$$

where D_l and D_s are, respectively, the maximum propagation delay of p_l and minimum propagation delay of p_s, and δ_c^s and δ_c^h are, respectively, the setup and hold time of the convergent register c. While delay padding may reduce the minimum clock period, this method requires finding all reconvergent paths within graph G. The complexity of finding a single simple path in a directed graph is $O(|V| + |E|)$ [337]. The number of simple paths can however be prohibitively large, up to $|V|!$ in a fully connected graph. Depending upon the complexity, an integrated circuit may contain hundreds of thousands of nodes, leading to an exorbitant number of simple paths. Delay insertion is therefore not practical for large circuits. An alternative approach, adapted from [604], is utilized in the algorithm presented here. The minimum clock period is determined by the delay uncertainty of the edges,

$$T_{ij}^{min} = \max_{(ij) \in E} (D_{ij}^{max} - d_{ij}^{min} + \delta_j^s + \delta_i^h), \tag{11.4}$$

where δ_i^h is the hold time of register i.

The minimum clock period of the overall system is

$$T_{min} = \max \left(\max_{(i,j) \in E} (D_{ij}^{max} - d_{ij}^{min} + \delta_j^s + \delta_i^h), \max_{i \in V} (T^i) \right). \tag{11.5}$$

This minimum clock period determines the target clock period in the clock skew scheduling process, as described in Subsection 11.1.3. Note that although setting the clock period to T_{min} maximizes the performance of the system, a higher clock period can be chosen to improve other metrics, such as robustness to parameter variations [239, 599].

11.1.3 Clock skew optimization

Once the minimum clock period is determined, clock skew optimization is performed in two steps. The permissible range (PR) [237, 239, 599] of the clock skew for each path is used to form an objective function. The basis cycles are determined within the graph to form a constraint function. The clock skew schedule is optimized for robustness to parameter variations.

The permissible range is the range of clock skew between sequentially-adjacent registers i and j that satisfy the setup and hold constraints of a circuit [230, 237], defined as

$$PR_{ij} = \left[-d_{ij}^{min} + \delta_i^h, T_{CP} - D_{ij}^{max} - \delta_j^s \right], \tag{11.6}$$

where T_{CP} is the target clock period. In vector form, the upper and lower bound of the permissible range for every combinational data path is expressed as vectors $\mathbf{s_{min}}, \mathbf{s_{max}} \in \mathbb{R}^{|E|}$. To maximize the robustness of the system, the clock skew of each data path is maintained at the center $\mathbf{s}^*$ of the PR,

$$\mathbf{s}^* = \frac{1}{2} (\mathbf{s_{min}} + s_{max}). \tag{11.7}$$

Clock skew deviations arising from parameter variations are therefore less likely to cause a setup or hold time violation. Note however that due to timing constraints, such as cycles, maintaining the clock skew at the center of the PR is often not possible [239]. The scheduling process therefore sets each target clock skew as close to the center of the PR while satisfying the local timing constraints.

Let $s \in \mathbb{R}^{|E|}$ be the vector of clock skews for each local data path. The clock skew scheduling optimization problem is expressed as

$$\underset{s}{\textbf{Minimize}} : ||s - s^*||^2 \tag{11.8}$$

$$\textbf{subject to}$$

$$s_{min}^i \le s^i \le s_{max}^i \forall i \in \mathbb{N}, i \le |E|, \tag{11.9}$$

$$Bs = 0, \tag{11.10}$$

where s_{min}^i, s^i, and s_{max}^i are the ith element of, respectively, s_{min}, s, and s_{max}; $0 \in \mathbb{R}^{|E|}$ is the zero vector; and $B \in \mathbb{R}^{(|E|-|V|+1)\times|E|}$ is the circuit connectivity matrix of graph G [231]. With (11.8), the clock skew of each data path is placed as close to the center of the permissible range as possible [230, 239]. Expression (11.9) requires the clock skew of each data path to be within the permissible range. Expression (11.10) requires the clock skew within a cycle to be zero. Each row b_i in B represents an independent cycle in G. The entry b_{ij} is equal to 1 or -1 if the edge, respectively, follows or opposes the direction of the cycle, and 0 if the edge does not belong to the cycle. An efficient solution of this problem can be achieved with quadratic programming (QP) in $O(|V|^3)$ time [241].

Once the final clock skew schedule is generated, a schedule of clock arrival times is produced. An arbitrary node x is marked as a reference node with a clock arrival time of 0. The clock arrival time at each register is determined using the fundamental equation of clock skew [230],

$$t_{skew} = \tau_i - \tau_f, \tag{11.11}$$

where τ_i and τ_f are, respectively, the clock arrival time at the initial and final register of a local data path. The arrival time τ_p of the register p preceding register x is

$$\tau_p = s_{px} + \tau_x, \tag{11.12}$$

where s_{px} is the clock skew of the edge (p, x) determined from the optimization process. Similarly, the arrival time τ_s of the successor s of register x is

$$\tau_s = \tau_x - s_{xs}, \tag{11.13}$$

where s_{px} is the clock skew of the edge (x, s). The process is repeated until the arrival time at each register is determined. The resulting schedule of arrival times is passed to the clock tree synthesis algorithm, as described in Section 11.2.

11.2 Clock tree synthesis

Once the clock arrival time of each logic gate is determined, the objective is to generate a clock network that satisfies these arrival times. A single external clock source is assumed in QuCTS. To distribute the clock signal from a single source to multiple sinks, a tree structure is utilized [368] due to area efficiency and adaptability of this structure as compared to symmetric H-tree or mesh topologies [232, 367]. Due to the limited fanout of RSFQ gates, splitters are required to distribute the clock signal to the many gates within a circuit. Standard splitters provide a fanout of two. Non-standard splitters with a higher fanout exist, although the bias margins are significantly degraded as compared to standard splitters [588, 591]. A binary clock tree is therefore assumed in QuCTS.

To distribute the clock signal to N gates, $N - 1$ splitters are required, forming a directed binary tree,

$$T = (V_T, E_T), \tag{11.14}$$

$$V_T = V_{SPL} \cup V_{sink}, \tag{11.15}$$

where V_{sink} is the set of clock sinks (logic gates), and V_{SPL} is the set of splitters. The leaf nodes within T (*i.e.*, nodes with zero fanout) correspond to the clock sinks. Other nodes correspond to splitters and have a fanout of two. The root node corresponds to the hierarchically topmost splitter, as shown in Fig. 11.3. The clock signal initially arrives at the root node within the clock tree and passes to the splitters corresponding to child nodes 0 and 1. At each successive node of tree T, the clock signal is split into multiple signals that eventually arrive at each sink within a cluster. The arrival time of the clock signal is set by the delay from the clock signal source (root node) to the clock sink. This delay is comprised of splitter delays, interconnect delay, and any intentional delay. By varying these components, the arrival time of the signal can be controlled to satisfy the timing requirements of each clock sink. The objective of the clock tree synthesis process in RSFQ is to produce a binary clock tree that delivers the clock signal at a precise time with minimum interconnect and junction area.

The first step in the clock tree synthesis process is to produce a binary tree. A common approach in binary tree synthesis is clustering [605], as illustrated in Fig. 11.3. Each gate is represented as a point in a two- or three-dimensional space. The location of each gate is represented by an X and Y coordinate, and the weighted clock signal wT serves as a third dimension. The importance of the clock arrival time is controlled by the weight parameter w. A greater weight groups gates that

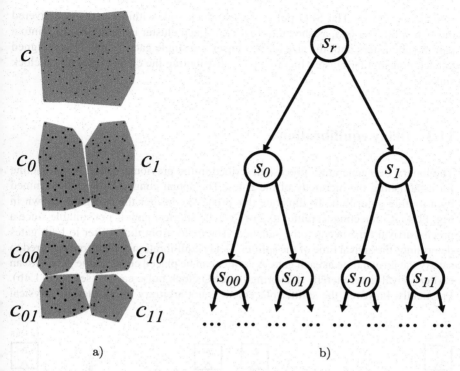

Fig. 11.3 Binary clock tree generation based on clustering. a) Hierarchical clustering of the gates based on location. All of the gates are initially within a single top level cluster c (top row). The gates are split into two clusters, c_0 and c_1 (middle row), which, in turn, are further divided into smaller clusters (bottom row), until the clusters contain only a single clock sink. b) Binary clock tree T with each node representing a splitter. The top level cluster corresponds to the root splitter s_r. Gates within c_0 and c_1 receive the clock signal from the two branches of the root splitter s_r. s_0 and s_1 are added to the binary clock tree as successors of the root splitter s_r to distribute the SFQ clock pulse from s_r to the corresponding clusters. The s_{00} and s_{01} (s_{10} and s_{11}) splitters therefore become the successors of splitter s_0 (s_1) to distribute the clock pulse to, respectively, c_{00} and c_{01} (c_{10} and c_{11}). Similarly, each successive clustering step adds two new successor splitters to the corresponding preceding node, resulting in a binary clock tree.

are not physically close but have a similar arrival time. In contrast, a smaller weight groups gates by physical proximity, disregarding any difference in arrival times. The gates are split into two clusters using a clustering algorithm, such as K-Means [606] and BIRCH [607]. The choice of clustering algorithm has a minor effect on the clock tree topology, affecting fewer than 1% of the clusters.

A binary clock tree is a rooted directed tree, where each node corresponds to a splitter. The topmost (root) splitter $s_r \in T$ receives a clock pulse from the external clock source. The SFQ pulse at each clock sink is delivered through the parent splitter. After the first clustering step, the gates are split into two groups, c_0 and c_1. The two SFQ output pulses of s_r are delivered to clusters c_0 and c_1 via corresponding splitters s_0 and s_1. The SFQ pulse at each clock sink within c_0 (c_1) is delivered through splitter s_0 (s_1), as shown in Fig. 11.3. Each cluster is iteratively split into a pair of subclusters until the size of the cluster is a single gate. A splitter is assigned to each nonsingular cluster, hierarchically distributing the clock signal to the clock sinks.

11.3   Delay equilibration

The binary tree generation process described in the previous section is a guideline for establishing the hierarchy of the gates. The actual connections are determined by a routing algorithm. To illustrate this process, consider the two gates shown in Fig. 11.4a. Connecting a splitter to gates via the shortest path is not suitable since a precise arrival time needs to be satisfied. The delay from the splitter to both gates determines the arrival time of the splitter. Delay equilibration is therefore required to satisfy the arrival time at each gate. A splitter can be placed closer to the gate with an earlier arrival time, thereby delivering the SFQ clock pulse earlier (see Fig. 11.4b). Practically, however, the splitter placement is not arbitrary but limited by physical

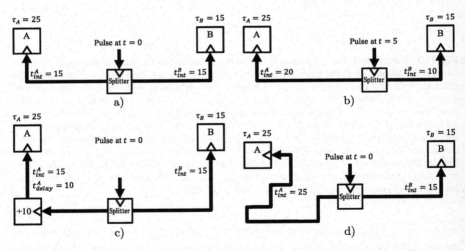

Fig. 11.4 Example of delay equilibration process. Two gates, A and B, require a clock pulse to arrive at, respectively, 25 and 15 time units. The clock signal initially arrives at the splitter, where two SFQ pulses for each gate are generated. a) An example of an invalid topology. While the delay requirement of B is satisfied, A receives the SFQ pulse too early, producing a timing violation. b) Strategic placement of the splitter closer to B reduces the delay from the splitter to B and increases the delay from the splitter to A. c) The wire connecting the splitter to A is intentionally lengthened to increase the delay. d) The delay element is placed between the splitter and A, thereby increasing the delay of the path.

layout constraints. In addition, if the difference in arrival time is large, the splitter placement may be insufficient to balance the arrival time of the clock signals.

In CMOS, a variety of techniques are available to adjust the wire delay, including wire snaking, wire sizing, dummy wire insertion, and active delay elements [375, 608, 609]. In RSFQ, passive transmission lines require impedance matching, complicating the wire sizing and dummy wire insertion process. The wire snaking technique, illustrated in Fig. 11.4c, is suitable for RSFQ, albeit requiring significant area for a modest increase in delay. A significantly larger delay with a relatively small area can be achieved with active delay elements. A JTL can be used as a delay element by controlling the bias current of the Josephson junctions [588]. JTLs, however, require dedicated space within the device layer. JTLs are therefore more suitable for providing large delays while PTL-based wire snaking can be used to tune the path delay.

Delay equilibration of a pair of gates requires the precise location of each gate. Since only the position of the clock sinks is initially known, the algorithm generates the clock tree layout in a reverse breadth first search order. The gates are processed in pairs, starting from the farthermost leaves (sinks) of the tree.

The embedding of the clock tree into the layout is accomplished in three steps. In the coarse embedding step presented in Subsection 11.3.1, the location of the splitter, JTL delay elements, and initial PTL routing for every pair of nodes in a binary tree is determined. The local portion of the layout is converted into a proxy graph where the potential location of the splitters and JTLs is determined. The graph is evaluated to determine the location and delay of the splitters and JTLs, satisfying the arrival time of the clock signal with minimum interconnect, as described in Subsection 11.3.2. Based on the location of the splitters, JTLs, and blockages, the layout is converted into a Hanan grid [144]. The approximate PTL layout is determined using a shortest path algorithm, such as the A-star shortest path algorithm [580]. The precise routing of the interconnect is determined during the fine routing stage, as described in Subsection 11.3.3. The delay of the wires is finely adjusted with wire snaking to satisfy the precise requirement of the clock arrival times.

11.3.1 Coarse routing

The coarse routing process for a pair of nodes A and B commences with identifying the cell location for the splitters and JTLs. The layout regions available for the JTLs and splitters are provided to QuCTS as a user input. Based on the cell dimensions and spacing information, these layout regions are converted into a set of points P describing a potential position of a cell (see Fig. 11.5).

Delay equilibrium can be achieved with wire snaking or delay insertion [375, 609]. Large delays with wire snaking however require prohibitively large area and increase the likelihood of routing congestion. Delay elements, in contrast, typically produce large delays, rendering them unsuitable if the delay difference is small. To

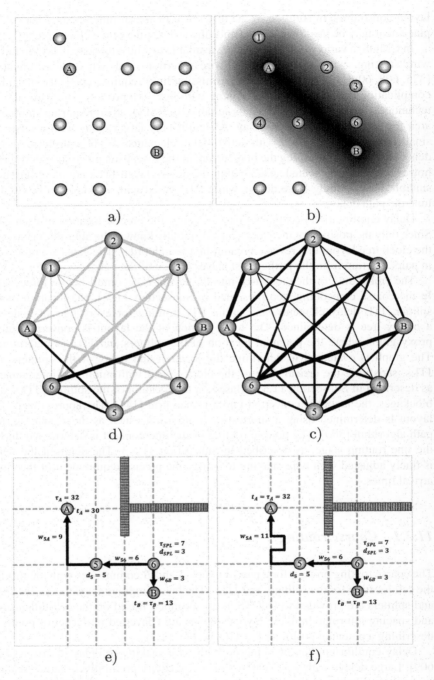

Fig. 11.5 (continued)

estimate the appropriate number of delay elements, a heuristic based on the delay difference n_{DE} is used,

$$n_{DE} = k \left\lfloor \frac{|\tau_A - \tau_B|}{t_{DE}} \right\rfloor, \tag{11.16}$$

where t_{DE} is the delay of the element, τ_A and τ_B are the required arrival time of the clock signal at, respectively, A and B, and k is an exploration parameter. To explore the layout space, additional gate cells are considered by setting the exploration parameter k above 1. A larger k produces a longer runtime at the cost of superior layout exploration. A good exploration-performance tradeoff is empirically achieved with $k \in [1.5, 2]$. According to (11.16), larger arrival time differences require additional delay elements. Note that n_{DE} provides an estimate of the number of elements. For example, fewer delay elements can be used if the elements are located far from the shortest path.

Those cells located close to the line connecting the nodes A and B form a subset of cells $P_{AB} \subset P$ suitable for routing. These gate cells combined with gates A and B form the set of proxy graph vertices $V_p = \{A, B\} \cup V_g$. Each pair of nodes in V_p except $\{A, B\}$ is connected with an edge. The weight of each edge is the Manhattan distance between the terminals. The edge weights therefore represent the length of the shortest rectilinear PTL connecting two points within a layout. For a proxy graph with $n_{DE} + 2$ nodes (two gates and n_{DE} gate cells), a total of $\frac{1}{2}(n_{DE} + 2)(n_{DE} + 1)$ edge weights is determined. The resulting undirected proxy graph is

$$\begin{aligned} G_p &= (V_p, E_p, w : E_p \to \mathbb{R}), \\ V_p &= \{A, B\} \cup V_g, \\ E_p &= \{\{a, b\} \in V_p^2 | a \neq b \wedge \{a, b\} \neq \{A, B\}\}, \\ w(a, b) &= |x_a - x_b| + |y_a - y_b|, \end{aligned} \tag{11.17}$$

where x_a and y_a are, respectively, x and y coordinates of node a. Note that the edge $\{A, B\}$ is explicitly excluded from the proxy graph since this proxy path

---

Fig. 11.5 Example of delay equilibration between gates A and B. τ_i is the arrival time of gate i. a) Initial layout. The empty circles represent vacant gate cells. b) Discovery of gate cells in proximity of the line connecting the two gates. The darker areas are closer to the line and are included in the proxy graph. c) Proxy graph containing six discovered gate cells and gates A and B. The thickness of the edges represents the closeness of the two nodes within a layout. d) A candidate proxy path $A - 5 - 6 - B$ is discovered in a proxy graph. e) The candidate proxy path transferred to the layout. w_{ij} is the delay of the path between nodes i and j, and d_i is the delay of the element at cell i. The splitter is therefore placed at node 6. The delay from the splitter to A relative to τ_A is smaller than the delay from the splitter to B relative to τ_B. The arrival time of the splitter is therefore based on the arrival time of B. Additional delay is required along the path to node A. f) Using wire snaking, additional delay is introduced along the path from the splitter to A. The arrival time is satisfied for both A and B.

does not include a necessary gate cell for a splitter. The paths between A and B model the connections in the layout. In this chapter, these paths are referred to as proxy paths. To determine the shortest proxy paths, the k-shortest path algorithm, described in [610], is used. This algorithm finds all loopless paths from source to target in increasing order of edge weight. By utilizing this algorithm, the proxy paths requiring the least interconnect resources are identified in the proxy graph.

Four crucial assumptions are made when producing the proxy graph G_p:

1. Each gate is equipped with a passive transmission line transmitter and receiver [611]. Including the PTL driver and receiver within each gate reduces the complexity of the routing process and enables a linear relationship between the length and delay of an interconnect [611].
2. The placement of splitters and delay elements is limited to certain areas of the layout. This assumption is consistent with a typical RSFQ IC layout where the placement of the cells is limited to narrow regions, such as the cell rows [612–614]. Only those nodes within the dedicated regions have a connection to the vacant gate cells. Other nodes are not connected to the device layer, preventing placement of the devices within prohibited zones. QuCTS can however handle arbitrary cell placement regions.
3. The size of the splitters and delay elements is assumed similar [615] and cells do not overlap. These assumptions simplify the placement of the splitters and delay elements, accelerating the clock tree synthesis process.
4. The orientation and pin configuration of the cells are assumed flexible, allowing the splitters and JTL elements to be arbitrarily oriented to satisfy routing needs.

The paths within a proxy graph model the connections in the layout. A shorter path corresponds to a PTL connection with a smaller interconnect length. To determine the shortest proxy paths, the k-shortest path algorithm, described in [610], is used. This algorithm finds all loopless paths from source to target in increasing order of edge weight. By utilizing this algorithm, the proxy paths requiring the least interconnect resources are identified.

11.3.2   Analysis of proxy path delay

If the proxy path contains more than one gate cell, the splitter placement is determined by the delay analysis described in this subsection. For example, consider the path $A - g_5 - g_6 - B$ shown in Fig. 11.5d. Placing a splitter at g_5 requires the SFQ clock pulse to arrive at the splitter at

$$\tau_{SPL|g_5} = \tau_A - w_{A,5} - d_{SPL}, \qquad (11.18)$$

where d_{SPL} is the splitter delay. The resulting clock arrival time at node B is

$$t_B = \tau_{SPL|g_5} + d_{SPL} + w_{5,6} + d_6 + w_{6,B} \gg \tau_B, \qquad (11.19)$$

where d_i is the delay of the element placed at node g_i, and $w_{i,j}$ for brevity is equivalent to $w(g_i, g_j)$. The resulting arrival time is significantly later than the required arrival time. Correcting this discrepancy with wire snaking requires significant area. If the splitter is instead placed at cell g_6, the SFQ clock pulse arrives at

$$\tau_{SPL|g_6} = \tau_A - w_{A,5} - d_5 - w_{5,6} - d_{SPL}, \tag{11.20}$$

yielding a clock arrival time at B,

$$t_B = \tau_{SPL|g_6} + d_{SPL} + w_{6,B} \approx \tau_B. \tag{11.21}$$

The discrepancy in arrival time is minimized and can be corrected with less area overhead using wire snaking.

To generalize this algorithm, consider the path $A - g_1 - \cdots - g_m - B$ with one splitter and $m - 1$ delay elements. Placing a splitter at cell g_k produces two paths,

$$q_A(g_k) = (A, g_1, \ldots, g_{k-1}, SPL), \tag{11.22}$$

$$q_B(g_k) = (SPL, g_k, g_{k+1}, \ldots, g_m, B). \tag{11.23}$$

The delay of each path is the sum of the splitter delay d_{SPL}, interconnect delay, and intentional delay,

$$d(q_A(g_k)) = W_{A,k} + S_{A,k} + d_{SPL}, \tag{11.24}$$

$$d(q_B(g_k)) = W_{B,k} + S_{B,k} + d_{SPL}, \tag{11.25}$$

where

$$W_{A,k} = w_{A,1} + \ldots + w_{k-1,k}, \tag{11.26}$$

$$W_{B,k} = w_{k,k+1} + \ldots + w_{m,B}, \tag{11.27}$$

$$S_{A,k} = \sum_{i=1}^{k-1} d_i, \tag{11.28}$$

$$S_{B,k} = \sum_{i=k+1}^{m} d_i. \tag{11.29}$$

Note that $d(g_k)$ is replaced with d_{SPL}.

To satisfy the arrival time at gate A, the SFQ clock pulse is required to arrive at the splitter at time

$$\tau_{SPL|g_k} = \tau_A - W_{A,k} - S_{A,k} - d_{SPL}. \tag{11.30}$$

The resulting arrival time at gate B is

$$t_B = \tau_A - W_{A,k} - S_{A,k} + W_{B,k} + S_{B,k}. \tag{11.31}$$

If the required arrival time at B is τ_B, the resulting mismatch in the clock arrival time is

$$\Delta(g_k) = \tau_A - \tau_B - W_{A,k} + W_{B,k} - S_{A,k} + S_{B,k}. \tag{11.32}$$

To minimize this mismatch, the splitter placement and delay of the delay elements are adjusted to minimize $|\Delta(g_k)|$. Ideally, $\Delta(g_k) = 0$, yielding

$$\tau_A - \tau_B = W_{A,k} + S_{A,k} - W_{B,k} - S_{B,k}. \tag{11.33}$$

Practically, however, a tolerance level $|\Delta(g_k)| < \varepsilon$ is set by the user that allows the proxy paths to be reasonably close to the target arrival time.

The intentional delay can be varied by choosing different delays from the set of possible delays, $D = d_1, d_2, ..., d_n | d_i < d_j \forall 1 < i < j < n$. The number of delay elements on each side of the splitter is, respectively, $k - 1$ and $m - k$. The total number of possible splitter locations is m, yielding a total number of delay combinations,

$$N = \sum_{k=1}^{n} \binom{k+n-2}{k-1} \binom{m-k+n-1}{m-k}. \tag{11.34}$$

To reduce the number of iterations, note that the gate with an earlier arrival time typically does not require a delay element. By varying the delay of the elements along the paths, the target arrival time can be achieved. In addition, a splitter is placed closer to the gate with a later arrival time, creating an unnecessary delay imbalance, requiring greater area. By restricting the splitter placement to $k \leq 2$, $i.e.$, no more than two nodes from the node with a later arrival time, the total number of combinations is reduced to

$$N = \binom{m+n-2}{m-1} + n \binom{m+n-3}{m-2}. \tag{11.35}$$

For $m = 10$ and $n = 5$, (11.35) yields 3,190 delay element combinations, as opposed to 48,620 by (11.34).

Many proxy paths are generated for further processing. Those proxy paths exhibiting a delay imbalance within a tolerance level are sorted by the number of delay elements and total interconnect length. The path tuning algorithm processes the least expensive paths first, yielding a significant savings in area.

11.3.3   Fine routing

In the fine routing stage, the proxy path selected in the previous section is converted into a layout. To determine a feasible placement for the interconnect, the routing is

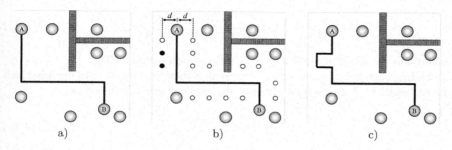

Fig. 11.6 Single iteration of the aura snaking process. a) Initial wire segment surrounded by vacant cells and blockages. b) Aura points generated within distance d from the wire. Two points near node A (filled) are selected for snaking. Note that the aura point is not generated within the blockage. c) The final extended segment.

based on a Hanan grid widely used in VLSI routing [616]. Hanan grid $H(S)$ is the set of points produced by drawing horizontal and vertical lines through each point in S. In QuCTS, the set of points for the Hanan grid consists of clocked gates, splitters, and JTL delay elements from the proxy graph, as well as bounds on the blockages, as illustrated in Fig. 11.5e. A graph $G_{H(S)}$ is based on points in $H(S)$. Two nodes in $G_{H(S)}$ are connected if the corresponding points are adjacent along any of the lines within the Hanan grid $H(S)$ and no blockage exists between the nodes. The weight of an edge is related to the propagation delay of the clock signal along the straight interconnect segment connecting the terminals of the edge.

The delay of the path generated in a Hanan grid graph is typically different from the estimate based on a proxy path. To adjust the delay and satisfy the arrival time requirements, the wire length is increased using wire snaking. A novel snaking method – aura snaking – is proposed here to increase the wire length, as illustrated in Fig. 11.6. The set of points Q within distance d from the interconnect segment is initially identified (see Fig. 11.6b). The set Q is referred to as an aura of the interconnect segment. The proximity metric of a point $q \in Q$ to other cells is defined as

$$p_q \equiv \sum_{p \in P_{AB} \ q} \frac{1}{||\vec{pq}||_s}, \qquad (11.36)$$

where $\vec{pq}$ is the vector connecting points p and q, and $||v_{pq}||_s$ is the s-norm of $\vec{pq}$. A point located closer to other cells has greater value of the proximity metric and can create congestion. Adjacent aura points with the smallest proximity metric are therefore chosen for snaking to minimize the likelihood of congestion. The aura points are evaluated for an intersection with blockages using polygon analysis algorithms, ensuring the feasibility of the wire snaking. Once the aura points are selected, the wire segment adjacent to the aura points is replaced with the snaking segment, as depicted in Fig. 11.6c. The interconnect is therefore extended by $2d$, increasing the wire delay by

$$\delta t = \frac{2d}{v}, \tag{11.37}$$

where v is the speed of the RSFQ pulse propagation within a PTL.

During each iteration, the delay of the path is increased by Δt until the mismatch is smaller than $\frac{2d}{v}$. In the final iteration, the aura distance is reduced to

$$d^* = \frac{v|t_A - t_B|}{2}, \tag{11.38}$$

where t_A and t_B denote the delay of the paths from the splitter to the corresponding gates. The last snaking operation therefore increases the delay of the extended path by exactly $|t_A - t_B|$, resulting in precise satisfaction of the clock arrival time.

Once a valid route for a pair of nodes is determined, several operations are necessary before the next pair is processed [65, 145]. The splitter, delay elements, and interconnect are placed into the layout. The corresponding points in P_{AB} are removed from the set P, preventing placement of additional gates in these locations. Interconnect is added to the blockages to ensure there is no intersection with any subsequent wires. The process described in this section is repeated for each pair of nodes within the circuit, thereby determining the position of the $N - 1$ splitters within a circuit with N clock sinks.

11.4   Case study

The validity of QuCTS is verified with the AMD2901 CPU Verilog model and the corresponding layout. 1,050 clocked gates are distributed within a 225 mm^2 IC. The maximum and minimum delay of each gate is known. The circuit topology is represented as a Verilog netlist. The PTL driver and receiver are embedded within each gate and splitter. The dimensions of each gate is 40 μm $\times$ 40 μm. Two layers of interconnect are dedicated to the clock distribution network. The vertical interconnects are placed in layer M2, and the horizontal interconnects are placed in layer M3. The gates are located in layer M5 and connected to layer M3 with vias. The interconnect pitch is 20 μm. The RSFQ pulse propagation speed in layers M2 and M3 is 6.25 μm/ps. The vertical connections between layers are established by the vias [72] and produce negligible delay.

The clock skew schedule is generated for a 154 ps clock period in less than one minute. The clock network layout is generated in 52.5 minutes and is shown in Fig. 11.7. A total of 2,290 gates are placed in the layout, 1,049 splitters and 1,241 delay elements. The total wire length is 1,027 mm occupying an area of 5.134 mm^2. 9,862 vias are placed between layers M2 and M3, and 6,676 vias are placed between layers M3 and M5. The maximum difference between the required and actual arrival times is 1.6 picoseconds.

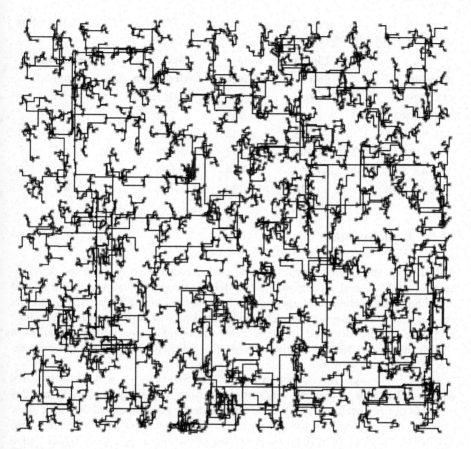

Fig. 11.7 Clock tree layout of AMD2901 synthesized with QuCTS.

The proposed tool has also been applied to a set of ISCAS'89 [617] and ITC'99 [618] benchmark circuits with high gate count. The cell placement for the benchmarks is generated with Synopsys IC design compiler [619]. The results are listed in Table 11.1. Note that the number of delay elements is linearly correlated with the number of clocked gates. For all six benchmarks, an average of 1.3 delay elements per splitter is required for the clock tree. This trend is explained by the clustering method used in QuCTS. Since the clock arrival time is considered during the routing process, gates with a similar arrival time are grouped together, resulting in a small delay imbalance, fewer delay elements, and less wire snaking. Despite the AMD2901 being composed of fewer gates than the S13207, the total wirelength is significantly larger. This trend is explained by the more compact placement of the cells in the S13207 as compared to the practical layout of AMD2901.

Table 11.1 Performance of QuCTS applied to AMD2901, ITC'99, and ISCAS'89 benchmark circuits with high gate count.

Circuit	Clocked gates	Delay elements	Total wirelength, mm	Runtime, minutes
AMD2901	1,049	1,241	1,027	53
ISCAS'89 S13207	1,636	2,405	272	73
ITC'99 B14	6,365	5,762	905	212
ISCAS'89 S38417	11,796	11,367	1,002	393
ISCAS'89 S35932	14,914	15,814	6,656	372
ITC'99 B18	45,710	71,090	31,736	2,309

11.5   Conclusions

Advances in RSFQ electronics over the past decades have enabled the development of sophisticated superconductive systems. Design methodologies and related algorithms and techniques targeting the large scale integration of RSFQ circuits are essential for managing the increasing complexity of superconductive systems. Elevating the performance of large scale superconductive systems requires a significant advancement in existing design capabilities, particularly the synchronous clock distribution network.

QuCTS — single flux Quantum Clock Tree Synthesis — is described in this chapter. This tool is the first clock tree synthesis capability for RSFQ circuits that also utilizes useful clock skew. Using quadratic programming, the clock skew schedule is optimized for robustness to parameter variations and converted into a schedule of clock arrival times. A binary clock tree is generated by recursive clustering of the clock sinks based on the physical location and, optionally, the clock arrival times. Splitters and delay elements are placed within the layout, and the paths are tuned to satisfy the schedule of arrival times. The tool is validated using the AMD2901 four bit microprocessor as well as ITC'99 and ISCAS'89 benchmark circuits. By exploring different topologies, QuCTS minimizes the number of delay elements and interconnect length. The clock arrival time schedule is precisely satisfied by using wire snaking.

Chapter 12
Conclusions

From its inception, the history of VLSI is characterized by the rapid rise in sophistication of integrated systems. Together with technology scaling and advancements in circuits and architecture, electronic design automation has greatly enhanced the complexity of VLSI systems, increasing the computational capabilities of humanity to unprecedented levels. Diverse technical expertise is required to produce modern high performance ICs, ranging from materials to software engineering. To facilitate collaboration among disparate fields, the VLSI system design process is divided into multiple abstraction layers. This approach concentrates the design effort on a specific level of an integrated system while assuming proper functionality at other abstraction layers.

A graph is a mathematical structure naturally suited for managing the complexity of large scale VLSI systems. By representing a complex system as an abstract network, the design effort can be concentrated on the key features of a system while discarding any extraneous information. Fewer details are considered, focusing on the key features of a system. Applications of graph theory are ubiquitous at every abstraction layer. At the register transfer layer, register allocation is often accomplished by graph coloring, minimizing communication between the CPU and memory. Ordered binary decision diagrams and AND-inverter graphs enable efficient graph-based processing of logic circuits. Graph-based techniques, such as random walks and network flow theory, facilitate the circuit analysis process of VLSI systems. Physical design is greatly enhanced by applying graph optimization algorithms to circuit partitioning, floorplanning, placement, and routing.

The complexity of modern VLSI systems necessitates the use of supporting infrastructural circuitry. Power, ground, and clock distribution networks are critical parts of an integrated system that ensure correct functionality and high levels of performance. A significant portion of the design resources is allocated to these supporting networks, such as dedicated on-chip layers and I/O pins.

R. Bairamkulov, E. G. Friedman, *Graphs in VLSI*,
https://doi.org/10.1007/978-3-031-11047-4_12

Maintaining correct functionality requires a nearly constant supply voltage despite highly volatile loads. Power delivery is a crucial part of a VLSI system, connecting the functional circuitry with an external power supply. Three major issues exist within the realm of power network design; namely, analysis, exploration, and synthesis. Graph theoretic methods for tackling these problems are presented in this book.

Analysis of the power delivery system requires an accurate estimate of the on-chip electrical characteristics. Conventional, general purpose analysis methods, such as modified nodal analysis and partial element equivalent circuit, enable accurate analysis of the behavior of power networks. The analysis of power delivery systems in modern integrated systems uses traditional methods; however, prohibitive computational time is required. Alternative graph-based approaches for circuit analysis have been developed. Domain decomposition, for example, utilizes graph cut algorithms to split a large circuit into multiple independent subcircuits to accelerate and parallelize the analysis of power grids. Geometric and algebraic multigrid techniques initially approximate a solution using a coarsened version of a grid. The final solution is subsequently determined by a smoothing process. Random walk-based methods exploit the duality between random walks in graphs and electrical circuits to evaluate the electrical behavior in linear time.

The size and regularity of power grids enable the use of compact models based on infinite grids. The advantage of an infinite grid model is the ability to estimate the effective resistance in constant time, i.e. the analysis runtime does not depend upon the size of the grid. This feature drastically accelerates the analysis of large power networks. The accuracy of an infinite grid model however significantly decreases near the boundaries of the grid. The image method is proposed to extend the infinite grid model to truncated infinite meshes, thereby maintaining the accuracy of the infinite grid model near the boundaries of the grid. The infinity mirror technique further extends the application of the infinite grid model to finite grids, greatly increasing the accuracy of the analysis. For a grid with ten billion nodes, a six orders of magnitude speedup with no degradation in accuracy is achieved as compared to modified nodal analysis.

Power network analysis is crucial for verifying the functionality of an IC. Precise analysis of an integrated system is only possible at the final stages of the design process when accurate system characteristics are known. Power integrity violations at these stages however require a massive system redesign, greatly increasing the design effort and time to market. To reduce the risk of violations and subsequent modification of the power network, system parameters, such as the number and characteristics of the voltage domains and decoupling capacitors, should be judiciously selected. The objective of power delivery exploration is to determine the electrical characteristics of an integrated system based on certain design parameters. With power delivery exploration, the effects of the design parameters on system performance can be determined, allowing informed decisions at early stages of the design process.

A versatile circuit level framework for power delivery exploration is presented. Based on certain design parameters, such as the total area of the decoupling capacitors or the number of voltage domains, a model of the power network can be generated. By analyzing this model, the effects of the electrical and non-electrical parameters on system performance can be evaluated. This framework is enhanced by constrained optimization, achieving a 15% reduction in decoupling capacitance and 38.6% reduction in power in an industrial case study.

While electrical modeling facilitates the development of power networks, further information can be extracted by considering the layout characteristics. Exploration of the physical power network is accomplished using SPROUT – the Smart Power ROUTing tool for prototyping board-level power networks. Based on the layout characteristics, such as the pattern of the ball grid array or the location of the decoupling capacitors, a prototype layout of the power network is generated. The electrical characteristics of a synthesized prototype are in close agreement with manually designed layouts. The effects of the physical design parameters on the electrical performance of a system can therefore be efficiently estimated during early stages of the design process.

Conventional power delivery networks utilize a single, board-level converter to generate and regulate on-chip supply voltages. Due to the large distance from the load, this method is sensitive to power noise due to fluctuations in load currents. In contrast, modern distributed power delivery systems utilize multiple on-chip converters placed near the load circuitry, effectively reducing the distance to the point-of-load. The number of on-chip regulators is however limited due to area constraints. A framework for placing on-chip voltage regulators is presented to minimize the voltage drop within the on-chip network with the fewest regulators. Using the fast grid analysis algorithm based on the infinity mirror technique, the computational runtime significantly accelerated, supporting grid-based power networks of arbitrary size. Practical scenarios related to the regulators are considered, such as restricted positions and limited current.

Since most high performance VLSI systems are synchronous, the clock distribution network is a vital part of the modern IC development process. The clock network is synthesized in three steps. The arrival time of the clock signal at each synchronous element is determined during clock skew scheduling. An abstract structure of a clock tree is determined during topological clock tree synthesis. A physical layout is produced during clock tree embedding. The position of the synchronous elements and wire lengths are adjusted to satisfy the required arrival times of the clock signal. Graph-based techniques are widely used during these processes, including cycle basis, spanning tree, Steiner minimal tree, and graph optimization.

The feature size of transistors in modern technology nodes is approaching atomic scales. Challenges in scaling of CMOS manufacturing technologies have motivated the development of alternative IC technologies and logic families. Rapid single flux quantum (RSFQ) is an emerging cryogenic superconductive technology promising a several orders of magnitude increase in speed and reduction in power as compared to CMOS. Several challenges however exist that prevent the widespread adoption of this technology. Most individual RSFQ logic gates require a clock signal, while a splitter gate is required to provide multiple fanout. Furthermore, special transmission lines are required for signaling. Due to these limitations, algorithms for synchronization of CMOS circuits need to be revised to support RSFQ circuits. QuCTS, an algorithm for clock distribution network synthesis in RSFQ circuits, is presented in this book. After producing a schedule of clock arrival times, a clock tree topology is determined by utilizing clustering algorithms. The location of the splitters is determined using the novel proxy graph technique, and the transmission lines are routed using a Hanan grid graph.

Graph theory plays an important role in facilitating the development of VLSI systems by providing powerful algorithms for design, analysis, and optimization. The relationship between graph theory and VLSI is however not unidirectional. While the applications of graph theory in VLSI greatly facilitate advancements in the VLSI system design process, diverse VLSI applications, in turn, motivate the development of novel graph algorithms. A virtuous cycle of theory and application therefore exists, advancing both graph theory and more powerful VLSI systems. The application side of this loop is explored in this book. The addressed issues highlight the vast potential of applying graph theory to the design of integrated systems, achieving orders of magnitude improvements in power, performance, and functionality. Exploring further applications of graph theory will likely bring enormous benefits to VLSI systems integration, greatly expanding the computational capabilities of humankind.

Appendix A
Green's function for a truncated grid

It is of interest to determine the lattice Green's function (LGF) for a truncated infinite mesh. The LGF is the response of a lattice to a unit perturbation at the origin,

$$\Delta_r G(x, y) = \delta(x, y); \, x, y \in \mathbb{Z}, \tag{A.1}$$

where $G(x, y)$ is the LGF, Δ_r is the discrete differential operator, and $\delta(x, y)$ is the Kronecker delta function, which is unity at the origin and zero elsewhere. The electrical form of (A.1) is obtained by applying KCL,

$$\Delta_r \phi(x, y) = r I_0 \delta(x, y); \, x, y \in \mathbb{Z}, \tag{A.2}$$

where

$$\phi(x, y) = r I_0 G(x, y) \tag{A.3}$$

is the potential distribution within the grid in response to a current I_0 injected at the origin. Combining (6.8) and (A.3) results in

$$R_{eff} = 2r \left(G(0, 0) - G(x - x_0, y - y_0) \right), \tag{A.4}$$

which is consistent with [479]. From [469], the LGF for an anisotropic infinite grid is

$$G(x, y) = \frac{k}{2\pi} \int_0^\pi \frac{e^{-|x|\alpha} \cos y\beta}{\sinh \alpha} d\beta. \tag{A.5}$$

To determine the LGF for a half-plane mesh, the following equation,

© The Author(s), under exclusive license to Springer Nature Switzerland AG 2023
R. Bairamkulov, E. G. Friedman, *Graphs in VLSI*,
https://doi.org/10.1007/978-3-031-11047-4

$$\Delta_r \phi_{half}(x, y) = r I_0 \delta(x, y); \, x \in \mathbb{N}_0, y \in \mathbb{Z}, \tag{A.6}$$

is solved by the image method. Expression (A.6) is transformed into

$$\Delta_r \phi_{half}(x, y) = r I_0 (\delta(x, y) + \delta(-x - 1, y)); \, x \in \mathbb{N}_0, y \in \mathbb{Z}. \tag{A.7}$$

Due to the linearity of Δ_r,

$$\Delta_r \phi_{half}(x, y) = \Delta_r \phi(x, y) + \Delta_r \phi(-x - 1, y); \, x \in \mathbb{N}_0, y \in \mathbb{Z}. \tag{A.8}$$

By the uniqueness theorem,

$$\phi_{half}(x, y) = \phi(x, y) + \phi(-x - 1, y); \, x \in \mathbb{N}_0, y \in \mathbb{Z}. \tag{A.9}$$

Using (6.22),

$$\phi(x, y) = \phi_0 - r I_0 \Omega_k(x, y); \, x, y \in \mathbb{Z}, \tag{A.10}$$

$$\phi_0 = \frac{k I_0 r}{2\pi} \int_0^\pi \frac{d\beta}{\sinh \alpha}. \tag{A.11}$$

Expression (A.9) reduces to

$$\phi_{half}(x, y) = 2\phi_0 - r I_0 (\Omega_k(x, y) + \Omega_k(-x - 1, y)); \, x \in \mathbb{N}_0, y \in \mathbb{Z}. \tag{A.12}$$

Based on (A.3), LGF for half-plane mesh is

$$G_{half}(x, y) = \frac{2\phi_0}{r I_0} - \Omega_k(x, y) - \Omega_k(-x - 1, y); \, x \in \mathbb{N}_0, y \in \mathbb{Z}. \tag{A.13}$$

Following similar steps for the quarter-plane mesh yields

$$\phi_{qt.}(x, y) = 4\phi_0 - r I_0 (\Omega_k(x, y) + \Omega_k(-x - 1, y) + \Omega_k(x, -y - 1) + \Omega_k(-x - 1, -y - 1));$$
$$x, y \in \mathbb{N}_0, \tag{A.14}$$

and

$$G_{qt.}(x, y) = \frac{4\phi_0}{r I_0} - \Omega_k(x, y) - \Omega_k(-x - 1, y) - \Omega_k(x, -y - 1) - \Omega_k(-x - 1, -y - 1));$$
$$x, y \in \mathbb{N}_0, \tag{A.15}$$

The effective resistance is determined in each case using (6.8).

Appendix B
Uniqueness based on boundary conditions

To demonstrate the validity of the method for a truncated mesh, it is proved here that the potentials within the circuit are uniquely determined by the boundary conditions. Thus, it is sufficient to maintain the same boundary conditions while modifying the topology to ensure the same electric potentials within a grid.

Consider the circuit shown in Fig. 6.2a. Boundary conditions $\phi_b(x, y)$ are imposed on a set of nodes $(x, y) \in S_v$. The arbitrary node (x_g, y_g) is connected to ground. The resulting boundary conditions of the system can be expressed as

$$\phi(x, y) = \phi_b(x, y), \text{ at } (x, y) \in S_v, \tag{B.1}$$

$$\phi(x_g, y_g) = 0. \tag{B.2}$$

Suppose current $I_{in}(x, y)$ is injected at specific nodes $(x, y) \in S_i$ such that

$$I(x, y) = \begin{cases} I_{in}(x, y), & \text{at } (x, y) \in S_i, & \text{(B.3a)} \\ 0 & \text{otherwise.} & \text{(B.3b)} \end{cases}$$

The uniqueness theorem states that the conditions described in (B.1) to (B.2) are sufficient to uniquely determine the potential $\phi(x, y)$ due to injected current $I(x, y)$. To prove this statement, assume this statement is incorrect and two distinct distributions of potentials exist that satisfy the boundary conditions:

$$\phi_1(x, y) \neq \phi_2(x, y). \tag{B.4}$$

Applying Kirchhoff's current law yields

$$I(x, y) = 4\phi_1(x, y) - \phi_1(x - 1, y) - \phi_1(x + 1, y) - \phi_1(x, y - 1) - \phi_1(x, y + 1), \tag{B.5}$$

© The Author(s), under exclusive license to Springer Nature Switzerland AG 2023   307
R. Bairamkulov, E. G. Friedman, *Graphs in VLSI*,
https://doi.org/10.1007/978-3-031-11047-4

$$I(x, y) = 4\phi_2(x, y) - \phi_2(x - 1, y) - \phi_2(x + 1, y) - \phi_2(x, y - 1) - \phi_2(x, y + 1).$$
$$(B.6)$$

Suppose that $\phi_3(x, y)$ is also a potential distribution such that

$$\phi_3(x, y) = \phi_1(x, y) - \phi_2(x, y). \tag{B.7}$$

From (B.5) and (B.6),

$$0 = 4\phi_3(x, y) - \phi_3(x - 1, y) - \phi_3(x + 1, y) - \phi_3(x, y - 1) - \phi_3(x, y + 1). \tag{B.8}$$

Expression (B.8) indicates that $\phi_3(x, y)$ is the potential distribution within a circuit without current injection. No currents, therefore, flow through the resistors and $\phi_3(x, y)$ is constant. Note that

$$\phi_3(x_1, y_1) = \phi_1(x_1, y_1) - \phi_2(x_1, y_1) = 0. \tag{B.9}$$

Therefore, since $\phi_3(x, y)$ is constant,

$$\phi_3(x, y) = \phi_3(x_1, y_1) = 0, \tag{B.10}$$

$$\phi_1(x, y) = \phi_2(x, y), \tag{B.11}$$

which contradicts (B.4), indicating that the conditions described in (B.1) to (B.2) uniquely determine the potential distribution in an infinite grid due to current injection $I(x, y)$.

Appendix C
Multilayer routing algorithm

If a routing path between terminals is not possible in a single layer due to the space being disjoint, a routing path can be allocated utilizing vias to connect the different layers. The routing process is decomposed into two parts. The layers through which a routing path are possible are initially determined. Due to the relatively high cost of the vias [620], the number of interlayer connections is also minimized. After placement of the vias, the routing process is decomposed into several single layer routing steps.

To determine the layers connecting the terminals, the routing process, described in Algorithm 7, is utilized. The available space for each layer is determined using Algorithm 1 (see Fig. C.1a). The available space within each layer is converted into an equivalent two-dimensional graph. The vertical edges connect the vertices within the adjacent layers through a via. This process produces a three-dimensional graph Γ_n^{3D}, as shown in Fig. C.1b. The vertical edges are assigned a higher cost, as compared to those edges within the same layer, to model the higher cost of the via.

Once a three-dimensional graph Γ_n^{3D} is generated, the shortest path between nodes in Θ_n is determined using a shortest path algorithm, such as Dijkstra [576] or Bellman-Ford [195]. After placing vias, the routing process is separately performed on each layer, from source to via, between vias, and from via to target. Those vias utilized during the routing process between nodes in Θ_n become a terminal on the respective layer (see Fig. C.1c). The multilayer routing process is thereby split into several two-dimensional routing steps.

R. Bairamkulov, E. G. Friedman, *Graphs in VLSI*,
https://doi.org/10.1007/978-3-031-11047-4

Algorithm 7 Determine least expensive multilayer path between nodes.

1: **procedure** MULTILAYER($\{A_n^1, A_n^2, ..., A_n^L\}, \{T_n = t_1^{l_1}, ..., t_k^{l_k}\}, r_{via}, w_{via}$)

2:     $\Gamma_n^{3D} = (V_n^{3D} = \varnothing, E_n^{3D} = \varnothing)$

3:     **for** $l = 1, 2, ..., L$ **do**

4:         $\Gamma_n^l = (V_n^l, E_n^l) \leftarrow space2graph(A_n^l, \Delta x = \Delta y = r_{via})$

5:         **for** each terminal $t_i^{l_i}$ in T_n **do**

6:             **if** $l_i = l$ **then**

7:                 $\Gamma_n^l, \Theta_l \leftarrow identifyTerminals(\Gamma_n^l, t_i^{l_i})$

8:         $V_n^{3D} \leftarrow V_n^{3D} \cup V_n^l$

9:         $E_n^{3D} \leftarrow E_n^{3D} \cup E_n^l$

10:        **for** each vertex v in Γ_n^l **do**

11:            **if** node v^{l-1} exists **then**

12:                $E_n^{3D} \leftarrow E_n^{3D} \cup \{v^l, v^{l-1}, w = w_{via}\}$

       $paths \leftarrow shortestpath(\Gamma_n, \theta_i, \{\theta_i + 1, ..., \theta_k\})$

13:    **for** $e = \{v_i, v_j\}$ in $paths$ **do**

14:        $\Theta_i \leftarrow v_i$

15:        $\Theta_j \leftarrow v_j$

16:    **return** $\Theta = \Theta_1, \Theta_2, ..., \Theta_L$

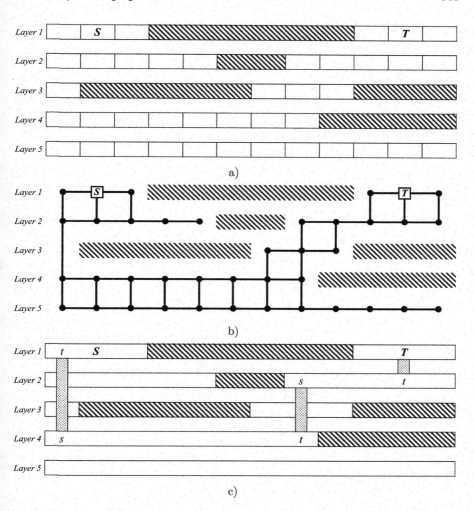

Fig. C.1 Cross-sectional view of the multilayer routing process. Prohibited areas are filled with a diagonal pattern. a) Available space is determined at each layer. Routing between the source (S) and target (T) is not possible within a single layer. b) Equivalent graph showing potential via locations. c) Via placement. The routing process is decomposed into three single layer routing steps between the local source s and target t

References

[1] S. F. Kennedy, D. J. Albers, G. L. Alexanderson, D. Dumbaugh, F. A. Faris, D. B. Haunsperger, and P. Zorn, *A Century of Advancing Mathematics*, The Mathematical Association of America, 2015.

[2] P. S. Rudman, *How Mathematics Happened: the First 50,000 Years*, Prometheus Books, 2009.

[3] B. Hopkins, *Resources for Teaching Discrete Mathematics: Classroom Projects, History Modules, and Articles*, Mathematical Association of America, 2009.

[4] L. Euler, "Solutio Problematis ad Geometriam Situs Pertinentis," *Commentarii Academiae Scientiarum Petropolitanae*, pp. 128–140, 1736.

[5] N. Biggs, E. K. Lloyd, and R. J. Wilson, *Graph Theory, 1736-1936*, Oxford University Press, 1986.

[6] A. Cayley, "On the Theory of the Analytical Forms Called Trees," *The London, Edinburgh, and Dublin Philosophical Magazine and Journal of Science*, Vol. 13, No. 85, pp. 172–176, March 1857.

[7] E. Frankland, *Lecture Notes for Chemical Students: Embracing Mineral and Organic Chemistry*, J. Van Voorst, 1866.

[8] J. J. Sylvester, "Chemistry and Algebra," *Nature*, Vol. 17, No. 432, pp. 284–284, February 1878.

[9] J. L. Moreno and H. H. Jennings, "Statistics of Social Configurations," *Sociometry*, Vol. 1, No. 3/4, pp. 342–374, April 1938.

[10] M. R. Garey and D. S. Johnson, "The Complexity of Near-Optimal Graph Coloring," *Journal of the ACM*, Vol. 23, No. 1, pp. 43–49, January 1976.

[11] B. W. Kernighan, *Some Graph Partitioning Problems Related to Program Segmentation.*, Ph.D. Thesis, Princeton University, 1969.

[12] J. S. Kilby, "Turning Potential into Realities: the Invention of the Integrated Circuit, Nobel Lecture," *ChemPhysChem*, Vol. 2, No. 8–9, pp. 482–489, September 2001.

[13] K. G. Beauchamp, *History of Telegraphy, Second Edition*, Institution of Engineering and Technology, 2001.

[14] S. Hemour and K. Wu, "Radio-Frequency Rectifier for Electromagnetic Energy Harvesting: Development Path and Future Outlook," *Proceedings of the IEEE*, Vol. 102, No. 11, pp. 1667–1691, November 2014.

[15] Digi-Key Electronics, *Omron Electronics G2R-1A-AC240*, [Online]. Available: https://www.digikey.com/en/products/detail/omron-electronics-inc-emc-div/G2R-1A-AC240/368714 [Accessed: June 24, 2021].

[16] Digi-Key Electronics, *Solen Électronique SI-12AX7B*, [Online]. Available: https://www.digikey.com/en/products/detail/solen/SI-12AX7B/10489816 [Accessed: June 24, 2021].

[17] P. E. Ceruzzi, "The Early Computers of Konrad Zuse, 1935 to 1945," *Annals of the History of Computing*, Vol. 3, No. 3, pp. 241–262, September 1981.

[18] H. H. Aiken and G. M. Hopper, "The Automatic Sequence Controlled Calculator – I," *Electrical Engineering*, Vol. 65, No. 8–9, pp. 384–391, September 1946.

[19] W. B. Fritz, "ENIAC – a Problem Solver," *IEEE Annals of the History of Computing*, Vol. 16, No. 1, pp. 25–45, March 1994.

[20] L. Hoddeson, "The Discovery of the Point-Contact Transistor," *Historical Studies in the Physical Sciences*, Vol. 12, No. 1, pp. 41–76, January 1981.

[21] P. R. Morris, *A History of the World Semiconductor Industry*, Peter Peregnius, 1990.

[22] A. E. Anderson, "Transistors in Switching Circuits," *Proceedings of the IRE*, Vol. 40, No. 11, pp. 1541–1558, November 1952.

[23] J. H. Felker, "Regenerative Amplifier for Digital Computer Applications," *Proceedings of the IRE*, Vol. 40, No. 11, pp. 1584–1596, November 1952.

[24] J. H. Felker, "Performance of TRADIC Transistor Digital Computer," *Proceedings of the Eastern Joint Computer Conference: Design and Application of Small Digital Computers*, pp. 46–49, December 1954.

[25] S. H. Lavington, *Early British Computers: the Story of Vintage Computers and the People Who Built Them*, Manchester University Press, 1980.

[26] T. Kilburn, R. Grimsdale, and D. Webb, "A Transistor Digital Computer with a Magnetic-Drum Store," *Proceedings of the IEE-Part B: Radio and Electronic Engineering*, Vol. 103, No. 3S, pp. 390–406, March 1956.

[27] L. C. Brown, "Flyable TRADIC: the First Airborne Transistorized Digital Computer," *IEEE Annals of the History of Computing*, Vol. 21, No. 4, pp. 55–61, December 1999.

[28] R. Herman, "Transistor Storage and Logic Circuits for Binary Data Processing," *Proceedings of the IEE-Part B: Electronic and Communication Engineering*, Vol. 106, No. 16S, pp. 663–674, May 1959.

[29] P. Cooper, "The U.S. Army Signal Corps' "Dick Tracy" Transistor Wrist Radio (1953)," *Proceedings of the IEEE*, Vol. 86, No. 1, pp. 163–169, January 1998.

[30] D. Burman, L. Fey, and D. Ingram, "Transistor Feedback Amplifiers in Carrier Telephony Systems," *Proceedings of the IEE-Part B: Electronic and Communication Engineering*, Vol. 106, No. 16S, pp. 587–595, May 1959.

[31] J. S. Kilby, "Invention of the Integrated Circuit," *IEEE Transactions on Electron Devices*, Vol. 23, No. 7, pp. 648–654, July 1976.

[32] W. F. Brinkman, D. E. Haggan, and W. W. Troutman, "A History of the Invention of the Transistor and Where It Will Lead Us," *IEEE Journal of Solid-State Circuits*, Vol. 32, No. 12, pp. 1858–1865, December 1997.

[33] K. Manchester, C. Sibley, and G. Alton, "Doping of Silicon by Ion Implantation," *Nuclear Instruments and Methods*, Vol. 38, pp. 169–174, December 1965.

[34] R. W. Bower and R. Dill, "Insulated Gate Field Effect Transistors Fabricated using the Gate as Source-Drain Mask," *Proceedings of the IEEE International Electron Devices Meeting*, pp. 102–104, October 1966.

[35] R. W. Bower, H. Dill, K. Aubuchon, and S. Thompson, "MOS Field Effect Transistors Formed by Gate Masked Ion Implantation," *IEEE Transactions on Electron Devices*, Vol. 15, No. 10, pp. 757–761, October 1968.

[36] C.-T. Sah, "Evolution of the MOS Transistor – from Conception to VLSI," *Proceedings of the IEEE*, Vol. 76, No. 10, pp. 1280–1326, October 1988.

[37] R. K. Booher, "MOS GP Computer," *Proceedings of the Fall Joint Computer Conference, Part I*, pp. 877–889, December 1968.

[38] A. G. Vacroux, "Microcomputers," *Scientific American*, Vol. 232, No. 5, pp. 32–41, May 1975.

[39] C. F. O'Donnell, "Engineering for Systems using Large Scale Integration," *Proceedings of the Fall Joint Computer Conference*, Vol. 1, pp. 867–876, December 1968.

[40] B. W. Kernighan and S. Lin, "An Efficient Heuristic Procedure for Partitioning Graphs," *The Bell System Technical Journal*, Vol. 49, No. 2, pp. 291–307, February 1970.

[41] C. Y. Lee, "An Algorithm for Path Connections and its Applications," *IRE Transactions on Electronic Computers*, Vol. EC-10, No. 3, pp. 346–365, September 1961.

[42] C. Smith, *The ZX Spectrum ULA: How to Design a Microcomputer*, ZX Design & Media, 2010.

[43] C. W. Beardsley, "Computer Aids for IC Design, Artwork, and Mask Generation," *IEEE Spectrum*, Vol. 8, No. 9, pp. 63–79, September 1971.

[44] W. Holt, "Computer Aided Design and New Manufacturing Methods for Electronic Materials," *Materials & Design*, Vol. 6, No. 1, pp. 42–45, March 1985.

[45] W. R. DeHaan, "The Bell Telephone Laboratories Automatic Graphic Schematic Drawing Program," *Proceedings of the SHARE Design Automation Project*, pp. 4.1 – 4.25, June 1966.

[46] C. Alaimo, "A Graphics Aided Drafting System (GRAD)," *Proceedings of the ACM/IEEE Design Automation Conference*, p. 4.1–4.23, January 1967.

[47] H. N. Lerman, "MADS - a Machine Aided Drafting System," *Proceedings of the SHARE Design Automation Project*, p. 10.1–10.31, January 1966.

[48] M. R. Davis and T. O. Ellis, "The RAND Tablet: a Man-Machine Graphical Communication Device," *Proceedings of the Fall Joint Computer Conference, Part I*, pp. 325–331, October 1964.

[49] P. LaCour, A. J. Reich, K. H. Nakagawa, S. F. Schulze, and L. Grodd, "New Stream Format: Progress Report on Containing Data Size Explosion," *Proceedings of the Design and Process Integration for Microelectronic Manufacturing*, Vol. 5042, pp. 214–221, July 2003.

[50] L. W. Nagel, "SPICE2: a Computer Program to Simulate Semiconductor Circuits," Memorandum No. ERL-M520, University of California, Berkeley, May 1975.

[51] W. Engl and D. Mlynski, "Topological Synthesis Procedure for Circuit Integration," *Proceedings of the IEEE International Solid-State Circuits Conference*, pp. 138–139, February 1969.

[52] E. Trischler, "An Integrated Design for Testability and Automatic Test Pattern Generation System: an Overview," *Proceedings of the ACM/IEEE Design Automation Conference*, pp. 209–215, June 1984.

[53] T. H. Bruggere and E. Hollomon, "Tools for Computer-Aided Engineering," *IEEE Computer Graphics and Applications*, Vol. 3, No. 9, pp. 48–53, December 1983.

[54] C. Sechen and A. Sangiovanni-Vincentelli, "The Timberwolf Placement and Routing Package," *IEEE Journal of Solid-State Circuits*, Vol. 20, No. 2, pp. 510–522, April 1985.

[55] E. G. Friedman, "Feedback in Silicon Compilers," *IEEE Circuits and Devices Magazine*, Vol. 1, No. 3, pp. 15–20, May 1985.

[56] B. M. Pangrle and D. D. Gajski, "Design Tools for Intelligent Silicon Compilation," *IEEE Transactions on Computer-Aided Design of Integrated Circuits and Systems*, Vol. 6, No. 6, pp. 1098–1112, November 1987.

[57] A. M. Prabhu, "Management Issues in EDA," *Proceedings of the ACM/IEEE Design Automation Conference*, pp. 41–47, June 1994.

[58] S. Schulz, G. Hinckley, G. Spirakis, K. Vahtra, J. Darringer, J. G. Janac, and H. Jones, "What Drives EDA Innovation?," *Proceedings of the ACM/IEEE Design Automation Conference*, pp. 790–791, June 2001.

[59] I. Bahar, A. K. Jones, S. Katkoori, P. H. Madden, D. Marculescu, and I. L. Markov, "'Scaling' the Impact of EDA Education: Preliminary Findings from the CCC Workshop Series on Extreme Scale Design Automation," *Proceedings of the IEEE International Conference on Microelectronic Systems Education*, pp. 64–67, June 2013.

[60] T. Rabuske, "Polymath: a Platform for Rapid Application Development of Modular EDA Tools," *Proceedings of the IEEE International Symposium on Circuits and Systems*, pp. 1–5, October 2020.

[61] L.-T. Wang, Y.-W. Chang, and K.-T. T. Cheng, *Electronic Design Automation: Synthesis, Verification, and Test*, Morgan Kaufmann, 2009.

[62] G. H. Mealy, "A Method for Synthesizing Sequential Circuits," *The Bell System Technical Journal*, Vol. 34, No. 5, pp. 1045–1079, September 1955.

[63] E. F. Moore, "Gedanken-Experiments on Sequential Machines," *Automata Studies.(AM-34), Volume 34*, pp. 129–154. April 1956.

[64] C. Ho, A. Ruehli, and P. Brennan, "The Modified Nodal Approach to Network Analysis," *IEEE Transactions on Circuits and Systems*, Vol. 22, No. 6, pp. 504–509, June 1975.

[65] R. Bairamkulov, A. Roy, M. Nagarajan, V. Srinivas, and E. G. Friedman, "SPROUT - Smart Power ROUting Tool for Board-Level Exploration and Prototyping," *Proceedings of the ACM/IEEE Design Automation Conference*, pp. 283–288, December 2021.

[66] R. Bairamkulov, T. Jabbari, E.G. Friedman, "QuCTS-single-flux quantum clock tree synthesis". IEEE Transactions on Computer-Aided Design of Integrated Circuits and Systems **41**(10), 3346–3358 (October 2022)

[67] C. Ababei, Y. Feng, B. Goplen, H. Mogal, T. Zhang, K. Bazargan, and S. Sapatnekar, "Placement and Routing in 3D Integrated Circuits," *IEEE Design & Test of Computers*, Vol. 22, No. 6, pp. 520–531, December 2005.

[68] M. Rostami, F. Koushanfar, and R. Karri, "A Primer on Hardware Security: Models, Methods, and Metrics," *Proceedings of the IEEE*, Vol. 102, No. 8, pp. 1283–1295, August 2014.

[69] M. Fyrbiak, S. Wallat, S. Reinhard, N. Bissantz, and C. Paar, "Graph Similarity and its Applications to Hardware Security," *IEEE Transactions on Computers*, Vol. 69, No. 4, pp. 505–519, April 2019.

[70] H. Qian, S. R. Nassif, and S. S. Sapatnekar, "Random Walks in a Supply Network," *Proceedings of the ACM/IEEE Design Automation Conference*, pp. 93–98, June 2003.

[71] R. Bairamkulov and E. G. Friedman, "Effective Resistance of Two-Dimensional Truncated Infinite Mesh Structures," *IEEE Transactions on Circuits and Systems I: Regular Papers*, Vol. 66, No. 11, pp. 4368–4376, November 2019.

[72] R. Bairamkulov and E. G. Friedman, "Effective Resistance of Finite Two-Dimensional Grids based on Infinity Mirror Technique," *IEEE Transactions on Circuits and Systems I: Regular Papers*, Vol. 67, No. 9, pp. 3224–3233, September 2020.

[73] T. Lei and S. Kumar, "A Two-Step Genetic Algorithm for Mapping Task Graphs to a Network on Chip Architecture," *Proceedings of the Euromicro Symposium on Digital System Design*, pp. 180–187, September 2003.

[74] A. E. Kiasari, Z. Lu, and A. Jantsch, "An Analytical Latency Model for Networks-on-Chip," *IEEE Transactions on Very Large Scale Integration (VLSI) Systems*, Vol. 21, No. 1, pp. 113–123, January 2013.

[75] R. Bairamkulov, K. Xu, M. Popovich, J. S. Ochoa, V. Srinivas, and E. G. Friedman, "Power Delivery Exploration Methodology based on Constrained Optimization," *IEEE Transactions on Computer-Aided Design of Integrated Circuits and Systems*, Vol. 39, No. 9, pp. 1916–1924, September 2020.

[76] J. A. Bondy and U. S. R. Murty, *Graph Theory with Applications*, Vol. 290, Macmillan London, 1976.

[77] D. B. West, *Introduction to Graph Theory*, Prentice Hall, 2001.

[78] R. J. Wilson, "What is a Graph?," *International Journal of Mathematical Education in Science and Technology*, Vol. 3, No. 2, pp. 107–115, June 1972.

[79] X. Ouvrard, J. M. Le Goff, and S. Marchand-Maillet, "On Adjacency and e-Adjacency in General Hypergraphs: Towards a New e-Adjacency Tensor," *Electronic Notes in Discrete Mathematics*, Vol. 70, pp. 71–76, December 2018.

[80] S. Sarkar and K. N. Sivarajan, "Hypergraph Models for Cellular Mobile Communication Systems," *IEEE Transactions on Vehicular Technology*, Vol. 47, No. 2, pp. 460–471, May 1998.

[81] T. F. Gonzalez, *Handbook of Approximation Algorithms and Metaheuristics: Contemporary and Emerging Applications, Vol. 2*, CRC Press, 2018.

[82] S. Klamt, U.-U. Haus, and F. Theis, "Hypergraphs and Cellular Networks," *PLoS Computational Biology*, Vol. 5, No. 5, pp. 1–6, May 2009.

[83] D. J. Galas, N. A. Sakhanenko, A. Skupin, and T. Ignac, "Describing the Complexity of Systems: Multivariable Set Complexity and the Information Basis of Systems Biology," *Journal of Computational Biology*, Vol. 21, No. 2, pp. 118–140, February 2014.

[84] H. Zhang, L. Song, and Z. Han, "Radio Resource Allocation for Device-to-Device Underlay Communication using Hypergraph Theory," *IEEE Transactions on Wireless Communications*, Vol. 15, No. 7, pp. 4852–4861, July 2016.

[85] J. Yu, D. Tao, and M. Wang, "Adaptive Hypergraph Learning and its Application in Image Classification," *IEEE Transactions on Image Processing*, Vol. 21, No. 7, pp. 3262–3272, July 2012.

[86] L. Li and T. Li, "News Recommendation via Hypergraph Learning: Encapsulation of User Behavior and News Content," *Proceedings of the ACM International Conference on Web Search and Data Mining*, pp. 305–314, February 2013.

[87] Y. M. Ponce, "Total and Local Quadratic Indices of the Molecular Pseudograph's Atom Adjacency Matrix: Applications to the Prediction of Physical Properties of Organic Compounds," *Molecules*, Vol. 8, No. 9, pp. 687–726, August 2003.

[88] H. Yang, Y. Gu, J. Zhu, K. Hu, and X. Zhang, "PGCN-TCA: Pseudo Graph Convolutional Network with Temporal and Channel-Wise Attention for Skeleton-Based Action Recognition," *IEEE Access*, Vol. 8, pp. 10040–10047, January 2020.

[89] M. Borowczak and R. Vemuri, "S* FSM: a Paradigm Shift for Attack Resistant FSM Designs and Encodings," *Proceedings of the ASE/IEEE International Conference on Biomedical Computing*, pp. 96–100, December 2012.

[90] Y. Sun, X. Yu, R. Bie, and H. Song, "Discovering Time-Dependent Shortest Path on Traffic Graph for Drivers Towards Green Driving," *Journal of Network and Computer Applications*, Vol. 83, pp. 204–212, April 2017.

[91] X. Xu, M. Niemeijer, Q. Song, M. Sonka, M. K. Garvin, J. M. Reinhardt, and M. D. Abramoff, "Vessel Boundary Delineation on Fundus Images using Graph-Based Approach," *IEEE Transactions on Medical Imaging*, Vol. 30, No. 6, pp. 1184–1191, June 2011.

[92] Y. Sun, T.-C. Wang, C. Wong, and C. Liu, "Routing for Symmetric FPGAs and FPICs," *Proceedings of the IEEE/ACM International Conference on Computer-Aided Design*, pp. 486–490, January 1993.

[93] W. Shi and B. Hong, "Resource Allocation with a Budget Constraint for Computing Independent Tasks in the Cloud," *Proceedings of the IEEE International Conference on Cloud Computing Technology and Science*, pp. 327–334, December 2010.

[94] D. E. Knuth and J. L. Szwarcfiter, "A Structured Program to Generate All Topological Sorting Arrangements," *Information Processing Letters*, Vol. 2, No. 6, pp. 153–157, February 1974.

[95] R. E. Bryant, "Graph-Based Algorithms for Boolean Function Manipulation," *IEEE Transactions on Computers*, Vol. C-35, No. 8, pp. 677–691, August 1986.

[96] F. Scarselli, M. Gori, A. C. Tsoi, M. Hagenbuchner, and G. Monfardini, "The Graph Neural Network Model," *IEEE Transactions on Neural Networks*, Vol. 20, No. 1, pp. 61–80, January 2009.

[97] H. Topcuoglu, S. Hariri, and M.-Y. Wu, "Performance-Effective and Low-Complexity Task Scheduling for Heterogeneous Computing," *IEEE Transactions on Parallel and Distributed Systems*, Vol. 13, No. 3, pp. 260–274, March 2002.

[98] W. Chen, Y. Yuan, and L. Zhang, "Scalable Influence Maximization in Social Networks under the Linear Threshold Model," *Proceedings of the IEEE International Conference on Data Mining*, pp. 88–97, December 2010.

[99] W. Shi and C. Su, "The Rectilinear Steiner Arborescence Problem is NP-Complete," *Proceedings of the ACM-SIAM Symposium on Discrete Algorithms*, pp. 780–787, February 2000.

[100] A. Artmeier, J. Haselmayr, M. Leucker, and M. Sachenbacher, "The Shortest Path Problem Revisited: Optimal Routing for Electric Vehicles," *Proceedings of the Conference on Artificial Intelligence*, pp. 309–316, September 2010.

[101] S. E. Dreyfus, "An Appraisal of Some Shortest-Path Algorithms," *Operations Research*, Vol. 17, No. 3, pp. 395–412, June 1969.

[102] T. J. Moser, "Shortest Path Calculation of Seismic Rays," *Geophysics*, Vol. 56, No. 1, pp. 59–67, January 1991.

[103] H. de Fraysseix, P. O. de Mendez, and P. Rosenstiehl, "Trémaux Trees and Planarity," *International Journal of Foundations of Computer Science*, Vol. 17, No. 5, pp. 1017–1029, April 2006.

[104] É. Lucas, *Récréations Mathématiques, Vol. 2*, Gauthier-Villars, 1883.

[105] R. Tarjan, "Depth-First Search and Linear Graph Algorithms," *SIAM Journal on Computing*, Vol. 1, No. 2, pp. 146–160, June 1972.

[106] V. N. Rao and V. Kumar, "Parallel Depth First Search. Part I. Implementation," *International Journal of Parallel Programming*, Vol. 16, No. 6, pp. 479–499, December 1987.

[107] T. H. Cormen, C. E. Leiserson, R. L. Rivest, and C. Stein, *Introduction to Algorithms*, MIT Press, 2009.

[108] R. E. Korf, "Depth-First Iterative-Deepening: An Optimal Admissible Tree Search," *Artificial Intelligence*, Vol. 27, No. 1, pp. 97–109, September 1985.

[109] E. F. Moore, "The Shortest Path through a Maze," *Proceedings of the International Symposium on Switching Theory*, pp. 285–292, April 1959.

[110] A. Shimbel, "Structure in Communication Nets," *Proceedings of the Symposium on Information Networks*, pp. 119–203, April 1954.

[111] L. R. Ford Jr, *Network Flow Theory*, RAND Corporation, Santa Monica, California, 1956.

[112] R. Bellman, "On a Routing Problem," *Quarterly of Applied Mathematics*, Vol. 16, No. 1, pp. 87–90, April 1958.

[113] E. W. Dijkstra, "A Note on Two Problems in Connexion with Graphs," *Numerische Mathematik*, Vol. 1, No. 1, pp. 269–271, December 1959.

[114] R. K. Ahuja, K. Mehlhorn, J. Orlin, and R. E. Tarjan, "Faster Algorithms for the Shortest Path Problem," *Journal of the ACM*, Vol. 37, No. 2, pp. 213–223, April 1990.

[115] D. B. Johnson, "Efficient Algorithms for Shortest Paths in Sparse Networks," *Journal of the ACM*, Vol. 24, No. 1, pp. 1–13, January 1977.

[116] M. L. Fredman and R. E. Tarjan, "Fibonacci Heaps and their Uses in Improved Network Optimization Algorithms," *Journal of the ACM*, Vol. 34, No. 3, pp. 596–615, July 1987.

[117] S. Hougardy, "The Floyd–Warshall Algorithm on Graphs with Negative Cycles," *Information Processing Letters*, Vol. 110, No. 8–9, pp. 279–281, April 2010.

[118] R. Dechter and J. Pearl, "Generalized Best-First Search Strategies and the Optimality of A*," *Journal of the ACM*, Vol. 32, No. 3, pp. 505–536, July 1985.

[119] N. Li, J. C. Hou, and L. Sha, "Design and Analysis of an MST-Based Topology Control Algorithm," *IEEE Transactions on Wireless Communications*, Vol. 4, No. 3, pp. 1195–1206, April 2005.

[120] D. Li, X. Jia, and H. Liu, "Energy Efficient Broadcast Routing in Static *Ad Hoc* Wireless Networks," *IEEE Transactions on Mobile Computing*, Vol. 3, No. 2, pp. 144–151, June 2004.

[121] Y. Tarabalka, J. Chanussot, and J. A. Benediktsson, "Segmentation and Classification of Hyperspectral Images using Minimum Spanning Forest Grown from Automatically Selected Markers," *IEEE Transactions on Systems, Man, and Cybernetics, Part B (Cybernetics)*, Vol. 40, No. 5, pp. 1267–1279, October 2009.

[122] W.-C. Tu, S. He, Q. Yang, and S.-Y. Chien, "Real-Time Salient Object Detection with a Minimum Spanning Tree," *Proceedings of the IEEE Conference on Computer Vision and Pattern Recognition*, pp. 2334–2342, June 2016.

[123] J. Cong, A. B. Kahng, G. Robins, M. Sarrafzadeh, and C.-K. Wong, "Provably Good Performance-Driven Global Routing," *IEEE Transactions on Computer-Aided Design of Integrated Circuits and Systems*, Vol. 11, No. 6, pp. 739–752, May 1992.

[124] O. Borůvka, "O Jistém Problému Minimálním," *Práce Moravské Přírodovědecké Společnosti*, Vol. 3, No. 3, pp. 37–58, January 1926.

[125] R. L. Graham and P. Hell, "On the History of the Minimum Spanning Tree Problem," *Annals of the History of Computing*, Vol. 7, No. 1, pp. 43–57, March 1985.

[126] G. Choquet, "'Etude de Certains R'eseaux de Routes," *Comptes Rendus Hebdomadaires des S'eances de L'Acad'emie des Sciences*, Vol. 206, pp. 310–313, July 1938.

[127] K. Florek, J. Łukaszewicz, J. Perkal, H. Steinhaus, and S. Zubrzycki, "Sur la Liaison et la Division des Points d'un Ensemble Fini," *Colloquium Mathematicum*, Vol. 2, No. 3–4, pp. 282–285, 1951.

[128] G. Zhou and M. Gen, "A Note on Genetic Algorithms for Degree-Constrained Spanning Tree Problems," *Networks: an International Journal*, Vol. 30, No. 2, pp. 91–95, December 1997.

[129] V. Jarník, "O Jistém Problému Minimálním. (z Dopisu Panu O. Borůvkovi)," *Práce Moravské Přírodovědecké Společnosti*, Vol. 6, No. 4, pp. 57–63, February 1930.

[130] B. M. Moret and H. D. Shapiro, "An Empirical Analysis of Algorithms for Constructing a Minimum Spanning Tree," *Proceedings of the Workshop on Algorithms and Data Structures*, pp. 400–411, August 1991.

[131] J. Kruskal, "On the Shortest Spanning Subtree of a Graph and the Traveling Salesman Problem," *Proceedings of the American Mathematical Society*, Vol. 7, No. 1, pp. 48–50, February 1956.

[132] F. Glover, D. Klingman, R. Krishnan, and R. Padman, "An In-Depth Empirical Investigation of Non-Greedy Approaches for the Minimum Spanning Tree Problem," *European Journal of Operational Research*, Vol. 56, No. 3, pp. 343–356, February 1992.

[133] B. Chazelle, "A Minimum Spanning Tree Algorithm with Inverse-Ackermann Type Complexity," *Journal of the ACM*, Vol. 47, No. 6, pp. 1028–1047, November 2000.

[134] S. Pettie, "An Inverse-Ackermann Style Lower Bound for the Online Minimum Spanning Tree Verification Problem," *Proceedings of the IEEE Symposium on Foundations of Computer Science*, pp. 155–163, November 2002.

[135] D. R. Karger, P. N. Klein, and R. E. Tarjan, "A Randomized Linear-Time Algorithm to Find Minimum Spanning Trees," *Journal of the ACM*, Vol. 42, No. 2, pp. 321–328, March 1995.

[136] J. Byrka, F. Grandoni, T. Rothvoß, and L. Sanita, "An Improved LP-Based Approximation for Steiner Tree," *Proceedings of the ACM Symposium on Theory of Computing*, pp. 583–592, June 2010.

[137] G. Robins and A. Zelikovsky, "Tighter Bounds for Graph Steiner Tree Approximation," *SIAM Journal on Discrete Mathematics*, Vol. 19, No. 1, pp. 122–134, May 2005.

[138] J. Byrka, F. Grandoni, T. Rothvoß, and L. Sanità, "Steiner Tree Approximation via Iterative Randomized Rounding," *Journal of the ACM*, Vol. 60, No. 1, pp. 1–33, February 2013.

[139] M. Brazil, R. L. Graham, D. A. Thomas, and M. Zachariasen, "On the History of the Euclidean Steiner Tree Problem," *Archive for History of Exact Sciences*, Vol. 68, No. 3, pp. 327–354, May 2014.

[140] R. Wilson and J. J. Watkins, *Combinatorics: Ancient & Modern*, Oxford University Press, 2013.

[141] J. Krarup and S. Vajda, "On Torricelli's Geometrical Solution to a Problem of Fermat," *IMA Journal of Management Mathematics*, Vol. 8, No. 3, pp. 215–224, July 1997.

[142] D. Du, P. M. Pardalos, and R. Graham, *Handbook of Combinatorial Optimization, Vol. 4*, Springer-Verlag, 1998.

[143] M. R. Garey, R. L. Graham, and D. S. Johnson, "The Complexity of Computing Steiner Minimal Trees," *SIAM Journal on Applied Mathematics*, Vol. 32, No. 4, pp. 835–859, June 1977.

[144] M. Hanan, "On Steiner's Problem with Rectilinear Distance," *SIAM Journal on Applied Mathematics*, Vol. 14, No. 2, pp. 255–265, March 1966.

[145] R. Bairamkulov, A. Roy, M. Nagarajan, V. Srinivas, and E. G. Friedman, "SPROUT - smart power routing tool for board-level exploration and prototyping," IEEE Transactions on Computer-Aided Design of Integrated Circuits and Systems, Vol. 41, No. 7, pp. 2263–2275, July 2022

[146] F. K. Hwang, D. S. Richards, and P. Winter, *The Steiner Tree Problem, Vol. 53*, Annals of Discrete Mathematics. Elsevier, 1992.

[147] J. M. Smith, *Algorithms for Generalized Steiner Network (GSN) Problems*, Ph.D. Thesis, University of Illinois at Urbana-Champaign, August 1978.

[148] G. Georgakopoulos and C. H. Papadimitriou, "The 1-Steiner Tree Problem," *Journal of Algorithms*, Vol. 8, No. 1, pp. 122–130, March 1987.

[149] A. B. Kahng and G. Robins, "A New Class of Steiner Trees Heuristics with Good Performance: the Iterated 1-Steiner-Approach.," *Proceedings of the IEEE/ACM International Conference on Computer-Aided Design*, pp. 428–431, November 1990.

[150] M. Borah, R. M. Owens, and M. J. Irwin, "An Edge-Based Heuristic for Steiner Routing," *IEEE Transactions on Computer-Aided Design of Integrated Circuits and Systems*, Vol. 13, No. 12, pp. 1563–1568, December 1994.

[151] M. Müller-Hannemann and S. Peyer, "Approximation of Rectilinear Steiner Trees with Length Restrictions on Obstacles," *Proceedings of the Workshop on Algorithms and Data Structures*, pp. 207–218, August 2003.

[152] H. Tang, G. Liu, X. Chen, and N. Xiong, "A Survey on Steiner Tree Construction and Global Routing for VLSI Design," *IEEE Access*, Vol. 8, pp. 68593–68622, April 2020.

[153] W. C. Elmore, "The Transient Response of Damped Linear Networks with Particular Regard to Wideband Amplifiers," *Journal of Applied Physics*, Vol. 19, No. 1, pp. 55–63, January 1948.

[154] G. Ajwani, C. Chu, and W.-K. Mak, "FOARS: FLUTE based Obstacle-Avoiding Rectilinear Steiner Tree Construction," *IEEE Transactions on Computer-Aided Design of Integrated Circuits and Systems*, Vol. 30, No. 2, pp. 194–204, February 2011.

[155] Y. Hu, T. Jing, X. Hong, Z. Feng, X. Hu, and G. Yan, "An-OARSMan: Obstacle-Avoiding Routing Tree Construction with Good Length Performance," *Proceedings of the ACM/IEEE Asia and South Pacific Design Automation Conference*, pp. 7–12, January 2005.

[156] C. Chiang and C.-S. Chiang, "Octilinear Steiner Tree Construction," *Proceedings of the IEEE Midwest Symposium on Circuits and Systems*, pp. I–603 – I–606, August 2002.

[157] G. Liu, G. Chen, and W. Guo, "DPSO based Octagonal Steiner Tree Algorithm for VLSI Routing," *Proceedings of the IEEE International Conference on Advanced Computational Intelligence*, pp. 383–387, October 2012.

[158] Y. Kanemoto, R. Sugawara, and M. Ohmura, "A Genetic Algorithm for the Rectilinear Steiner Tree in 3-D VLSI Layout Design," *Proceedings of the IEEE Midwest Symposium on Circuits and Systems*, Vol. 1, pp. I–465, July 2004.

[159] S. Tayu and S. Ueno, "On the Complexity of Three-Dimensional Channel Routing," *Proceedings of the IEEE International Symposium on Circuits and Systems*, pp. 3399–3402, May 2007.

[160] J.-T. Yan, Z.-W. Chen, and D.-H. Hu, "Timing-Driven Steiner Tree Construction for Three-Dimensional ICs," *Proceedings of the Joint International IEEE Northeast Workshop on Circuits and Systems and TAISA Conference*, pp. 335–338, June 2008.

[161] P. Maritz and S. Mouton, "Francis Guthrie: a Colourful Life," *The Mathematical Intelligencer*, Vol. 34, No. 3, pp. 67–75, July 2012.

[162] D. A. MacKenzie, *Mechanizing Proof: Computing, Risk, and Trust*, MIT Press, 2001.

[163] P. G. Tait, "On the Colouring of Maps," *Proceedings of the Royal Society of Edinburgh*, Vol. 10, pp. 501–503, November 1880.

[164] P. J. Heawood, "Map Color Theorems," *Quarterly Journal of Mathematics*, Vol. 24, pp. 332–338, December 1890.

[165] K. A. Appel and W. Haken, "Every Planar Map is Four Colorable. I. Discharging," *Illinois Journal of Mathematics*, Vol. 21, pp. 429–490, September 1977.

[166] K. A. Appel, W. Haken, and J. Koch, "Every Planar Map is Four Colorable. II. Reducibility," *Illinois Journal of Mathematics*, Vol. 21, pp. 491–567, September 1977.

[167] P. Formanowicz and K. Tanaś, "A Survey of Graph Coloring – its Types, Methods and Applications," *Foundations of Computing and Decision Sciences*, Vol. 37, No. 3, pp. 223–238, October 2012.

[168] H. Lu, M. Halappanavar, D. Chavarría-Miranda, A. H. Gebremedhin, A. Panyala, and A. Kalyanaraman, "Algorithms for Balanced Graph Colorings with Applications in Parallel Computing," *IEEE Transactions on Parallel and Distributed Systems*, Vol. 28, No. 5, pp. 1240–1256, May 2017.

[169] S. A. Taleb, H. Slimani, and M. E. Khanouche, "A Routing Approach based on (N, p)-Equitable b-Coloring of Graphs for Wireless Sensor Networks," *Proceedings of the IEEE International Conference on Smart Communications in Network Technologies*, pp. 90–95, October 2018.

[170] V. G. Vizing, "On Evaluation of the Chromatic Class of a p-Graph," *Discrete Analysis: Compilation of the Scientific Works*, Vol. 3, pp. 25–30, 1964.

[171] V. G. Vizing, "The Chromatic Class of a Multigraph," *Cybernetics*, Vol. 1, No. 3, pp. 32–41, May 1965.

[172] G. Raeisi and M. Gholami, "Edge Coloring of Graphs with Applications in Coding Theory," *China Communications*, Vol. 18, No. 1, pp. 181–195, January 2021.

[173] S. Gandham, M. Dawande, and R. Prakash, "Link Scheduling in Sensor Networks: Distributed Edge Coloring Revisited," *Proceedings of the Joint Conference of the IEEE Computer and Communications Societies.*, Vol. 4, pp. 2492–2501, March 2005.

[174] H. Li, P. Shenoy, and K. Ramamritham, "Scheduling Communication in Real-Time Sensor Applications," *Proceedings of the IEEE Real-Time and Embedded Technology and Applications Symposium*, pp. 10–18, May 2004.

[175] B. Awerbuch and Y. Azar, "Local Optimization of Global Objectives: Competitive Distributed Deadlock Resolution and Resource Allocation," *Proceedings of the IEEE Symposium on Foundations of Computer Science*, pp. 240–249, November 1994.

[176] A. B. Kahn, "Topological Sorting of Large Networks," *Communications of the ACM*, Vol. 5, No. 11, pp. 558–562, November 1962.

[177] J. Barnat, L. Brim, and P. Ročkai, "Parallel Partial Order Reduction with Topological Sort Proviso," *Proceedings of the IEEE International Conference on Software Engineering and Formal Methods*, pp. 222–231, September 2010.

[178] C. Mead and L. Conway, *Introduction to VLSI Systems*, Addison-Wesley, 1980.

[179] E. Gray and D. Tall, "Abstraction as a Natural Process of Mental Compression," *Mathematics Education Research Journal*, Vol. 19, No. 2, pp. 23–40, September 2007.

[180] C. Panara and M. R. Varney, *Local Government in Europe: the 'Fourth Level' in the EU Multi-Layered System of Governance*, Routledge, 2013.

[181] Council of State Governments, *The Book of the States*, Council of State Governments, 2019.

[182] Y. Wang, *Software Engineering Foundations: a Software Science Perspective*, CRC Press, 2007.

[183] B. Leiner, R. Cole, J. Postel, and D. Mills, "The DARPA Internet Protocol Suite," *IEEE Communications Magazine*, Vol. 23, No. 3, pp. 29–34, March 1985.

[184] Y. LeCun, Y. Bengio, and G. Hinton, "Deep Learning," *Nature*, Vol. 521, No. 7553, pp. 436–444, May 2015.

[185] N. H. E. Weste and K. Eshraghian, *Principles of CMOS VLSI Design: a Systems Perspective*, Addison-Wesley Publishing, 1985.

[186] M.-B. Lin, *Introduction to VLSI Systems: a Logic, Circuit, and System Perspective*, CRC Press, 2011.

[187] C. H. Sequin, "Managing VLSI Complexity: an Outlook," *Proceedings of the IEEE*, Vol. 71, No. 1, pp. 149–166, January 1983.

[188] A. Hashimoto and J. Stevens, "Wire Routing by Optimizing Channel Assignment within Large Apertures," *Proceedings of the ACM/IEEE Design Automation Conference*, pp. 155–169, June 1971.

[189] D. W. Hightower, "A Solution to Line-Routing Problems on the Continuous Plane," *Proceedings of the ACM/IEEE Design Automation Conference*, pp. 1–24, January 1969.

[190] K. Mikami, "A Computer Program for Optimal Routing of Printed Circuit Connectors," *Proceedings of the IFIPS*, pp. 1475–1478, November 1968.

[191] R. B. Hitchcock, "Cellular Wiring and the Cellular Modeling Technique," *Proceedings of the ACM/IEEE Design Automation Conference*, pp. 25–41, January 1969.

[192] J. D. Lesser and J. J. Shedletsky, "An Experimental Delay Test Generator for LSI Logic," *IEEE Computer Architecture Letters*, Vol. 29, No. 3, pp. 235–248, March 1980.

[193] Y. Crouzet and C. Landrault, "Design of Self-Checking MOS-LSI Circuits: Application to a Four-Bit Microprocessor," *IEEE Transactions on Computers*, Vol. C-29, No. 6, pp. 532–537, June 1980.

[194] A. J. Carlan, *State-of-the-Art Assessment of Testing and Testability of Custom LSI/VLSI Circuits, Vol. 2*, Aerospace Corporation, El Segundo, California, 1982.

[195] S. H. Gerez, *Algorithms for VLSI Design Automation*, Wiley, 1998.

[196] T. C. Bartee, I. L. Lebow, and I. S. Reed, *Theory and Design of Digital Machines*, McGraw-Hill, 1962.

[197] W. A. Clark, "Macromodular Computer Systems," *Proceedings of the Spring Joint Computer Conference*, pp. 335–336, April 1967.

[198] S. M. Ornstein, M. J. Stucki, and W. A. Clark, "A Functional Description of Macromodules," *Proceedings of the Spring Joint Computer Conference*, pp. 337–355, April 1967.

[199] M. J. Stucki, S. M. Ornstein, and W. A. Clark, "Logical Design of Macromodules," *Proceedings of the Spring Joint Computer Conference*, pp. 357–364, April 1967.

[200] J. R. Cox Jr., "Economy of Scale and Specialization in Large Computing Systems," *Computer Design*, Vol. 77, No. 11, pp. 77–80, November 1968.

[201] C. G. Bell and J. Grason, "The Register Transfer Module Design Concept," *Computer Design*, Vol. 10, No. 5, pp. 87–94, May 1971.

[202] S. H. Fuller, D. P. Siewiorek, and R. J. Swan, "Computer Modules: an Architecture for Large Digital Modules," *ACM SIGARCH Computer Architecture News*, Vol. 2, No. 4, pp. 231.-237, December 1973.

[203] J. Grason, C. G. Bell, and J. Eggert, "The Commercialization of Register Transfer Modules," *Computer*, Vol. 6, No. 10, pp. 23–28, October 1973.

[204] D. E. Thomas, E. D. Lagnese, R. A. Walker, J. V. Rajan, R. L. Blackburn, and J. A. Nestor, *Algorithmic and Register-Transfer Level Synthesis: the System Architect's Workbench*, Springer Science & Business Media, 1989.

[205] A. K. Bose, "Progress in VLSI Design Automation," *Proceedings of the European Solid-State Circuits Conference*, pp. 103–104, September 1984.

[206] A. Dewey, "VHSIC Hardware Description (VHDL) Development Program," *Proceedings of the ACM/IEEE Design Automation Conference*, pp. 625–628, June 1983.

[207] C.-L. Huang and S. Y. H. Su, "Approaches for Computer-Aided Logic/System Design using Hardware Description Language," *Proceedings of the International Computer Symposium*, pp. 772–790, December 1980.

[208] G. De Micheli, *Synthesis and Optimization of Digital Circuits*, McGraw-Hill Higher Education, 1994.

[209] D. D. Gajski, S. Abdi, A. Gerstlauer, and G. Schirner, *Embedded System Design: Modeling, Synthesis and Verification*, Springer Science & Business Media, 2009.

[210] M. R. Zargham, *Computer Architecture: Single and Parallel Systems*, Prentice-Hall, 1996.

[211] P. Gray, "On the Arithmometer of M. Thomas (de Colmar), and its Application to the Construction of Life Contingency Tables," *Journal of the Institute of Actuaries and Assurance Magazine*, Vol. 17, No. 4, pp. 249–266, October 1873.

[212] J. von Neumann, "First Draft of a Report on the EDVAC," *IEEE Annals of the History of Computing*, Vol. 15, No. 4, pp. 27–75, April 1993.

[213] R. Nair, "Evolution of Memory Architecture," *Proceedings of the IEEE*, Vol. 103, No. 8, pp. 1331–1345, August 2015.

[214] D. Comer, *Essentials of Computer Architecture*, CRC Press, 2017.

[215] V. Herdt, D. Große, P. Pieper, and R. Drechsler, "RISC-V based Virtual Prototype: an Extensible and Configurable Platform for the System-Level," *Journal of Systems Architecture*, Vol. 109, pp. 101756, October 2020.

[216] A. Waterman and K. Asanović, *The RISC-V Instruction Set Manual, Volume I: User-Level ISA*, RISC-V Foundation, 2019.

[217] J. L. Hennessy and D. A. Patterson, *Computer Architecture: a Quantitative Approach*, Elsevier, 2011.

[218] P. Briggs, K. D. Cooper, and L. Torczon, "Improvements to Graph Coloring Register Allocation," *ACM Transactions on Programming Languages and Systems*, Vol. 16, No. 3, pp. 428–455, May 1994.

[219] G. J. Chaitin, "Register Allocation and Spilling via Graph Coloring," *ACM SIGPLAN Notices*, Vol. 17, No. 6, pp. 98–101, June 1982.

[220] D. Watson, *A Practical Approach to Compiler Construction*, Springer, 2017.

[221] D. J. A. Welsh and M. B. Powell, "An Upper Bound for the Chromatic Number of a Graph and its Application to Timetabling Problems," *The Computer Journal*, Vol. 10, No. 1, pp. 85–86, January 1967.

[222] M. Poletto and V. Sarkar, "Linear Scan Register Allocation," *ACM Transactions on Programming Languages and Systems*, Vol. 21, No. 5, pp. 895–913, September 1999.

[223] J. Schneider and R. Wattenhofer, "A New Technique for Distributed Symmetry Breaking," *Proceedings of the ACM SIGACT-SIGOPS Symposium on Principles of Distributed Computing*, pp. 257–266, July 2010.

[224] F. Kuhn, "Weak Graph Colorings: Distributed Algorithms and Applications," *Proceedings of the Symposium on Parallelism in Algorithms and Architectures*, pp. 138–144, August 2009.

[225] N. Alon, L. Babai, and A. Itai, "A Fast and Simple Randomized Parallel Algorithm for the Maximal Independent Set Problem," *Journal of Algorithms*, Vol. 7, No. 4, pp. 567–583, December 1986.

[226] M. Luby, "A Simple Parallel Algorithm for the Maximal Independent Set Problem," *SIAM Journal on Computing*, Vol. 15, No. 4, pp. 1036–1053, November 1986.

[227] M. Rigo, *Advanced Graph Theory and Combinatorics*, John Wiley & Sons, 2016.

[228] A. Y. Zomaya and Y.-C. Lee, *Energy-Efficient Distributed Computing Systems*, Wiley, 2012.

[229] P. Marwedel, *Embedded System Design: Embedded Systems, Foundations of Cyber-Physical Systems, and the Internet of Things*, Springer Nature, 2021.

[230] E. G. Friedman, "Clock Distribution Networks in Synchronous Digital Integrated Circuits," *Proceedings of the IEEE*, Vol. 89, No. 5, pp. 665–692, May 2001.

[231] I. S. Kourtev, B. Taskin, and E. G. Friedman, *Timing Optimization through Clock Skew Scheduling*, Springer Science & Business Media, 2008.

[232] E. Salman and E. G. Friedman, *High Performance Integrated Circuit Design*, McGraw-Hill Professional, 2012.

[233] J. P. Fishburn, "Clock Skew Optimization," *IEEE Transactions on Computers*, Vol. 39, No. 7, pp. 945–951, July 1990.

[234] T. M. McWilliams, "Verification of Timing Constraints on Large Digital Systems," *Proceedings of the ACM/IEEE Design Automation Conference*, pp. 139–147, November 1980.

[235] T.-H. Chao, Y.-C. Hsu, J.-M. Ho, and A. B. Kahng, "Zero Skew Clock Routing with Minimum Wirelength," *IEEE Transactions on Circuits and Systems II: Analog and Digital Signal Processing*, Vol. 39, No. 11, pp. 799–814, September 1992.

[236] H. B. Bakoglu, "A Symmetric Clock Distribution Tree and Optimized High-Speed Interconnections for Reduced Clock Skew in ULSI & WSI Circuits," *Proceedings of the IEEE International Conference on Computer Design: VLSI in Computers*, pp. 118–122, October 1986.

[237] J. L. Neves and E. G. Friedman, "Design Methodology for Synthesizing Clock Distribution Networks Exploiting Nonzero Localized Clock Skew," *IEEE Transactions on Very Large Scale Integration (VLSI) Systems*, Vol. 4, No. 2, pp. 286–291, June 1996.

[238] E. G. Friedman, *Performance Limitations in Synchronous Digital Systems*, Ph.D. Thesis, University of California, Irvine, 1989.

[239] J. L. Neves and E. G. Friedman, "Optimal Clock Skew Scheduling Tolerant to Process Variations," *Proceedings of the ACM/IEEE Design Automation Conference*, pp. 623–628, June 1996.

[240] I. S. Kourtev and E. G. Friedman, "Simultaneous Clock Scheduling and Buffered Clock Tree Synthesis," *Proceedings of the IEEE International Symposium on Circuits and Systems*, Vol. 3, pp. 1812–1815, June 1997.

[241] I. S. Kourtev and E. G. Friedman, "Clock Skew Scheduling for Improved Reliability via Quadratic Programming," *Proceedings of the IEEE/ACM International Conference on Computer-Aided Design*, pp. 239–243, November 1999.

[242] A. V. Mezhiba and E. G. Friedman, *Power Distribution Networks in High Speed Integrated Circuits*, Springer Science & Business Media, 2004.

[243] J.-P. Queille and J. Sifakis, "Specification and Verification of Concurrent Systems in CESAR," *Proceedings of the International Symposium on Programming*, pp. 337–351, April 1982.

[244] E. M. Clarke and E. A. Emerson II, "Design and Synthesis of Synchronization Skeletons using Branching Time Temporal Logic," *Proceedings of the Logics of Programs Workshop*, pp. 52–71, May 1982.

[245] K. L. McMillan, *Symbolic Model Checking*, Springer, 1993.

[246] D. K. Pradhan and I. G. Harris, *Practical Design Verification*, Cambridge University Press, 2009.

[247] A. Mishchenko, S. Chatterjee, R. Jiang, and R. K. Brayton, *FRAIGs: a Unifying Representation for Logic Synthesis and Verification*, University of California, Berkeley, 2005.

[248] M. K. Ganai and A. Kuehlmann, "On-the-Fly Compression of Logical Circuits," *Proceedings of the International Workshop on Logic Synthesis*, pp. 1–7, June 2000.

[249] A. Kuehlmann, V. Paruthi, F. Krohm, and M. K. Ganai, "Robust Boolean Reasoning for Equivalence Checking and Functional Property Verification," *IEEE Transactions on Computer-Aided Design of Integrated Circuits and Systems*, Vol. 21, No. 12, pp. 1377–1394, December 2002.

[250] A. Kuehlmann and F. Krohm, "Equivalence Checking using Cuts and Heaps," *Proceedings of the ACM/IEEE Design Automation Conference*, pp. 263–268, June 1997.

[251] A. C. Ling, J. Zhu, and S. D. Brown, "Delay Driven AIG Restructuring using Slack Budget Management," *Proceedings of the ACM Great Lakes Symposium on VLSI*, pp. 163–166, May 2008.

[252] G. Pasandi and M. Pedram, "A Dynamic Programming-Based, Path Balancing Technology Mapping Algorithm Targeting Area Minimization," *Proceedings of the IEEE/ACM International Conference on Computer-Aided Design*, pp. 1–8, November 2019.

[253] A. Mishchenko, J. S. Zhang, S. Sinha, J. R. Burch, R. Brayton, and M. Chrzanowska-Jeske, "Using Simulation and Satisfiability to Compute Flexibilities in Boolean Networks," *IEEE Transactions on Computer-Aided Design of Integrated Circuits and Systems*, Vol. 25, No. 5, pp. 743–755, May 2006.

[254] M. Backes, J. M. Matos, R. Ribas, and A. Reis, "Reviewing AIG Equivalence Checking Approaches," *Proceedings of the ACM Microelectronics Student Forum*, pp. 1–4, September 2014.

[255] K. T. S. Oldham, *The Doctrine of Description: Gustav Kirchhoff, Classical Physics, and the "Purpose of all Science" in 19th-Century Germany*, Ph.D. Thesis, University of California, Berkeley, 2008.

[256] G. R. Kirchhoff, "Ueber die Auflösung der Gleichungen, auf welche man bei der Untersuchung der linearen Vertheilung galvanischer Ströme geführt wird," *Annalen der Physik*, Vol. 148, No. 12, pp. 497–508, October 1847.

[257] P. Appell and J. Drach, *Oeuvres de Henri Poincaré Tome I*, Gauthier-Villars, 1928.

[258] C. A. Desoer and E. S. Kuh, *Basic Circuit Theory*, McGraw-Hill, 1969.

[259] R. Merris, "Laplacian Matrices of Graphs: a Survey," *Linear Algebra and its Applications*, Vol. 197–198, pp. 143–176, February 1994.

[260] L. Nagel and R. Rohrer, "Computer Analysis of Nonlinear Circuits, Excluding Radiation (CANCER)," *IEEE Journal of Solid-State Circuits*, Vol. 6, No. 4, pp. 166–182, August 1971.

[261] W. J. McCalla and W. G. Howard, "BIAS-3-A Program for the Nonlinear D.C. Analysis of Bipolar Transistor Circuits," *IEEE Journal of Solid-State Circuits*, Vol. 6, No. 1, pp. 14–19, February 1971.

[262] A. Hemani, "Charting the EDA Roadmap," *IEEE Circuits and Devices Magazine*, Vol. 20, No. 6, pp. 5–10, December 2004.

[263] A. B. Kahng, J. Lienig, I. L. Markov, and J. Hu, *VLSI Physical Design: from Graph Partitioning to Timing Closure*, Springer Science & Business Media, 2011.

[264] T. Lengauer, *Combinatorial Algorithms for Integrated Circuit Layout*, Springer Science & Business Media, 2012.

[265] C. J. Alpert, D. P. Mehta, and S. S. Sapatnekar, *Handbook of Algorithms for Physical Design Automation*, CRC Press, 2008.

[266] N. Cserhalmi, O. Lowenschuss, and B. Scheff, "Efficient Partitioning for the Batch-Fabricated Fourth Generation Computer," *Proceedings of the Fall Joint Computer Conference*, Vol. 1, pp. 857–865, December 1968.

[267] H. Beelitz, H. Müller, R. Linhardt, and R. Sidnam, "Partitioning for Large-Scale Integration," *Proceedings of the IEEE International Solid-State Circuits Conference*, Vol. 10, pp. 50–51, February 1967.

[268] M. Armbruster, M. Fügenschuh, C. Helmberg, and A. Martin, "A Comparative Study of Linear and Semidefinite Branch-and-Cut Methods for Solving the Minimum Graph Bisection Problem," *Integer Programming and Combinatorial Optimization*, pp. 112–124, May 2008.

[269] G. Karypis, R. Aggarwal, V. Kumar, and S. Shekhar, "Multilevel Hypergraph Partitioning: Applications in VLSI Domain," *IEEE Transactions on Very Large Scale Integration (VLSI) Systems*, Vol. 7, No. 1, pp. 69–79, March 1999.

[270] C. M. Fiduccia and R. M. Mattheyses, "A Linear-Time Heuristic for Improving Network Partitions," *Proceedings of the ACM/IEEE Design Automation Conference*, pp. 175–181, June 1982.

[271] B. Krishnamurthy, "An Improved Min-Cut Algorithm for Partitioning VLSI Networks," *IEEE Transactions on Computers*, Vol. C-33, No. 5, pp. 438–446, May 1984.

[272] L. A. Sanchis, "Multiple-Way Network Partitioning," *IEEE Transactions on Computers*, Vol. 38, No. 1, pp. 62–81, January 1989.

[273] L.-T. Liu, M.-T. Kuo, S.-C. Huang, and C.-K. Cheng, "A Gradient Method on the Initial Partition of Fiduccia-Mattheyses Algorithm," *Proceedings of the IEEE/ACM International Conference on Computer-Aided Design*, pp. 229–234, November 1995.

[274] G. Karypis and V. Kumar, *METIS – Unstructured Graph Partitioning and Sparse Matrix Ordering System, Version 2.0*, University of Minnesota, Minneapolis, Minnesota, 1995.

[275] C. J. Alpert, L. W. Hagen, and A. B. Kahng, "A Hybrid Multilevel/Genetic Approach for Circuit Partitioning," *Proceedings of the IEEE Asia Pacific Conference on Circuits and Systems*, pp. 298–301, November 1996.

[276] G. Wang, W. Gong, and R. Kastner, "Application Partitioning on Programmable Platforms using the Ant Colony Optimization," *Journal of Embedded Computing*, Vol. 2, No. 1, pp. 119–136, September 2006.

[277] M. Abdelhalim, A. Salama, and S.-D. Habib, "Hardware Software Partitioning using Particle Swarm Optimization Technique," *Proceedings of the International Workshop on System on Chip for Real Time Applications*, pp. 189–194, December 2006.

[278] H. Murata, K. Fujiyoshi, S. Nakatake, and Y. Kajitani, "Rectangle-Packing-Based Module Placement," *Proceedings of the IEEE/ACM International Conference on Computer-Aided Design*, pp. 472–479, November 1995.

[279] M. Tang and X. Yao, "A Memetic Algorithm for VLSI Floorplanning," *IEEE Transactions on Systems, Man, and Cybernetics, Part B (Cybernetics)*, Vol. 37, No. 1, pp. 62–69, January 2007.

[280] I. L. Markov, J. Hu, and M.-C. Kim, "Progress and Challenges in VLSI Placement Research," *Proceedings of the IEEE*, Vol. 103, No. 11, pp. 1985–2003, November 2015.

[281] X. Hong, Sheqin D., Gang H., Y. Cai, C.-K. Cheng, and J. Gu, "Corner Block List Representation and its Application to Floorplan Optimization," *IEEE Transactions on Circuits and Systems II: Express Briefs*, Vol. 51, No. 5, pp. 228–233, May 2004.

[282] P.-N. Guo, C.-K. Cheng, and T. Yoshimura, "An O-Tree Representation of Non-Slicing Floorplan and its Applications," *Proceedings of the ACM/IEEE Design Automation Conference*, pp. 268–273, June 1999.

[283] F. Mao, N. Xu, and Y. Ma, "Hybrid Algorithm for Floorplanning using B^*-Tree Representation," *Proceedings of the IEEE International Symposium on Intelligent Information Technology Application*, Vol. 3, pp. 228–231, November 2009.

[284] C. Chu and Y.-C. Wong, "FLUTE: Fast Lookup Table based Rectilinear Steiner Minimal Tree Algorithm for VLSI Design," *IEEE Transactions on Computer-Aided Design of Integrated Circuits and Systems*, Vol. 27, No. 1, pp. 70–83, January 2008.

[285] R. H. J. M. Otten and R. K. Brayton, "Performance Planning," *Integration, the VLSI Journal*, Vol. 29, No. 1, pp. 1–24, March 2000.

[286] H. Ren, D. Z. Pan, C. J. Alpert, G.-J. Nam, and P. Villarrubia, "Hippocrates: First-Do-No-Harm Detailed Placement," *Proceedings of the ACM/IEEE Asia and South Pacific Design Automation Conference*, pp. 141–146, January 2007.

[287] J. Westra and P. Groeneveld, "Is Probabilistic Congestion Estimation Worthwhile?," *Proceedings of the ACM International Workshop on System Level Interconnect Prediction*, pp. 99–106, April 2005.

[288] C.-K. Koh and P. H. Madden, "Manhattan or Non-Manhattan? a Study of Alternative VLSI Routing Architectures," *Proceedings of the ACM Great Lakes Symposium on VLSI*, pp. 47–52, March 2000.

[289] M. Pathak and S. K. Lim, "Performance and Thermal-Aware Steiner Routing for 3-D Stacked ICs," *IEEE Transactions on Computer-Aided Design of Integrated Circuits and Systems*, Vol. 28, No. 9, pp. 1373–1386, September 2009.

[290] N. J. Nilsson, *The Quest for Artificial Intelligence*, Cambridge University Press, 2009.

[291] T. Yoshimura and E. S. Kuh, "Efficient Algorithms for Channel Routing," *IEEE Transactions on Computer-Aided Design of Integrated Circuits and Systems*, Vol. 1, No. 1, pp. 25–35, January 1982.

[292] T. G. Szymanski, "Dogleg Channel Routing is NP-Complete," *IEEE Transactions on Computer-Aided Design of Integrated Circuits and Systems*, Vol. 4, No. 1, pp. 31–41, January 1985.

[293] P. M. Kogge and H. S. Stone, "A Parallel Algorithm for the Efficient Solution of a General Class of Recurrence Equations," *IEEE Transactions on Computers*, Vol. C-22, No. 8, pp. 786–793, August 1973.

[294] R. P. Brent and H. T. Kung, "A Regular Layout for Parallel Adders," *IEEE Transactions on Computers*, Vol. C-31, No. 3, pp. 260–264, March 1982.

[295] A. E. Shapiro, F. Atallah, K. Kim, J. Jeong, J. Fischer, and E. G. Friedman, "Adaptive Power Gating of 32-bit Kogge Stone Adder," *Integration, the VLSI Journal*, Vol. 53, pp. 80–87, March 2016.

[296] R. Rojas, "Konrad Zuse's Legacy: the Architecture of the Z1 and Z3," *IEEE Annals of the History of Computing*, Vol. 19, No. 2, pp. 5–16, April/June 1997.

[297] D. A. Huffman, "The Synthesis of Sequential Switching Circuits," *Journal of the Franklin Institute*, Vol. 257, No. 4, pp. 275–303, March 1954.

[298] S. J. Mason, "Feedback Theory – Some Properties of Signal Flow Graphs," *Proceedings of the IRE*, Vol. 41, No. 9, pp. 1144–1156, September 1953.

[299] J. A. Brzozowski and E. J. McCluskey, "Signal Flow Graph Techniques for Sequential Circuit State Diagrams," *IEEE Transactions on Electronic Computers*, Vol. EC-12, No. 2, pp. 67–76, April 1963.

[300] M. C. Paull and S. H. Unger, "Minimizing the Number of States in Incompletely Specified Sequential Switching Functions," *IRE Transactions on Electronic Computers*, Vol. EC-8, No. 3, pp. 356–367, September 1959.

[301] B. Gilchrist, J. H. Pomerene, and S. Y. Wong, "Fast Carry Logic for Digital Computers," *IRE Transactions on Electronic Computers*, Vol. EC-4, No. 4, pp. 133–136, December 1955.

[302] J. Sklansky, "Conditional-Sum Addition Logic," *IRE Transactions on Electronic Computers*, Vol. EC-9, No. 2, pp. 226–231, June 1960.

[303] R. Bellman, *The Theory of Dynamic Programming*, RAND Corporation, Santa Monica, California, 1954.

[304] T. Haigh, M. Priestley, and C. Rope, "Engineering 'The Miracle of the ENIAC': Implementing the Modern Code Paradigm," *IEEE Annals of the History of Computing*, Vol. 36, No. 2, pp. 41–59, June 2014.

[305] G. H. Barnes, R. M. Brown, M. Kato, D. J. Kuck, D. L. Slotnick, and R. A. Stokes, "The ILLIAC IV Computer," *IEEE Transactions on Computers*, Vol. C-17, No. 8, pp. 746–757, August 1968.

[306] J. Cortadella and S. S Sapatnekar, "Static Timing Analysis," *Electronic Design Automation for IC Implementation, Circuit Design, and Process Technology: Circuit Design, and Process Technology*, pp. 133–154. January 2016.

[307] D. G. Malcolm, J. H. Roseboom, C. E. Clark, and W. Fazar, "Application of a Technique for Research and Development Program Evaluation," *Operations Research*, Vol. 7, No. 5, pp. 646–669, October 1959.

[308] M. Engwall, "PERT, Polaris, and the Realities of Project Execution," *International Journal of Managing Projects in Business*, pp. 595–616, September 2012.

[309] T. I. Kirkpatrick and N. R. Clark, "PERT as an Aid to Logic Design," *IBM Journal of Research and Development*, Vol. 10, No. 2, pp. 135–141, March 1966.

[310] L. W. Cotten, "Circuit Implementation of High-Speed Pipeline Systems," *Proceedings of the Fall Joint Computer Conference*, Vol. 1, pp. 489–504, November 1965.

[311] H. H. Loomis and M. R. McCoy, "A Theory of High-Speed Clocked Logic," *Proceedings of the Symposium on Switching Circuit Theory and Logical Design*, pp. 150–161, October 1965.

[312] W. H. Howe, "High-Speed Logic Circuit Considerations," *Proceedings of the Fall Joint Computer Conference*, Vol. 1, pp. 505–510, November 1965.

[313] L. W. Cotten, "Maximum-Rate Pipeline Systems," *Proceedings of the AFIPS Spring Joint Computer Conference*, pp. 581–586, May 1969.

[314] S. Harting and P. Verma, "Universal Time Frame: a New Network Feature for Delay Minimization," *IEEE Transactions on Communications*, Vol. 23, No. 11, pp. 1339–1342, November 1975.

[315] S. D. Rosenbaum and J. T. Caves, "8192-Bit Block Addressable CCD Memory," *IEEE Journal of Solid-State Circuits*, Vol. 10, No. 5, pp. 273–280, October 1975.

[316] G. P. Hyatt and G. Ohlberg, "Electrically Alterable Digital Differential Analyzer," *Proceedings of the Spring Joint Computer Conference*, pp. 161–169, April 1968.

[317] S. Dhar, M. A. Franklin, and D. F. Wan, "Reduction of Clock Delays in VLSI Structures," *Proceedings of the IEEE International Conference on Computer Design: VLSI in Computers*, pp. 778–783, October 1984.

[318] J. W. Goodman, F. J. Leonberger, S.-Y. Kung, and R. A. Athale, "Optical Interconnections for VLSI Systems," *Proceedings of the IEEE*, Vol. 72, No. 7, pp. 850–866, July 1984.

[319] A. J. Martin and M. Nystrom, "Asynchronous Techniques for System-on-Chip Design," *Proceedings of the IEEE*, Vol. 94, No. 6, pp. 1089–1120, June 2006.

[320] L. C. Bening, T. A. Lane, C. R. Alexander, and J. E. Smith, "Developments in Logic Network Path Delay Analysis," *Proceedings of the ACM/IEEE Design Automation Conference*, pp. 605–615, June 1982.

[321] A. E. Dunlop, V. D. Agrawal, D. N. Deutsch, M. F. Jukl, P. Kozak, and M. Wiesel, "Chip Layout Optimization using Critical Path Weighting," *Proceedings of the ACM/IEEE Design Automation Conference*, pp. 133–136, June 1984.

[322] J. K. Ousterhout, "Crystal: a Timing Analyzer for nMOS VLSI Circuits," *Proceedings of the Caltech Conference on Very Large Scale Integration*, pp. 57–69, March 1983.

[323] G. Martin, J. Berrie, T. Little, D. Mackay, J. McVean, D. Tomsett, and L. Weston, "An Integrated LSI Design Aids System," *Microelectronics Journal*, Vol. 12, No. 4, pp. 18–22, July 1981.

[324] R.-S. Tsay and I. Lin, "Robin Hood: a System Timing Verifier for Multi-Phase Level-Sensitive Clock Designs," *Proceedings of the IEEE International ASIC Conference and Exhibit*, pp. 516–519, September 1992.

[325] K. A. Sakallah, T. N. Mudge, and O. A. Olukotun, "CheckT_c and minT_c: Timing Verification and Optimal Clocking of Synchronous Digital Circuits," *Proceedings of the IEEE/ACM International Conference on Computer-Aided Design*, pp. 552–555, November 1990.

[326] T. G. Szymanski, "Computing Optimal Clock Schedules," *Proceedings of the ACM/IEEE Design Automation Conference*, pp. 399–404, June 1992.

[327] J. Rosenfeld and E. G. Friedman, "Design Methodology for Global Resonant H-Tree Clock Distribution Networks," *IEEE Transactions on Very Large Scale Integration (VLSI) Systems*, Vol. 15, No. 2, pp. 135–148, April 2007.

[328] C. A. Tsao and C. Koh, "UST/DME: a Clock Tree Router for General Skew Constraints," *ACM Transactions on Design Automation of Electronic Systems*, Vol. 7, No. 3, pp. 359–379, July 2002.

[329] K. D. Boese and A. B. Kahng, "Zero-Skew Clock Routing Trees with Minimum Wirelength," *Proceedings of the IEEE International ASIC Conference and Exhibit*, pp. 17–21, September 1992.

[330] N. Shenoy, R. K. Brayton, and Sangiovanni-Vincentelli A. L., "Graph Algorithms for Clock Schedule Optimization," *Proceedings of the IEEE/ACM International Conference on Computer-Aided Design*, pp. 132–136, November 1992.

[331] R. B. Deokar and S. S. Sapatnekar, "A Graph-Theoretic Approach to Clock Skew Optimization," *Proceedings of the IEEE International Symposium on Circuits and Systems*, Vol. 1, pp. 407–410, May 1994.

[332] C.-T. Tsai and W.-Y. Yip, "An Experimental Technique for Full Package Inductance Matrix Characterization," *IEEE Transactions on Components, Packaging, and Manufacturing Technology: Part B*, Vol. 19, No. 2, pp. 338–343, May 1996.

[333] C.-H. Park and B. Kim, "A Low-Noise, 900-MHz VCO in 0.6-μm CMOS," *IEEE Journal of Solid-State Circuits*, Vol. 34, No. 5, pp. 586–591, May 1999.

[334] J. Lee and S. Cho, "A 10MHz 80μW 67 ppm/°C CMOS Reference Clock Oscillator with a Temperature Compensated Feedback Loop in 0.18μm CMOS," *Proceedings of the IEEE Symposium on VLSI Circuits*, pp. 226–227, June 2009.

[335] U. Denier, "Analysis and Design of an Ultralow-Power CMOS Relaxation Oscillator," *IEEE Transactions on Circuits and Systems I: Regular Papers*, Vol. 57, No. 8, pp. 1973–1982, August 2010.

[336] J. T. Santos and R. G. Meyer, "A One-Pin Crystal Oscillator for VLSI Circuits," *IEEE Journal of Solid-State Circuits*, Vol. 19, No. 2, pp. 228–236, April 1984.

[337] R. Sedgewick, *Algorithms in C, Part 5: Graph Algorithms, Third Edition*, Addison-Wesley Professional, 2001.

[338] D. B. Johnson, "Finding All the Elementary Circuits of a Directed Graph," *SIAM Journal on Computing*, Vol. 4, No. 1, pp. 77–84, March 1975.

[339] D. M. Chapiro, *Globally Asynchronous Locally Synchronous Circuits*, Ph.D. Thesis, Stanford University, 1984.

[340] I. Lin, J. A. Ludwig, and K. Eng, "Analyzing Cycle Stealing on Synchronous Circuits with Level-Sensitive Latches," *Proceedings of the ACM/IEEE Design Automation Conference*, pp. 393–398, June 1992.

[341] T.-Y. Cheung, "Graph Traversal Techniques and the Maximum Flow Problem in Distributed Computation," *IEEE Transactions on Software Engineering*, Vol. SE-9, No. 4, pp. 504–512, July 1983.

[342] E. Korach and Z. Ostfeld, "DFS Tree Construction: Algorithms and Characterizations," *Proceedings of the International Workshop on Graph-Theoretic Concepts in Computer Science*, pp. 87–106, June 1988.

[343] J.-L. Tsai, D. Baik, C. C.-P. Chen, and K. K. Saluja, "A Yield Improvement Methodology using Pre- and Post-Silicon Statistical Clock Scheduling," *Proceedings of the IEEE/ACM International Conference on Computer-Aided Design*, pp. 611–618, November 2004.

[344] J.-J. Liou, A. Krstic, L.-C. Wang, and K.-T. Cheng, "False-Path-Aware Statistical Timing Analysis and Efficient Path Selection for Delay Testing and Timing Validation," *Proceedings of the ACM/IEEE Design Automation Conference*, pp. 566–569, June 2002.

[345] D. Blaauw, K. Chopra, A. Srivastava, and L. Scheffer, "Statistical Timing Analysis: from Basic Principles to State of the Art," *IEEE Transactions on Computer-Aided Design of Integrated Circuits and Systems*, Vol. 27, No. 4, pp. 589–607, April 2008.

[346] Y. Wang, W.-S. Luk, X. Zeng, J. Tao, C. Yan, J. Tong, W. Cai, and J. Ni, "Timing Yield Driven Clock Skew Scheduling considering Non-Gaussian Distributions of Critical Path Delays," *Proceedings of the ACM/IEEE Design Automation Conference*, pp. 223–226, June 2008.

[347] S.-H. Huang, C.-H. Cheng, C.-M. Chang, and Y.-T. Nieh, "Clock Period Minimization with Minimum Delay Insertion," *Proceedings of the ACM/IEEE Design Automation Conference*, pp. 970—975, June 2007.

[348] S.-H. Huang and Y.-T. Nieh, "Clock Period Minimization of Non-Zero Clock Skew Circuits," *Proceedings of the IEEE/ACM International Conference on Computer-Aided Design*, pp. 809–812, November 2003.

[349] B. Taskin and I. S. Kourtev, "Delay Insertion Method in Clock Skew Scheduling," *IEEE Transactions on Computer-Aided Design of Integrated Circuits and Systems*, Vol. 25, No. 4, pp. 651–663, April 2006.

[350] W. Burleson, M. Ciesielski, F. Klass, and W. Liu, "Wave-Pipelining: a Tutorial and Research Survey," *IEEE Transactions on Very Large Scale Integration (VLSI) Systems*, Vol. 6, No. 3, pp. 464–474, September 1998.

[351] C. T. Gray, W. Liu, and R. K. Cavin, III, *Wave Pipelining: Theory and CMOS Implementation*, Springer Science & Business Media, 1994.

[352] D. Velenis, K. T. Tang, I. S. Kourtev, V. Adler, F. Baez, and E. G. Friedman, "Demonstration of Speed and Power Enhancements through Application of Nonzero Clock Skew Scheduling," *Proceedings of the ACM/IEEE International Workshop on Timing Issues in the Specification and Synthesis of Digital Systems*, pp. 58–63, December 2000.

[353] D. Velenis, K. T. Tang, I. S. Kourtev, V. Adler, F. Baez, and E. G. Friedman, "Demonstration of Speed and Power Enhancements on an Industrial Circuit through Application of Clock Skew Scheduling," *Journal of Circuits, Systems, and Computers*, Vol. 11, No. 03, pp. 231–245, June 2002.

[354] M. El-Moursy and E. Friedman, "Power Characteristics of Inductive Interconnect," *IEEE Transactions on Very Large Scale Integration (VLSI) Systems*, Vol. 12, No. 12, pp. 1295–1306, December 2004.

[355] K. M. Cao, W.-C. Lee, W. Liu, X. Jin, P. Su, S. K. H. Fung, J. X. An, B. Yu, and C. Hu, "BSIM4 Gate Leakage Model Including Source-Drain Partition," *Proceedings of the IEEE International Electron Devices Meeting*, pp. 815–818, December 2000.

[356] S.-M. Kang and Y. Leblebici, *CMOS Digital Integrated Circuits*, McGraw-Hill Education, 2003.

[357] K. Usami, M. Igarashi, F. Minami, T. Ishikawa, M. Kanzawa, M. Ichida, and K. Nogami, "Automated Low-Power Technique Exploiting Multiple Supply Voltages Applied to a Media Processor," *IEEE Journal of Solid-State Circuits*, Vol. 33, No. 3, pp. 463–472, March 1998.

[358] S. Köse, S. Tam, S. Pinzon, B. McDermott, and E. G. Friedman, "Active Filter-Based Hybrid On-Chip DC–DC Converter for Point-of-Load Voltage Regulation," *IEEE Transactions on Very Large Scale Integration (VLSI) Systems*, Vol. 21, No. 4, pp. 680–691, April 2013.

[359] R. Bairamkulov and E. G. Friedman, "On-Chip Voltage Regulator Distribution," *Proceedings of the IEEE International Conference on Computer-Aided Design*, November 2022.

[360] V. Kursun and E. G. Friedman, "Low Swing Dual Threshold Voltage Domino Logic," *Proceedings of the ACM Great Lakes Symposium on VLSI*, pp. 47–52, April 2002.

[361] V. Kursun and E. G. Friedman, *Multi-Voltage CMOS Circuit Design*, John Wiley & Sons, 2006.

[362] O. Coudert, "Gate Sizing for Constrained Delay/Power/Area Optimization," *IEEE Transactions on Very Large Scale Integration (VLSI) Systems*, Vol. 5, No. 4, pp. 465–472, December 1997.

[363] M. Ni and S. O. Memik, "Leakage Power-Aware Clock Skew Scheduling: Converting Stolen Time into Leakage Power Reduction," *Proceedings of the ACM/IEEE Design Automation Conference*, pp. 610–613, June 2008.

[364] L. Li, J. Sun, Y. Lu, H. Zhou, and X. Zeng, "Low Power Discrete Voltage Assignment under Clock Skew Scheduling," *Proceedings of the ACM/IEEE Asia and South Pacific Design Automation Conference*, pp. 515–520, January 2011.

[365] L. H. Chen, M. Marek-Sadowska, and F. Brewer, "Buffer Delay Change in the Presence of Power and Ground Noise," *IEEE Transactions on Very Large Scale Integration (VLSI) Systems*, Vol. 11, No. 3, pp. 461–473, August 2003.

[366] V. F. Pavlidis, I. Savidis, and E. G. Friedman, "Clock Distribution Networks for 3-D Integrated Circuits," *IEEE Custom Integrated Circuits Conference*, pp. 651–654, September 2008.

[367] V. F. Pavlidis, I. Savidis, and E. G. Friedman, "Clock Distribution Networks in 3-D Integrated Systems," *IEEE Transactions on Very Large Scale Integration (VLSI) Systems*, Vol. 19, No. 12, pp. 2256–2266, December 2011.

[368] J. L. Neves and E. G. Friedman, "Topological Design of Clock Distribution Networks based on Non-Zero Clock Skew Specifications," *Proceedings of the IEEE Midwest Symposium on Circuits and Systems*, Vol. 1, pp. 468–471, August 1993.

[369] J. Cong, L. He, C.-K. Koh, and P. H. Madden, "Performance Optimization of VLSI Interconnect Layout," *Integration, the VLSI Journal*, Vol. 21, No. 1, pp. 1–94, November 1996.

[370] J. G. Xi and W. W.-M. Dai, "Useful-Skew Clock Routing with Gate Sizing for Low Power Design," *Proceedings of the ACM/IEEE Design Automation Conference*, pp. 383–388, June 1996.

[371] A. Kahng, J. Cong, and G. Robins, "High-Performance Clock Routing based on Recursive Geometric Matching," *Proceedings of the ACM/IEEE Design Automation Conference*, pp. 322–327, June 1991.

[372] W. Khan and N. Sherwani, "Zero Skew Clock Routing Algorithm for High Performance ASIC Systems," *Proceedings of the IEEE International ASIC Conference and Exhibit*, pp. 79–82, September 1993.

[373] M. Edahiro, "A Clustering-Based Optimization Algorithm in Zero-Skew Routings," *Proceedings of the ACM/IEEE Design Automation Conference*, pp. 612–616, July 1993.

[374] M. Edahiro, "An Efficient Zero-Skew Routing Algorithm," *Proceedings of the ACM/IEEE Design Automation Conference*, pp. 375–380, June 1994.

[375] R. Chaturvedi and J. Hu, "Buffered Clock Tree for High Quality IC Design," *Proceedings of the IEEE International Symposium on Signals, Circuits and Systems*, pp. 381–386, July 2004.

[376] M. A. B. Jackson, A. Srinivasan, and E. S. Kuh, "Clock Routing for High-Performance ICs," *Proceedings of the ACM/IEEE Design Automation Conference*, pp. 573–579, January 1990.

[377] T.-H. Chao, Y.-C. Hsu, and J.-M. Ho, "Zero Skew Clock Net Routing," *Proceedings of the ACM/IEEE Design Automation Conference*, pp. 518–523, June 1992.

[378] A. B. Kahng and C.-W. Albert Tsao, "More Practical Bounded-Skew Clock Routing," *Proceedings of the ACM/IEEE Design Automation Conference*, pp. 594–599, June 1997.

[379] P. Penfield, Jr. and J. Rubinstein, "Signal Delay in RC Tree Networks," *Proceedings of the ACM/IEEE Design Automation Conference*, pp. 613–617, June 1981.

[380] T. Sakurai, "Closed-Form Expressions for Interconnection Delay, Coupling, and Crosstalk in VLSIs," *IEEE Transactions on Electron Devices*, Vol. 40, No. 1, pp. 118–124, January 1993.

[381] I. Partin-Vaisband, R. Jakushokas, M. Popovich, A. V. Mezhiba, S. Köse, and E. G. Friedman, *On-Chip Power Delivery and Management, Fourth Edition*, Springer International Publishing, 2016.

[382] K. Xu, B. Vaisband, G. Sizikov, X. Li, and E. G. Friedman, "Power Noise and Near-Field EMI of High-Current System-in-Package with VR Top and Bottom Placements," *IEEE Transactions on Components, Packaging and Manufacturing Technology*, Vol. 9, No. 4, pp. 712–718, April 2019.

[383] J. Zhang and E. Friedman, "Crosstalk Modeling for Coupled RLC Interconnects with Application to Shield Insertion," *IEEE Transactions on Very Large Scale Integration (VLSI) Systems*, Vol. 14, No. 6, pp. 641–646, June 2006.

[384] R. Bairamkulov, A. Roy, M. Nagarajan, V. Srinivas, and E. G. Friedman, "Graph-Based Power Network Routing for Board-Level High Performance Systems," *Proceedings of the IEEE International Symposium on Circuits and Systems*, October 2020.

[385] R. Bairamkulov, A. Ruderman, and Y. L. Familiant, "Time Domain Optimization of Voltage and Current THD for a Three-Phase Cascaded H-Bridge Inverter," *Proceedings of the IEEE International Power Electronics and Motion Control Conference*, September 2016.

[386] V. Adler and E. Friedman, "Uniform Repeater Insertion in RC Trees," *IEEE Transactions on Circuits and Systems I: Fundamental Theory and Applications*, Vol. 47, No. 10, pp. 1515–1523, October 2000.

[387] R.-S. Tsay, "An Exact Zero-Skew Clock Routing Algorithm," *IEEE Transactions on Computer-Aided Design of Integrated Circuits and Systems*, Vol. 12, No. 2, pp. 242–249, February 1993.

[388] M. Edahiro, "Delay Minimization for Zero-Skew Routing," *Proceedings of the IEEE/ACM International Conference on Computer-Aided Design*, pp. 563–566, November 1993.

[389] J. Cong and C.-K. Koh, "Minimum-Cost Bounded-Skew Clock Routing," *Proceedings of the IEEE International Symposium on Circuits and Systems*, Vol. 1, pp. 215–218, April 1995.

[390] D. J. H. Huang, A. B. Kahng, and C.-W. A. Tsao, "On the Bounded-Skew Clock and Steiner Routing Problems," *Proceedings of the ACM/IEEE Design Automation Conference*, pp. 508–513, June 1995.

[391] J. Oh, I. Pyo, and M. Pedram, "Constructing Minimal Spanning Trees with Lower and Upper Bounded Path Delays," *Proceedings of the IEEE International Symposium on Circuits and Systems*, Vol. 4, pp. 416–419, April 1996.

[392] J. Cong, A. B. Kahng, C.-K. Koh, and C.-W. A. Tsao, "Bounded-Skew Clock and Steiner Routing," *ACM Transactions on Design Automation of Electronic Systems*, Vol. 3, No. 3, pp. 341–388, July 1998.

[393] E. G Friedman, "Clock Distribution Design in VLSI Circuits – an Overview," *Proceedings of the IEEE International Symposium on Circuits and Systems*, pp. 1475–1478, May 1993.

[394] G. S. Ohm, *Die galvanische Kette, mathematisch bearbeitet*, Bei T. H. Reimann, Berlin, 1827.

[395] J. C. Maxwell, *A Treatise on Electricity and Magnetism*, Clarendon Press, 1873.

[396] J. R. Carson, "A Theoretical Study of the Three-Element Vacuum Tube," *Proceedings of the Institute of Radio Engineers*, Vol. 7, No. 2, pp. 187–200, April 1919.

[397] G. Kron, "Non-Riemannian Dynamics of Rotating Electrical Machinery," *Journal of Mathematics and Physics*, Vol. 13, No. 1–4, pp. 103–194, April 1934.

[398] F. H. Branin, "D-C and Transient Analysis of Networks using a Digital Computer," *Proceedings of the SHARE Design Automation Workshop*, p. 4.1–4.23, January 1964.

[399] D. O. Pederson, "A Historical Review of Circuit Simulation," *IEEE Transactions on Circuits and Systems*, Vol. 31, No. 1, pp. 103–111, January 1984.

[400] A. F. Malmberg and F. L. Cornwell, *NET-1 Network Analysis Program*, Los Alamos Scientific Laboratory of the University of California, Los Alamos, New Mexico, 1963.

[401] H. W. Mathers, S. R. Sedore, and J. R. Sents, *Automated Digital Computer Program for Determining Responses of Electronic Circuits to Transient Nuclear Radiation (SCEPTRE)*, IBM Federal Systems Division, Electronics Systems Center, Oswego, New York, 1967.

[402] H. Shichman, "Computation of DC Solutions for Bipolar Transistor Networks," *IEEE Transactions on Circuit Theory*, Vol. 16, No. 4, pp. 460–466, November 1969.

[403] H. Shichman, "Integration System of a Nonlinear Transient Network-Analysis Program," *IEEE Transactions on Circuit Theory*, Vol. 17, No. 3, pp. 378–386, August 1970.

[404] G. Hachtel, R. Brayton, and F. Gustavson, "The Sparse Tableau Approach to Network Analysis and Design," *IEEE Transactions on Circuit Theory*, Vol. 18, No. 1, pp. 101–113, January 1971.

[405] A. R. Newton and D. O. Pederson, "Analysis Time, Accuracy and Memory Requirement Tradeoffs in Spice2," *Proceedings of the IEEE Asilomar Conference on Circuits, Systems and Computers*, pp. 6–9, November 1977.

[406] T.-H. Chen, C. Luk, and C. C.-P. Chen, "INDUCTWISE: Inductance-Wise Interconnect Simulator and Extractor," *IEEE Transactions on Computer-Aided Design of Integrated Circuits and Systems*, Vol. 22, No. 7, pp. 884–894, July 2003.

[407] A. B. Bhattacharyya, *Compact MOSFET Models for VLSI Design*, John Wiley & Sons (Asia), 2009.

[408] Y. Zhang, X. Wang, Y. Li, and E. G. Friedman, "Memristive Model for Synaptic Circuits," *IEEE Transactions on Circuits and Systems II: Express Briefs*, Vol. 64, No. 7, pp. 767–771, July 2017.

[409] C. A. Floudas and P. M. Pardalos, *Encyclopedia of Optimization*, Springer Science & Business Media, 2008.

[410] M. R. Hestenes and E. Stiefel, "Methods of Conjugate Gradients for Solving Linear Systems," *Journal of Research of the National Bureau of Standards*, Vol. 49, No. 6, pp. 409–435, December 1952.

[411] A. Buttari, J. Langou, J. Kurzak, and J. Dongarra, "A Class of Parallel Tiled Linear Algebra Algorithms for Multicore Architectures," *Parallel Computing*, Vol. 35, No. 1, pp. 38–53, January 2009.

[412] T.-H. Chen and C. C.-P. Chen, "Efficient Large-Scale Power Grid Analysis based on Preconditioned Krylov-Subspace Iterative Methods," *Proceedings of the ACM/IEEE Design Automation Conference*, pp. 559–562, June 2001.

[413] Y. Saad and M. H. Schultz, "GMRES: a Generalized Minimal Residual Algorithm for Solving Nonsymmetric Linear Systems," *SIAM Journal on Scientific and Statistical Computing*, Vol. 7, No. 3, pp. 856–869, July 1986.

[414] R. Barrett, M. Berry, T. F. Chan, J. Demmel, J. Donato, J. Dongarra, V. Eijkhout, R. Pozo, C. Romine, and H. Van der Vorst, *Templates for the Solution of Linear Systems: Building Blocks for Iterative Methods*, SIAM, 1994.

[415] Y. Saad, *Iterative Methods for Sparse Linear Systems, Second Edition*, PWS Publishing, 1996.

[416] D. S. Watkins, *Fundamentals of Matrix Computations*, Vol. 64, John Wiley & Sons, 2004.

[417] D. Mandic, "A Generalized Normalized Gradient Descent Algorithm," *IEEE Signal Processing Letters*, Vol. 11, No. 2, pp. 115–118, February 2004.

[418] C. F. Van Loan and G. Golub, *Matrix Computations*, The Johns Hopkins University Press, 1996.

[419] M. J. Grote and T. Huckle, "Parallel Preconditioning with Sparse Approximate Inverses," *SIAM Journal on Scientific Computing*, Vol. 18, No. 3, pp. 838–853, May 1997.

[420] M. Benzi and M. Tůma, "A Comparative Study of Sparse Approximate Inverse Preconditioners," *Applied Numerical Mathematics*, Vol. 30, No. 2, pp. 305–340, June 1999.

[421] W. Shi and F. Yu, "A Divide-and-Conquer Algorithm for 3-D Capacitance Extraction," *IEEE Transactions on Computer-Aided Design of Integrated Circuits and Systems*, Vol. 23, No. 8, pp. 1157–1163, July 2004.

[422] K. Sun, Q. Zhou, K. Mohanram, and D. C. Sorensen, "Parallel Domain Decomposition for Simulation of Large-Scale Power Grids," *Proceedings of the IEEE/ACM International Conference on Computer-Aided Design*, pp. 54–59, November 2007.

[423] U. Kleis, O. Wallat, U. Wever, and Q. Zheng, "Domain Decomposition Methods for Circuit Simulation," *Proceedings of the Workshop on Parallel and Distributed Simulation*, pp. 183–186, July 1994.

[424] Y. Zhong and M. D. F. Wong, "Fast Block-Iterative Domain Decomposition Algorithm for IR Drop Analysis in Large Power Grid," *Proceedings of the IEEE International Symposium on Quality Electronic Design*, pp. 277–283, March 2010.

[425] D. E. Keyes, A. Sameh, and V. Venkatakrishnan, *Parallel Numerical Algorithms, Vol. 4*, Springer Science & Business Media, 2012.

[426] T. Yu, Z. Xiao, and M. D. F. Wong, "Efficient Parallel Power Grid Analysis via Additive Schwarz Method," *Proceedings of the IEEE/ACM International Conference on Computer-Aided Design*, pp. 399–406, November 2012.

[427] S. Le Borne and L. Grasedyck, "$\mathcal{H}$-Matrix Preconditioners in Convection-Dominated Problems," *SIAM Journal on Matrix Analysis and Applications*, Vol. 27, No. 4, pp. 1172–1183, December 2006.

[428] M. L. Zitzmann, *Fast and Efficient Methods for Circuit-Based Automotive EMC Simulation*, Ph.D. Thesis, University of Erlangen-Nuremberg, 2007.

[429] H.-B. Chen, Y.-C. Li, S. X.-D. Tan, X. Huang, H. Wang, and N. Wong, "$\mathcal{H}$-Matrix-Based Finite-Element-Based Thermal Analysis for 3D ICs," *ACM Transactions on Design Automation of Electronic Systems*, Vol. 20, No. 4, September 2015.

[430] R. P. Fedorenko, "A Relaxation Method for Solving Elliptic Difference Equations," *USSR Computational Mathematics and Mathematical Physics*, Vol. 1, No. 4, pp. 1092–1096, September 1962.

[431] A. Brandt, "Multi-Level Adaptive Solutions to Boundary-Value Problems," *Mathematics of Computation*, Vol. 31, No. 138, pp. 333–390, April 1977.

[432] W. Hackbusch, "On the Multi-Grid Method Applied to Difference Equations," *Computing*, Vol. 20, No. 4, pp. 291–306, December 1978.

[433] A. H. Baker, E. R. Jessup, and T. Manteuffel, "A Technique for Accelerating the Convergence of Restarted GMRES," *SIAM Journal on Matrix Analysis and Applications*, Vol. 26, No. 4, pp. 962–984, 2005.
;

[434] J. Mandel and S. V. Parter, "On the Multigrid F-Cycle," *Applied Mathematics and Computation*, Vol. 37, No. 1, pp. 19–36, May 1990.

[435] S. R. Nassif and J. N. Kozhaya, "Fast Power Grid Simulation," *Proceedings of the ACM/IEEE Design Automation Conference*, pp. 156–161, June 2000.

[436] S. Williams, D. D. Kalamkar, A. Singh, A. M. Deshpande, B. Van Straalen, M. Smelyanskiy, A. Almgren, P. Dubey, J. Shalf, and L. Oliker, "Optimization of Geometric Multigrid for Emerging Multi- and Manycore Processors," *Proceedings of the International Conference on High Performance Computing, Networking, Storage and Analysis*, pp. 1–11, November 2012.

[437] R. D. Falgout, *An Introduction to Algebraic Multigrid*, Lawrence Livermore National Lab, Livermore, California, 2006.

[438] J. N. Kozhaya, S. R. Nassif, and F. N. Najm, "A Multigrid-Like Technique for Power Grid Analysis," *IEEE Transactions on Computer-Aided Design of Integrated Circuits and Systems*, Vol. 21, No. 10, pp. 1148–1160, October 2002.

[439] J. Yang, Z. Li, Y. Cai, and Q. Zhou, "PowerRush: a Linear Simulator for Power Grid," *Proceedings of the IEEE/ACM International Conference on Computer-Aided Design*, pp. 482–487, December 2011.

[440] J. Yang, Z. Li, Y. Cai, and Q. Zhou, "PowerRush: Efficient Transient Simulation for Power Grid Analysis," *Proceedings of the IEEE/ACM International Conference on Computer-Aided Design*, pp. 653–659, November 2012.

[441] C. Zhuo, J. Hu, M. Zhao, and K. Chen, "Fast Decap Allocation based on Algebraic Multigrid," *Proceedings of the IEEE/ACM International Conference on Computer-Aided Design*, pp. 107–111, November 2006.

[442] Z. Feng and P. Li, "Multigrid on GPU: Tackling Power Grid Analysis on Parallel SIMT Platforms," *Proceedings of the IEEE/ACM International Conference on Computer-Aided Design*, pp. 647–654, November 2008.

[443] Z. Feng and Z. Zeng, "Parallel Multigrid Preconditioning on Graphics Processing Units (GPUs) for Robust Power Grid Analysis," *Proceedings of the ACM/IEEE Design Automation Conference*, pp. 661–666, June 2010.

[444] V. V. Rao and I. Savidis, "Parameter Biasing Obfuscation for Analog IP Protection," *Proceedings of the IEEE International Symposium on Hardware Oriented Security and Trust*, pp. 161–161, May 2017.

[445] R. S. Chakraborty and S. Bhunia, "HARPOON: an Obfuscation-Based SoC Design Methodology for Hardware Protection," *IEEE Transactions on Computer-Aided Design of Integrated Circuits and Systems*, Vol. 28, No. 10, pp. 1493–1502, October 2009.

[446] T. Reveyrand, "Multiport Conversions between S, Z, Y, h, $ABCD$, and T Parameters," *Proceedings of the International Workshop on Integrated Nonlinear Microwave and Millimetre-Wave Circuits*, pp. 1–3, July 2018.

[447] D. Schreurs and J. Verspecht, "Large-Signal Modelling and Measuring Go Hand-in-Hand: Accurate Alternatives to Indirect S-Parameter Methods," *International Journal of RF and Microwave Computer-Aided Engineering*, Vol. 10, No. 1, pp. 6–18, December 2000.

[448] G. F. Lawler and V. Limic, *Random Walk: a Modern Introduction*, Cambridge University Press, 2010.

[449] B. B. Mandelbrot, *The Fractal Geometry of Nature*, W. H. Freeman, New York, 1982.

[450] Lord Rayleigh F. R. S., "XII. On the Resultant of a Large Number of Vibrations of the Same Pitch and of Arbitrary Phase," *The London, Edinburgh, and Dublin Philosophical Magazine and Journal of Science*, Vol. 10, No. 60, pp. 73–78, August 1880.

[451] D. Abbott, B. R. Davis, N. J. Phillips, and K. Eshraghian, "Simple Derivation of the Thermal Noise Formula using Window-Limited Fourier Transforms and Other Conundrums," *IEEE Transactions on Education*, Vol. 39, No. 1, pp. 1–13, February 1996.

[452] D. Ben-Avraham and S. Havlin, *Diffusion and Reactions in Fractals and Disordered Systems*, Cambridge University Press, 2000.

[453] J. E. Neigel and J. C. Avise, "Application of a Random Walk Model to Geographic Distributions of Animal Mitochondrial DNA Variation," *Genetics*, Vol. 135, No. 4, pp. 1209–1220, December 1993.

[454] S. J. Hardiman and L. Katzir, "Estimating Clustering Coefficients and Size of Social Networks via Random Walk," *Proceedings of the International Conference on World Wide Web*, pp. 539–550, May 2013.

[455] A. Dasgupta, G. Das, and H. Mannila, "A Random Walk Approach to Sampling Hidden Databases," *Proceedings of the International Conference on Management of Data*, pp. 629–640, June 2007.

[456] C. St. J. A. Nash-Williams, "Random Walk and Electric Currents in Networks," *Mathematical Proceedings of the Cambridge Philosophical Society*, Vol. 55, pp. 181–194, April 1959.

[457] W. Guo, S. X.-D. Tan, Z. Luo, and X. Hong, "Partial Random Walk for Large Linear Network Analysis," *Proceedings of the IEEE International Symposium on Circuits and Systems*, Vol. 5, pp. 173–176, May 2004.

[458] Z. Yu and M. D. F. Wong, "Fast Algorithms for IR Drop Analysis in Large Power Grid," *Proceedings of the IEEE/ACM International Conference on Computer-Aided Design*, pp. 351–357, November 2005.

[459] H. Qian, S. R. Nassif, and S. S. Sapatnekar, "Power Grid Analysis using Random Walks," *IEEE Transactions on Computer-Aided Design of Integrated Circuits and Systems*, Vol. 24, No. 8, pp. 1204–1224, August 2005.

[460] B. Boghrati and S. S. Sapatnekar, "Incremental Analysis of Power Grids using Backward Random Walks," *ACM Transactions on Design Automation of Electronic Systems*, Vol. 19, No. 3, pp. 1–29, June 2014.

[461] T. Miyakawa, K. Yamanaga, H. Tsutsui, H. Ochi, and T. Sato, "Acceleration of Random-Walk Based Linear Circuit Analysis using Importance Sampling," *Proceedings of the ACM Great Lakes Symposium on VLSI*, pp. 211–216, May 2011.

[462] P. Li, "Variational Analysis of Large Power Grids by Exploring Statistical Sampling Sharing and Spatial Locality," *Proceedings of the IEEE/ACM International Conference on Computer-Aided Design*, pp. 645–651, November 2005.

[463] J. Wang, "Deterministic Random Walk Preconditioning for Power Grid Analysis," *Proceedings of the IEEE/ACM International Conference on Computer-Aided Design*, pp. 392–398, November 2012.

[464] Y. Liang, W. Yu, and H. Qian, "A Hybrid Random Walk Algorithm for 3-D Thermal Analysis of Integrated Circuits," *Proceedings of the ACM/IEEE Asia and South Pacific Design Automation Conference*, pp. 849–854, January 2014.

[465] Y. Cai, L. Kang, J. Shi, X. Hong, and S. X.-D. Tan, "Random Walk Guided Decap Embedding for Power/Ground Network Optimization," *IEEE Transactions on Circuits and Systems II: Express Briefs*, Vol. 55, No. 1, pp. 36–40, January 2008.

[466] D.-A. Li, M. Marek-Sadowska, and S. R. Nassif, "T-VEMA: a Temperature- and Variation-Aware Electromigration Power Grid Analysis Tool," *IEEE Transactions on Very Large Scale Integration (VLSI) Systems*, Vol. 23, No. 10, pp. 2327–2331, October 2015.

[467] A. Çiprut, *Grids in Very Large Scale Integration Systems*, Ph.D. Thesis, University of Rochester, May 2019.

[468] A. V. Mezhiba and E. G. Friedman, "Inductive Properties of High-Performance Power Distribution Grids," *IEEE Transactions on Very Large Scale Integration (VLSI) Systems*, Vol. 10, No. 6, pp. 762–776, December 2002.

[469] S. Köse and E. G. Friedman, "Effective Resistance of a Two Layer Mesh," *IEEE Transactions on Circuits and Systems II: Express Briefs*, Vol. 58, No. 11, pp. 739–743, November 2011.

[470] W. H. McCrea, "A Problem on Random Paths," *The Mathematical Gazette*, Vol. 20, No. 241, pp. 311–317, December 1936.

[471] W. H. McCrea and F. J. W. Whipple, "Random Paths in Two and Three Dimensions," *Proceedings of the Royal Society of Edinburgh*, Vol. 60, No. 3, pp. 281–298, January 1940.

[472] H. Flanders, "Infinite Networks: II–Resistance in an Infinite Grid," *Journal of Mathematical Analysis and Applications*, Vol. 40, No. 1, pp. 30–35, October 1972.

[473] B. van der Pol and H. Bremmer, *Operational Calculus based on the Two-Sided Laplace Integral*, Cambridge University Press, 1950.

[474] R. J. Duffin, "Basic Properties of Discrete Analytic Functions," *Duke Mathematical Journal*, Vol. 23, No. 2, pp. 335–363, June 1956.

[475] A. Stöhr, "Über einige lineare partielle Differenzengleichungen mit konstanten Koeffizienten," *Mathematische Nachrichten*, Vol. 3, No. 6, pp. 330–357, September 1950.

[476] G. Venezian, "On the Resistance between Two Points on a Grid," *American Journal of Physics*, Vol. 62, No. 11, pp. 1000–1004, November 1994.

[477] D. Atkinson and F. J. Van Steenwijk, "Infinite Resistive Lattices," *American Journal of Physics*, Vol. 67, No. 6, pp. 486–492, June 1999.

[478] F. Spitzer, *Principles of Random Walk*, Vol. 34, Springer Science & Business Media, 2013.

[479] J. Cserti, "Application of the Lattice Green's Function for Calculating the Resistance of an Infinite Network of Resistors," *American Journal of Physics*, Vol. 68, No. 10, pp. 896–906, October 2000.

[480] S. Köse and E. G. Friedman, "Efficient Algorithms for Fast IR Drop Analysis Exploiting Locality," *Integration, the VLSI Journal*, Vol. 45, No. 2, pp. 149–161, March 2012.

[481] Y. Ogasahara, M. Hashimoto, T. Kanamoto, and T. Onoye, "Measurement of Supply Noise Suppression by Substrate and Deep N-well in 90nm Process," *Proceedings of the IEEE Asian Solid-State Circuits Conference*, pp. 397–400, November 2008.

[482] R. E. Aitchison, "Resistance between Adjacent Points of Liebman Mesh," *American Journal of Physics*, Vol. 32, pp. 566–566, July 1964.

[483] E. M. Purcell, *Electricity and Magnetism*, McGraw-Hill, 1963.

[484] P. G. Doyle and J. L. Snell, *Random Walks and Electric Networks*, Vol. 22, Mathematical Association of America, 1984.

[485] K. Brown, "Infinite Grid of Resistors," Available at https://www.mathpages.com/home/kmath668/kmath668.htm.

[486] J. T. Moody, *Efficient Methods for Calculating Equivalent Resistance between Nodes of a Highly Symmetric Resistor Network*, Major Qualifying Project, Worcester Polytechnic Institute, March 2013.

[487] M. Jeng, "Random Walks and Effective Resistances on Toroidal and Cylindrical Grids," *American Journal of Physics*, Vol. 68, No. 1, pp. 37–40, January 2000.

[488] "SciPy: Reference Guide," 2014, [Online; accessed April 15, 2019].

[489] *MATLAB 2019a Documentation*, MathWorks, February 2019.

[490] F. Dorfler and F. Bullo, "Kron Reduction of Graphs with Applications to Electrical Networks," *IEEE Transactions on Circuits and Systems I: Regular Papers*, Vol. 60, No. 1, pp. 150–163, January 2013.

[491] K. Chang, K. Acharya, S. Sinha, B. Cline, G. Yeric, and S. K. Lim, "Power Benefit Study of Monolithic 3D IC at the 7nm Technology Node," *Proceedings of the IEEE/ACM International Symposium on Low Power Electronics and Design*, pp. 201–206, July 2015.

[492] E. Liu and E. Li, "Fast Voltage Drop Modeling of Power Grid with Application to Silicon Interposer Analysis," *Proceedings of the IEEE Electronic Components and Technology Conference*, pp. 1109–1114, May 2013.

[493] M. Popovich, M. Sotman, A. Kolodny, and E. G. Friedman, "Effective Radii of On-Chip Decoupling Capacitors," *IEEE Transactions on Very Large Scale Integration (VLSI) Systems*, Vol. 16, No. 7, pp. 894–907, July 2008.

[494] S. Zhao, K. Roy, and C.-K. Koh, "Decoupling Capacitance Allocation and its Application to Power-Supply Noise-Aware Floorplanning," *IEEE Transactions on Computer-Aided Design of Integrated Circuits and Systems*, Vol. 21, No. 1, pp. 81–92, January 2002.

[495] S. Köse and E. G. Friedman, "Distributed On-Chip Power Delivery," *IEEE Journal on Emerging and Selected Topics in Circuits and Systems*, Vol. 2, No. 4, pp. 704–713, December 2012.

[496] J. M. Kosterlitz and D. J. Thouless, "Ordering, Metastability and Phase Transitions in Two-Dimensional Systems," *Journal of Physics C: Solid-State Physics*, Vol. 6, No. 7, pp. 1181, April 1973.

[497] A. K. Chandra, P. Raghavan, W. L. Ruzzo, R. Smolensky, and P. Tiwari, "The Electrical Resistance of a Graph Captures its Commute and Cover Times," *Computational Complexity*, Vol. 6, No. 4, pp. 312–340, December 1996.

[498] A. H. Zemanian, "Infinite Electrical Networks: a Reprise," *IEEE Transactions on Circuits and Systems*, Vol. 35, No. 11, pp. 1346–1358, November 1988.

[499] A. Carmona, A. M. Encinas, and M. Mitjana, "Kirchhoff Index of Periodic Linear Chains," *Journal of Mathematical Chemistry*, Vol. 53, No. 5, pp. 1195–1206, May 2015.

[500] H. Zhang and Y. Yang, "Resistance Distance and Kirchhoff Index in Circulant Graphs," *International Journal of Quantum Chemistry*, Vol. 107, No. 2, pp. 330–339, November 2007.

[501] F.-Y. Wu, "Theory of Resistor Networks: the Two-Point Resistance," *Journal of Physics A: Mathematical and General*, Vol. 37, No. 26, pp. 6653–6673, July 2004.

[502] R. Jakushokas and E. G. Friedman, "Power Network Optimization based on Link Breaking Methodology," *IEEE Transactions on Very Large Scale Integration (VLSI) Systems*, Vol. 21, No. 5, pp. 983–987, May 2013.

[503] C.-J. Lee, J. K. Park, S. Kim, and J.-H. Chun, "A Study on a Lattice Resistance Mesh Model of Display Cathode Electrodes for Capacitive Touch Screen Panel Sensors," *Procedia Engineering*, Vol. 168, pp. 884 – 887, September 2016.

[504] W. Xu and E. G. Friedman, "Clock Feedthrough in CMOS Analog Transmission Gate Switches," *IEEE International ASIC/SOC Conference*, pp. 181–185, September 2002.

[505] R. M. Secareanu, S. Warner, S. Seabridge, C. Burke, J. Becerra, T. E. Watrobski, C. Morton, W. Staub, T. Tellier, I. S. Kourtev, and E. G. Friedman, "Substrate Coupling in Digital Circuits in Mixed-Signal Smart-Power Systems," *IEEE Transactions on Very Large Scale Integration (VLSI) Systems*, Vol. 12, No. 1, pp. 67–78, January 2004.

[506] J. Xie, Y. Jia, and M. Miao, "High Sensitivity Knitted Fabric Bi-Directional Pressure Sensor based on Conductive Blended Yarn," *Smart Materials and Structures*, Vol. 28, No. 3, pp. 035017, February 2019.

[507] J. Xie and H. Long, "Equivalent Resistance Calculation of Knitting Sensor under Strip Biaxial Elongation," *Sensors and Actuators A: Physical*, Vol. 220, pp. 118–125, December 2014.

[508] E. Bendito, A. Carmona, A. M. Encinas, and J. M. Gesto, "Characterization of Symmetric M-Matrices as Resistive Inverses," *Linear Algebra and its Applications*, Vol. 430, No. 4, pp. 1336–1349, February 2009.

[509] S. J. Kirkland and M. Neumann, "The M-matrix Group Generalized Inverse Problem for Weighted Trees," *SIAM Journal on Matrix Analysis and Applications*, Vol. 19, No. 1, pp. 226–234, January 1998.

[510] R. Bairamkulov, K. Xu, E. G. Friedman, M. Popovich, J. Ochoa, and V. Srinivas, "Versatile Framework for Power Delivery Exploration," *Proceedings of the IEEE International Symposium on Circuits and Systems*, pp. 1–5, May 2018.

[511] I. Vaisband and E. G. Friedman, "Heterogeneous Methodology for Energy Efficient Distribution of On-Chip Power Supplies," *IEEE Transactions on Power Electronics*, Vol. 28, No. 9, pp. 4267–4280, September 2013.

[512] I. Vaisband, B. Price, S. Köse, Y. Kolla, E. G. Friedman, and J. Fischer, "Distributed LDO Regulators in a 28 nm Power Delivery System," *Analog Integrated Circuits and Signal Processing*, Vol. 83, No. 3, pp. 295–309, June 2015.

[513] S. K. Khatamifard, L. Wang, W. Yu, S. Köse, and U. R. Karpuzcu, "ThermoGater: Thermally-Aware On-Chip Voltage Regulation," *Proceedings of the ACM/IEEE International Symposium on Computer Architecture*, pp. 120–132, December 2017.

[514] X. Zhan, J. Riad, P. Li, and E. Sánchez, "Design Space Exploration of Distributed On-Chip Voltage Regulation under Stability Constraint," *IEEE Transactions on Very Large Scale Integration (VLSI) Systems*, Vol. 26, No. 8, pp. 1580–1584, August 2018.

[515] A. Çiprut and E. G. Friedman, "Stability of On-Chip Power Delivery Systems with Multiple Low-Dropout Regulators," *IEEE Transactions on Very Large Scale Integration (VLSI) Systems*, Vol. 27, No. 8, pp. 1779–1789, August 2019.

[516] L. Wang, S. K. Khatamifard, U. R. Karpuzcu, and S. Köse, "Exploiting Algorithmic Noise Tolerance for Scalable On-Chip Voltage Regulation," *IEEE Transactions on Very Large Scale Integration (VLSI) Systems*, Vol. 27, No. 1, pp. 229–242, January 2019.

[517] M. Zhao, Y. Fu, V. Zolotov, S. Sundareswaran, and R. Panda, "Optimal Placement of Power-Supply Pads and Pins," *IEEE Transactions on Computer-Aided Design of Integrated Circuits and Systems*, Vol. 25, No. 1, pp. 144–154, January 2006.

[518] J. Kennedy and R. Eberhart, "Particle Swarm Optimization," *Proceedings of the IEEE International Conference on Neural Networks*, Vol. 4, pp. 1942–1948, November 1995.

[519] E. A. Burton, G. Schrom, F. Paillet, J. Douglas, W. J. Lambert, K. Radhakrishnan, and M. J. Hill, "FIVR – Fully Integrated Voltage Regulators on 4Th Generation Intel Core SoCs," *Proceedings of the IEEE Applied Power Electronics Conference and Exposition*, pp. 432–439, March 2014.

[520] B. Amelifard and M. Pedram, "Optimal Design of the Power-Delivery Network for Multiple Voltage-Island System-on-Chips," *IEEE Transactions on Computer-Aided Design of Integrated Circuits and Systems*, Vol. 28, No. 6, pp. 888–900, May 2009.

[521] W. Kim, M. S. Gupta, G.-Y. Wei, and D. Brooks, "System Level Analysis of Fast, Per-Core DVFS using On-Chip Switching Regulators," *Proceedings of the IEEE International Symposium on High Performance Computer Architecture*, pp. 123–134, February 2008.

[522] O. A. Uzun and S. Köse, "Converter-Gating: A Power Efficient and Secure On-Chip Power Delivery System," *IEEE Journal on Emerging and Selected Topics in Circuits and Systems*, Vol. 4, No. 2, pp. 169–179, June 2014.

[523] Z. Feng, Z. Zeng, and P. Li, "Parallel On-Chip Power Distribution Network Analysis on Multi-Core-Multi-GPU Platforms," *IEEE Transactions on Very Large Scale Integration (VLSI) Systems*, Vol. 19, No. 10, pp. 1823–1836, October 2011.

[524] S. R. Nassif, "Power Grid Analysis Benchmarks," *Proceedings of the ACM/IEEE Asia and South Pacific Design Automation Conference*, pp. 376–381, March 2008.

[525] K. Wang and M. Marek-Sadowska, "On-Chip Power-Supply Network Optimization using Multigrid-Based Technique," *IEEE Transactions on Computer-Aided Design of Integrated Circuits and Systems*, Vol. 24, No. 3, pp. 407–417, March 2005.

[526] D. J. Wales and J. P. Doye, "Global Optimization by Basin-Hopping and the Lowest Energy Structures of Lennard-Jones Clusters Containing up to 110 Atoms," *The Journal of Physical Chemistry A*, Vol. 101, No. 28, pp. 5111–5116, July 1997.

[527] D. Pham and D. Karaboga, *Intelligent Optimisation Techniques: Genetic Algorithms, Tabu Search, Simulated Annealing and Neural Networks*, Springer Science and Business Media, 2012.

[528] K. Xu, R. Patel, P. Raghavan, and E. G. Friedman, "Exploratory Design of On-Chip Power Delivery for 14, 10, and 7 nm and Beyond FinFET ICs," *Integration, the VLSI Journal*, Vol. 61, pp. 11–19, March 2018.

[529] M. Popovich, E. G. Friedman, R. M. Secareanu, and O. L. Hartin, "Efficient Placement of Distributed On-Chip Decoupling Capacitors in Nanoscale ICs," *Proceedings of the IEEE/ACM International Conference on Computer-Aided Design*, pp. 811–816, November 2007.

[530] W. Liao, L. He, and K. M. Lepak, "Temperature and Supply Voltage-Aware Performance and Power Modeling at Microarchitecture Level," *IEEE Transactions on Computer-Aided Design of Integrated Circuits and Systems*, Vol. 24, No. 7, pp. 1042–1053, July 2005.

[531] S. Dropsho, V. Kursun, D. H. Albonesi, S. Dwarkadas, and E. G. Friedman, "Managing Static Leakage Energy in Microprocessor Functional Units," *Proceedings of the IEEE/ACM International Symposium on Microarchitecture*, pp. 321–332, February 2002.

[532] E. Salman, E. G Friedman, R. M. Secareanu, and O. L. Hartin, "Worst Case Power/Ground Noise Estimation using an Equivalent Transition Time for Resonance," *IEEE Transactions on Circuits and Systems I: Regular Papers*, Vol. 56, No. 5, pp. 997–1004, May 2009.

[533] E. Rotem, A. Naveh, A. Ananthakrishnan, E. Weissmann, and D. Rajwan, "Power-Management Architecture of the Intel Microarchitecture Code-Named Sandy Bridge," *IEEE Micro*, Vol. 32, No. 2, pp. 20–27, March 2012.

[534] A. E. Engin, "Efficient Sensitivity Calculations for Optimization of Power Delivery Network Impedance," *IEEE Transactions on Electromagnetic Compatibility*, Vol. 52, No. 2, pp. 332–339, May 2010.

[535] A. Orlandi, "Differential Evolutionary Multiple-Objective Sequential Optimization of a Power Delivery Network," *IEEE Transactions on Electromagnetic Compatibility*, Vol. 60, No. 3, pp. 754–760, June 2018.

[536] B. Ko, J. Kim, J. Ryoo, C. Hwang, J. Song, and S. W. Kim, "Simplified Chip Power Modeling Methodology without Netlist Information in Early Stage of SoC Design Process," *IEEE Transactions on Components, Packaging and Manufacturing Technology*, Vol. 6, No. 10, pp. 1513–1521, October 2016.

[537] S. Köse, E. G. Friedman, R. M. Secareanu, and O. Hartin, "Current Profile of a Microcontroller to Determine Electromagnetic Emissions," *Proceedings of the IEEE International Symposium on Circuits and Systems*, pp. 2650–2653, May 2013.

[538] X. Huang, T. Yu, V. Sukharev, and S. X. D. Tan, "Physics-Based Electromigration Assessment for Power Grid Networks," *Proceedings of the ACM/IEEE Design Automation Conference*, pp. 1–6, June 2014.

[539] P. Salome, C. Leroux, P. Crevel, and J. P. Chante, "Investigations on the Thermal Behavior of Interconnects under ESD Transients using a Simplified Thermal RC Network," *Proceedings of the Electrical Overstress / Electrostatic Discharge Symposium*, pp. 187–198, October 1998.

[540] *HSPICE Quick Reference*, Synopsys, March 2017.

[541] H. Su, K. H. Gala, and S. S. Sapatnekar, "Analysis and Optimization of Structured Power/Ground Networks," *IEEE Transactions on Computer-Aided Design of Integrated Circuits and Systems*, Vol. 22, No. 11, pp. 1533–1544, November 2003.

[542] A. J. Fleming, S. Behrens, and S. O. Reza Moheimani, "Optimization and Implementation of Multimode Piezoelectric Shunt Damping Systems," *IEEE/ASME Transactions on Mechatronics*, Vol. 7, No. 1, pp. 87–94, March 2002.

[543] M. Rewienski and J. White, "A Trajectory Piecewise-Linear Approach to Model Order Reduction and Fast Simulation of Nonlinear Circuits and Micromachined Devices," *IEEE Transactions on Computer-Aided Design of Integrated Circuits and Systems*, Vol. 22, No. 2, pp. 155–170, February 2003.

[544] P. Moreno and A. Ramirez, "Implementation of the Numerical Laplace Transform: a Review," *IEEE Transactions on Power Delivery*, Vol. 23, No. 4, pp. 2599–2609, October 2008.

[545] E. I. Verriest, "Linear Systems over the Perspective Field as a Class of Nonlinear Systems for which a "Laplace" Transform can be Defined," *Proceedings of the IEEE International Conference on Control and Automation*, pp. 271–276, June 2016.

[546] *Verilog-AMS Language Reference Manual Analog and Mixed-Signal Extensions to Verilog-HDL*, Accellera, June 2009.

[547] E. Anderson *et al.*, *LAPACK Users' Guide*, Society for Industrial and Applied Mathematics, 1999.

[548] *MATLAB Control Systems Toolbox 10.3*, MathWorks, Inc., March 2018.

[549] H. Iwai, "Future of Integrated Devices," *Proceedings of the International Workshop on Junction Technology*, p. 43, June 2015.

[550] J. Kim, "Power Integrity of SiP (System in Package)," *IEEE Video Distinguished Lecturer Program*, August 2010.

[551] B. Archambeault, "Effective Power/Ground Plane Decoupling for PCB," *IEEE Video Distinguished Lecturer Program*, October 2007.

[552] Z. Or-Bach, "Moore's Law Has Stopped at 28nm," March 2014. [Online]. Available: http://electroiq.com/blog/2014/03/moores-law-has-stopped-at-28nm/, [Accessed: 2018-01-25].

[553] E. Esteve, "Why SOI is the Future Technology of Semiconductor," January 2014. [Online]. Available: https://www.semiwiki.com/forum/content/3077-why-soi-future-technology-semiconductor.html, [Accessed: 2018-01-25].

[554] Anysilicon.com, "IC Package Price Estimator," [Online]. Available: http://anysilicon.com/package-price-estimator/, [Accessed: 2018-01-25].

[555] Pcbshopper.com, "A Price Comparison Site for Printed Circuit Boards," [Online]. Available: https://pcbshopper.com/, [Accessed: 2018-01-25].

[556] Pcbcart.com, "Printed Circuit Board Calculator," [Online]. Available: https://www.pcbcart.com/quote, [Accessed: 2018-01-25].

[557] *Optimization Toolbox User's Guide*, MathWorks, Inc., September 2017.

[558] *Qualcomm Snapdragon 600E Processor APQ8064E*, Qualcomm Technologies, October 2017.

[559] M. S. Tanaka, M. Toyama, R. Mori, H. Nakashima, M. Haida, and I. Ooshima, "Early Stage Chip/Package/Board Co-Design Techniques for System-on-Chip," *Proceedings of the IEEE Conference on Electrical Performance of Electronic Packaging and Systems*, pp. 21–24, October 2011.

[560] T. Tseng, C. Lin, C. Lee, Y. Chou, and D. Kwai, "A Power Delivery Network (PDN) Engineering Change Order (ECO) Approach for Repairing IR-Drop Failures after the Routing Stage," *Proceedings of the IEEE Design, Automation and Test in Europe Conference and Exhibition*, pp. 1–4, April 2014.

[561] K. Shringarpure, B. Zhao, L. Wei, B. Archambeault, A. Ruehli, M. Cracraft, M. Cocchini, E. Wheeler, J. Fan, and J. Drewniak, "On Finding the Optimal Number of Decoupling Capacitors by Minimizing the Equivalent Inductance of the PCB PDN," *Proceedings of the IEEE International Symposium on Electromagnetic Compatibility*, pp. 218–223, September 2014.

[562] C. M. Smutzer, C. K. White, C. R. Haider, and B. K. Gilbert, "Power Delivery Network Pre-Layout Design Planning and Analysis through Automated Scripting," *Proceedings of the IEEE Workshop on Signal and Power Integrity*, pp. 1–4, June 2019.

[563] J. Mohamed, T. Michalka, S. Ozbayat, and G. R. Luevano, "PDN Design and Sensitivity Analysis using Synthesized Models in DDR SI/PI Co-Simulations," *Proceedings of the IEEE Electrical Design of Advanced Packaging and Systems Symposium*, pp. 1–3, December 2018.

[564] S. Yang, Y. S. Cao, H. Ma, J. Cho, A. E. Ruehli, J. L. Drewniak, and E. Li, "PCB PDN Prelayout Library for Top-Layer Inductance and the Equivalent Model for Decoupling Capacitors," *IEEE Transactions on Electromagnetic Compatibility*, Vol. 60, No. 6, pp. 1898–1906, August 2017.

[565] Y. S. Cao, T. Makharashvili, J. Cho, S. Bai, S. Connor, B. Archambeault, L. Jiang, A. E. Ruehli, J. Fan, and J. L. Drewniak, "Inductance Extraction for PCB Prelayout Power Integrity using PMSR Method," *IEEE Transactions on Electromagnetic Compatibility*, Vol. 59, No. 4, pp. 1339–1346, August 2017.

[566] B. Zhao, C. Huang, K. Shringarpure, J. Fan, B. Archambeault, B. Achkir, S. Connor, M. Cracraft, M. Cocchini, A. Ruehli, and J. Drewniak, "Analytical PDN Voltage Ripple Calculation using Simplified Equivalent Circuit Model of PCB PDN," *Proceedings of the IEEE Symposium on Electromagnetic Compatibility and Signal Integrity*, pp. 133–138, March 2015.

[567] S. Sun, D. Pommerenke, J. L. Drewniak, K. Xiao, S.-T. Chen, and T.-L. Wu, "Characterizing Package/PCB PDN Interactions from a Full-Wave Finite-Difference Formulation," *Proceedings of the IEEE International Symposium on Electromagnetic Compatibility*, Vol. 2, pp. 550–555, August 2006.

[568] B. Archambeault, M. Cocchini, G. Selli, J. Fan, J. L. Knighten, S. Connor, A. Orlandi, and J. Drewniak, "Design Methodology for PDN Synthesis on Multilayer PCBs," *Proceedings of the IEEE International Symposium on Electromagnetic Compatibility*, pp. 1–6, August 2008.

[569] J. Fan, J. L. Drewniak, J. L. Knighten, N. W. Smith, A. Orlandi, T. P. Van Doren, T. H. Hubing, and R. E. DuBroff, "Quantifying SMT Decoupling Capacitor Placement in DC Power-Bus Design for Multilayer PCBs," *IEEE Transactions on Electromagnetic Compatibility*, Vol. 43, No. 4, pp. 588–599, August 2001.

[570] T.-L. Wu, H.-H. Chuang, and T.-K. Wang, "Overview of Power Integrity Solutions on Package and PCB: Decoupling and EBG Isolation," *IEEE Transactions on Electromagnetic Compatibility*, Vol. 52, No. 2, pp. 346–356, July 2010.

[571] T. Hubing, "PCB EMC Design Guidelines: a Brief Annotated List," *Proceedings of the IEEE Symposium on Electromagnetic Compatibility*, pp. 34–36, August 2003.

[572] M. Moganti, F. Ercal, C. H. Dagli, and S. Tsunekawa, "Automatic PCB Inspection Algorithms: a Survey," *Computer Vision and Image Understanding*, Vol. 63, No. 2, pp. 287–313, March 1996.

[573] G. Greiner and K. Hormann, "Efficient Clipping of Arbitrary Polygons," *ACM Transactions on Graphics*, Vol. 17, No. 2, pp. 71–83, April 1998.

[574] B. R. Vatti, "A Generic Solution to Polygon Clipping," *Communications of the ACM*, Vol. 35, No. 7, pp. 56–63, July 1992.

[575] I. Savidis and E. G. Friedman, "Electrical Modeling and Characterization of 3-D Vias," *IEEE International Symposium on Circuits and Systems*, pp. 784–787, May 2008.

[576] S. Peyer, D. Rautenbach, and J. Vygen, "A Generalization of Dijkstra's Shortest Path Algorithm with Applications to VLSI Routing," *Journal of Discrete Algorithms*, Vol. 7, No. 4, pp. 377–390, December 2009.

[577] P. J. van Laarhoven and E. H. Aarts, *Simulated Annealing: Theory and Applications*, Vol. 37, Springer Science & Business Media, 2013.

[578] Y. K. Liu, X. Q. Wang, S. Z. Bao, M. Gomboši, and B. Žalik, "An Algorithm for Polygon Clipping, and for Determining Polygon Intersections and Unions," *Computers and Geosciences*, Vol. 33, No. 5, pp. 589 – 598, May 2007.

[579] Z. Tang, J. Zhu, F. He, L. Feng, G. Yang, and G. Han, "Adaptive Polygon Simplification basing on Delaunay Triangulation and its Application in High Speed PCBs and IC Packages Simulation," *Proceedings of the IEEE International Conference on Microwave Technology & Computational Electromagnetics*, pp. 253–256, May 2011.

[580] W. Zeng and R. L. Church, "Finding Shortest Paths on Real Road Networks: the Case for A-Star," *International Journal of Geographical Information Science*, Vol. 23, No. 4, pp. 531–543, June 2009.

[581] A. George and E. Ng, "On the Complexity of Sparse QR and LU Factorization of Finite-Element Matrices," *SIAM Journal on Scientific and Statistical Computing*, Vol. 9, No. 5, pp. 849–861, September 1988.

[582] V. K. Semenov, Y. A. Polyakov, and S. K. Tolpygo, "New AC-Powered SFQ Digital Circuits," *IEEE Transactions on Applied Superconductivity*, Vol. 25, No. 3, pp. 1–7, June 2014.

[583] Y. Ando, R. Sato, M. Tanaka, K. Takagi, N. Takagi, and A. Fujimaki, "Design and Demonstration of an 8-bit Bit-Serial RSFQ Microprocessor: CORE e4," *IEEE Transactions on Applied Superconductivity*, Vol. 26, No. 5, pp. 1–5, August 2016.

[584] K. Gaj, Q. P. Herr, V. Adler, A. Krasniewski, E. G. Friedman, and M. J Feldman, "Tools for the Computer-Aided Design of Multigigahertz Superconducting Digital Circuits," *IEEE Transactions on Applied Superconductivity*, Vol. 9, No. 1, pp. 18–38, March 1999.

[585] C. J. Fourie, "Digital Superconducting Electronics Design Tools–Status and Roadmap," *IEEE Transactions on Applied Superconductivity*, Vol. 28, No. 5, pp. 1–12, August 2018.

[586] K. Gaj, E. G. Friedman, and M. J. Feldman, "Timing of Multi-Gigahertz Rapid Single Flux Quantum Digital Circuits," *Journal of VLSI Signal Processing Systems for Signal, Image and Video Technology*, Vol. 16, No. 2, pp. 247–276, June 1997.

[587] K. K. Likharev and V. K. Semenov, "RSFQ Logic/Memory Family: a New Josephson-Junction Technology for Sub-Terahertz-Clock-Frequency Digital Systems," *IEEE Transactions on Applied Superconductivity*, Vol. 1, No. 1, pp. 3–28, March 1991.

[588] T. Jabbari, G. Krylov, S. Whiteley, E. Mlinar, J. Kawa, and E. G. Friedman, "Interconnect Routing for Large-Scale RSFQ Circuits," *IEEE Transactions on Applied Superconductivity*, Vol. 29, No. 5, pp. 1–5, August 2019.

[589] T. Jabbari, G. Krylov, S. Whiteley, J. Kawa, and E. G. Friedman, "Repeater Insertion in SFQ Interconnect," *IEEE Transactions on Applied Superconductivity*, Vol. 30, No. 8, pp. 1–8, December 2020.

[590] T. Jabbari and E. G. Friedman, "Global Interconnects in VLSI Complexity Single Flux Quantum Systems," *Proceedings of the ACM/IEEE International Workshop on System-Level Interconnect Problems and Pathfinding*, November 2020.

[591] T. Jabbari, G. Krylov, J. Kawa, and E. G. Friedman, "Splitter Trees in Single Flux Quantum Circuits," *IEEE Transactions on Applied Superconductivity*, Vol. 31, No. 5, pp. 1–6, August 2021.

[592] Z. J. Deng, N. Yoshikawa, S. R. Whiteley, and T. Van Duzer, "Data-Driven Self-Timed RSFQ Digital Integrated Circuit and System," *IEEE Transactions on Applied Superconductivity*, Vol. 7, No. 2, pp. 3634–3637, June 1997.

[593] H. R. Gerber, C. J. Fourie, W. J. Perold, and L. C. Muller, "Design of an Asynchronous Microprocessor using RSFQ-AT," *IEEE Transactions on Applied Superconductivity*, Vol. 17, No. 2, pp. 490–493, June 2007.

[594] T. V. Filippov, A. Sahu, A. F. Kirichenko, I. V. Vernik, M. Dorojevets, C. L. Ayala, and O. A. Mukhanov, "20 GHz Operation of an Asynchronous Wave-Pipelined RSFQ Arithmetic-Logic Unit," *Physics Procedia*, Vol. 36, pp. 59–65, January 2012.

[595] Y. Nobumori, T. Nishigai, K. Nakamiya, N. Yoshikawa, A. Fujimaki, H. Terai, and S. Yorozu, "Design and Implementation of a Fully Asynchronous SFQ Microprocessor: SCRAM2," *IEEE Transactions on Applied Superconductivity*, Vol. 17, No. 2, pp. 478–481, June 2007.

[596] Z. J. Deng, N. Yoshikawa, J. A. Tierno, A. R. Whiteley, and T. Van Duzer, "Asynchronous Circuits and Systems in Superconducting RSFQ Digital Technology," *Proceedings of the IEEE International Symposium on Advanced Research in Asynchronous Circuits and Systems*, pp. 274–285, April 1998.

[597] R. N. Tadros and P. A. Beerel, "A Robust and Self-Adaptive Clocking Technique for RSFQ Circuits – the Architecture," *Proceedings of the IEEE International Symposium on Circuits and Systems*, pp. 1–5, May 2018.

[598] S. N. Shahsavani and M. Pedram, "A Minimum-Skew Clock Tree Synthesis Algorithm for Single Flux Quantum Logic Circuits," *IEEE Transactions on Applied Superconductivity*, Vol. 29, No. 8, pp. 1–13, December 2019.

[599] J. L. Neves and E. G. Friedman, "Buffered Clock Tree Synthesis with Non-Zero Clock Skew Scheduling for Increased Tolerance to Process Parameter Variations," *Journal of VLSI Signal Processing Systems for Signal, Image, and Video Technology*, Vol. 16, No. 2, pp. 149–161, June 1997.

[600] E. G. Friedman, "The Application of Localized Clock Distribution Design to Improving the Performance of Retimed Sequential Circuits," *Proceedings of the IEEE Asia-Pacific Conference on Circuits and Systems*, pp. 12–17, December 1992.

[601] L. Xiao, Z. Xiao, Z. Qian, Y. Jiang, T. Huang, H. Tian, and E. F. Y. Young, "Local Clock Skew Minimization using Blockage-Aware Mixed Tree-Mesh Clock Network," *Proceedings of the IEEE/ACM International Conference on Computer-Aided Design*, pp. 458–462, November 2010.

[602] Y.-S. Su, W.-K. Hon, C.-C. Yang, S.-C. Chang, and Y.-J. Chang, "Value Assignment of Adjustable Delay Buffers for Clock Skew Minimization in Multi-Voltage Mode Designs," *Proceedings of the IEEE/ACM International Conference on Computer-Aided Design*, pp. 535–538, November 2009.

[603] A. L. Pankratov and B. Spagnolo, "Suppression of Timing Errors in Short Overdamped Josephson Junctions," *Physical Review Letters*, Vol. 93, No. 17, pp. 177001, October 2004.

[604] C. Lin and H. Zhou, "Clock Skew Scheduling with Delay Padding for Prescribed Skew Domains," *Proceedings of the ACM/IEEE Asia and South Pacific Design Automation Conference*, pp. 541–546, January 2007.

[605] K. Han, A. B. Kahng, and J. Li, "Optimal Generalized H-Tree Topology and Buffering for High-Performance and Low-Power Clock Distribution," *IEEE Transactions on Computer-Aided Design of Integrated Circuits and Systems*, Vol. 39, No. 2, pp. 478–491, February 2020.

[606] T. Kanungo, D. M. Mount, N. S. Netanyahu, C. D. Piatko, R. Silverman, and A. Y. Wu, "An Efficient k-Means Clustering Algorithm: Analysis and Implementation," *IEEE Transactions on Pattern Analysis and Machine Intelligence*, Vol. 24, No. 7, pp. 881–892, July 2002.

[607] T. Zhang, R. Ramakrishnan, and M. Livny, "BIRCH: an Efficient Data Clustering Method for Very Large Databases," *ACM SIGMOD Record*, Vol. 25, No. 2, pp. 103–114, June 1996.

[608] A. Balatsos, *Clock Buffer IC with Dynamic Impedance Matching and Skew Compensation*, Ph.D. Thesis, University of Toronto, 1998.

[609] J.-L. Tsai, T.-H. Chen, and C. C.-P. Chen, "Zero Skew Clock-Tree Optimization with Buffer Insertion/Sizing and Wire Sizing," *IEEE Transactions on Computer-Aided Design of Integrated Circuits and Systems*, Vol. 23, No. 4, pp. 565–572, April 2004.

[610] J. Y. Yen, "Finding the k Shortest Loopless Paths in a Network," *Management Science*, Vol. 17, No. 11, pp. 712–716, June 1971.

[611] M. Kou, P.-Y. Cheng, J. Zeng, T.-Y. Ho, K. Takagi, and H. Yao, "Splitter-Aware Multiterminal Routing With Length-Matching Constraint for RSFQ Circuits," *IEEE Transactions on Computer-Aided Design of Integrated Circuits and Systems*, Vol. 40, No. 11, pp. 2251–2264, November 2021.

[612] N. Kito, K. Takagi, and N. Takagi, "A Fast Wire-Routing Method and an Automatic Layout Tool for RSFQ Digital Circuits considering Wire-Length Matching," *IEEE Transactions on Applied Superconductivity*, Vol. 28, No. 4, pp. 1–5, June 2018.

[613] S. N. Shahsavani, T. Lin, A. Shafaei, C. J. Fourie, and M. Pedram, "An Integrated Row-Based Cell Placement and Interconnect Synthesis Tool for Large SFQ Logic Circuits," *IEEE Transactions on Applied Superconductivity*, Vol. 27, No. 4, pp. 1–8, June 2017.

[614] C. J. Fourie, C. L. Ayala, L. Schindler, T. Tanaka, and N. Yoshikawa, "Design and Characterization of Track Routing Architecture for RSFQ and AQFP Circuits in a Multilayer Process," *IEEE Transactions on Applied Superconductivity*, Vol. 30, No. 6, pp. 1–9, September 2020.

[615] R. S. Bakolo and C. J. Fourie, "Development of a RSFQ Cell Library for the University of Stellenbosch," *Proceedings of the IEEE AFRICON*, pp. 1–5, September 2011.

[616] M. Zachariasen, "A Catalog of Hanan Grid Problems," *Networks*, Vol. 38, pp. 200–1, September 2000.

[617] F. Brglez, D. Bryan, and K. Kozminski, "Combinational Profiles of Sequential Benchmark Circuits," *Proceedings of the IEEE International Symposium on Circuits and Systems,*, Vol. 3, pp. 1929–1934, May 1989.

[618] S. Davidson, "ITC'99 Benchmark Circuits - Preliminary Results," *Proceedings of the IEEE International Test Conference*, pp. 1125–1125, September 1999.

[619] *Design Compiler and IC Compiler Physical Guidance Technology Application Note*, Synopsys, Version G-2012.06, June 2012.

[620] Y. Xu, Y. Zhang, and C. Chu, "FastRoute 4.0: Global Router with Efficient Via Minimization," *Proceedings of the ACM/IEEE Asia and South Pacific Design Automation Conference*, pp. 576–581, January 2009.

Index